Valorization of Biomass: Current Trends, Challenges and Future Prospects

Valorization of Biomass: Current Trends, Challenges and Future Prospects

Editor: Nina Goldman

STATES
ACADEMIC PRESS
www.statesacademicpress.com

States Academic Press,
109 South 5th Street,
Brooklyn, NY 11249, USA

Visit us on the World Wide Web at:
www.statesacademicpress.com

ISBN: 978-1-63989-554-0 (Hardback)

Cataloging-in-Publication Data

Valorization of biomass : current trends, challenges and future prospects / edited by Nina Goldman.
p. cm.
Includes bibliographical references and index.
ISBN 978-1-63989-554-0
1. Biomass. 2. Biomass energy. 3. Biomass conversion. I. Goldman, Nina.
TP339 .V35 2022
662.88--dc23

Table of Contents

Preface

It is often said that books are a boon to mankind. They document every progress and pass on the knowledge from one generation to the other. They play a crucial role in our lives. Thus I was both excited and nervous while editing this book. I was pleased by the thought of being able to make a mark but I was also nervous to do it right because the future of students depends upon it. Hence, I took a few months to research further into the discipline, revise my knowledge and also explore some more aspects. Post this process, I begun with the editing of this book.

Any plant or animal material that can be used as a fuel for the production of heat and electricity is known as biomass. Some common examples of biomass are wood, energy crops, farm waste, etc. The main components of any biomass are carbohydrates, lipids, protein and lignin. Valorization is the process of converting biomass into energy, fuels and other useful resources. Some common valorization techniques are pyrolysis and liquefaction. Parameters affecting the valorization processes are temperature, pressure, catalysts, residence time and feedstock. This book brings forth some of the most innovative concepts and elucidates the unexplored aspects of valorization of biomass. From theories to research to practical applications, case studies related to all contemporary topics of relevance to this field have been included herein. This book is a vital tool for all researching or studying biomass valorization as it gives incredible insights into emerging trends and concepts.

I thank my publisher with all my heart for considering me worthy of this unparalleled opportunity and for showing unwavering faith in my skills. I would also like to thank the editorial team who worked closely with me at every step and contributed immensely towards the successful completion of this book. Last but not the least, I wish to thank my friends and colleagues for their support.

Editor

1

Sustainability of the Biowaste Utilization for Energy Production

Thorsten Ahrens, Silvia Drescher-Hartung and Olga Anne

Abstract

This article presents strategies for the development and practical modelling of biogas processes under economy market conditions. Herewith, anaerobic digestion results out of practical tests in different scales (lab to pilot) and different substrate mixtures were taken into account. Two lab-scale reactors on the one hand and pilot-scale examinations of chosen substrate mixtures on the other hand led to workable conclusions such as mixture suitability for biogas production, gas amounts and technical demands for full-scale implementation under market economy conditions. A comparison of both laboratory and pilot system performance with a full-scale biogas system has been done; herewith, the suitability of the corresponding practical process upscaling simulation has been proven. On the basis of the results, calculations regarding the necessary full-scale fermenter sizes and the required substrate amounts as well as the disposable (reusable) fermentation residues were made. The conclusion of outputs on biogas technology particularly with regard to the demand-driven production of electricity (500 + 250 kW flexible) as a special request for energy from renewable sources is given. As a further result, a general outlook and estimation for the economical implementation on a common Baltic Sea region country basis have been developed.

Keywords: biowaste, biogas, pilot scale, full scale, circular economy, waste-to-energy, sustainability, process simulation

1. Introduction

The European Union (EU) calls on all parties to the efficient use of energy and accelerates the integration of energy recovery technologies from wastes that emit less carbon dioxide [1].

Biogas production by anaerobic digestion is a well-known method of energy recovery from biomass. Nowadays, such technologies are widely implemented in accordance with the EU policies for green and sustainable energy supply, for example, in all Baltic Sea region countries, but in different percentages. As an example, in Lithuania, mostly sewage sludge was used for biogas production till year 2012. Half of the Lithuanian energy demand in 2009 was covered by fossil fuels, 16% by renewable sources [2]. As different publications of FNR (Fachagentur Nachwachsende Rohstoffe, Germany) show that the anaerobic digestion is well investigated, but due to the technology development, there is need of research, especially in terms of market economy conditions, because funding for such processes will end in the near future [3]. Funding strategies should be taken into account in order to allow the development and pilot implementation of new technologies. In any case, new technologies need to position themselves in market economy, which means that a long-lasting funding (e.g. for more than 20 years) should not be taken into account. In terms of biogas processes, long-lasting funding took place and market demands were not considered in most cases, corresponding solutions for the future technology development shall be addressed in this article. For example, in order to identify substrate potentials (especially from waste materials) for biogas production in connection with suitable operational strategies for different parts of the Baltic Sea region, two EU financed BSR programme research projects were performed—regional mobilizing of sustainable waste-to-energy production (REMOWE) and ABOWE (implementing advanced strategies for biological utilization of waste) [4, 5]. The technological and economic efficiency analyses of the biogas development were fulfilled within above-mentioned projects. Therefore—besides to operational data—costs (investment and operating costs) and expected revenues from the products were considered, because efficiency analyses constitute the basis for the process and evaluation of decision-making process [6]. A standard method for the execution of company evaluations is the calculation of discounted cash flow via assessing the value of future cash flows via discounting to the valuation date and deduction of the initial net investment [7]. The energy demand, which is presented by load profiles in the article, shows the difficulty of the energy management from different renewable energy sources like wind and solar power. From the authors point of view, an adapted feeding and demand-driven operation of biogas plant provides a possibility to solve this problem.

2. Renewable energy production: initial situation

The production of renewable energy currently plays an important role in the European Union and will gain importance in the future. Renewable energy sources are wind power, solar power, hydroelectric power, tidal power, geothermal energy, biomass and renewable parts of waste. Between 2003 and 2013, the production of renewable energy within the European Union increased by 84.4%. Thereby, the most important source was biomass and renewable waste. The renewable electric energy production had a share of 25.4% of the gross electricity consumption in the EU. Here the main growth could be registered in three renewable energy sources, which are wind turbines, solar power and biomass [8]. In Germany, the production of renewable energies had a share of 30% of the gross electricity production; in Lithuania, a rate of 16% is reached. Until 2025 the share, for example, in Germany should be increased up to 40–45% and until 2035 up to 55–60% [9].

2.1. Load profiles vs. renewable energy from wind and solar power

The term load profile relating to electricity describes a detailed listing of the yearly energy consumption or demand [10]. That means the load profile provides a pattern of the electricity usage of one or more and also different customers. Herewith, the current collection will be projected at 15-min intervals. Unless the energy consumption is not being measured directly, there are standard load profiles available, which will be used for forecasting calculations [11]. Based on measured data or standard load profiles, the yearly electricity demand of a certain region or customer group can be determined. An example of an annual load profile is being shown in **Figure 1**. The figure constitutes a simplified version. The curve was straightened and therefore shows no 15-min-pattern.

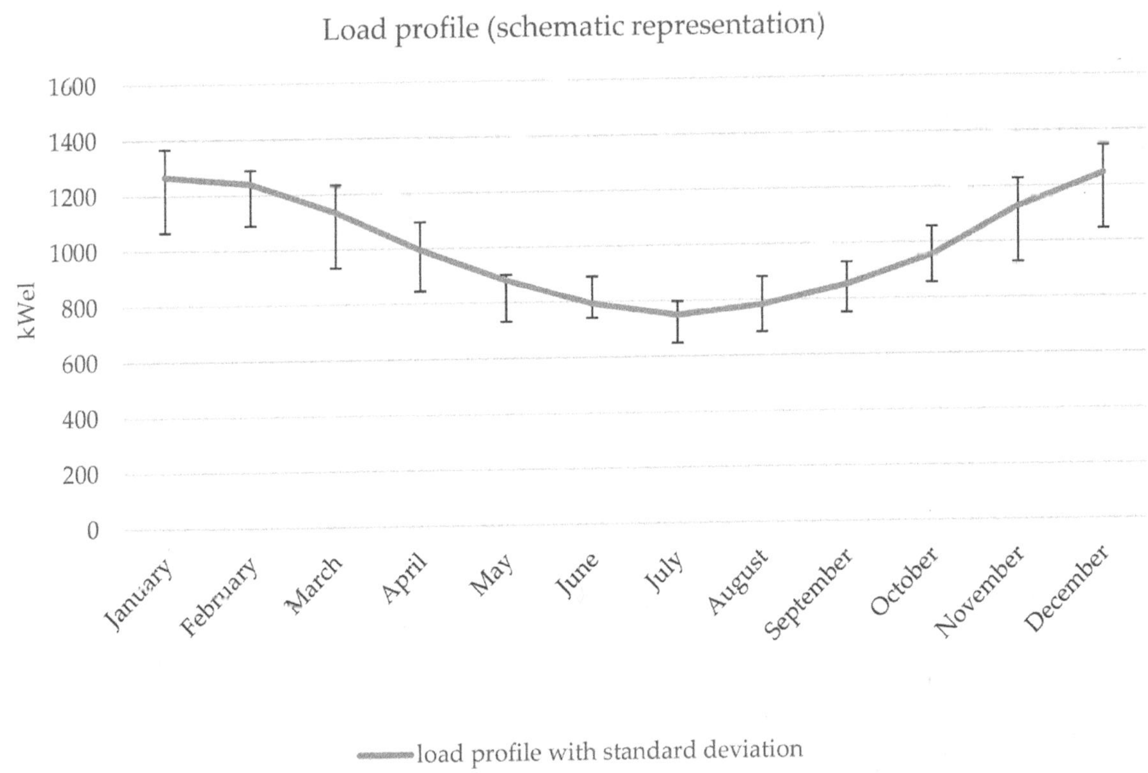

Figure 1. Load profile electricity demand.

As shown in **Figure 1**, the electricity demand is not evenly during the year. In current energy management concepts, the fluctuating demand is being compensated by the quick ramp-up of natural gas power plants [12]. The production of electricity from renewable energy sources had a high growth in between the last year. The relative share of electricity in the total quantity of electricity which was produced from biomass (incl. renewable waste) rose to 17.8%, the share of wind power to 26.5% and the share of solar power to 9.6% [8]. Anyhow wind and solar power are fluctuating producing electricity systems. Energy storage (for wind and solar power) is one of the possibilities to solve this problem [13]. Because of the increasing rate of fluctuating energies, it becomes more important to modify the electricity system, respectively, to find possibilities for energy storage or adapted energy production. **Figure 2** shows in an exemplary presentation how the covered amount of electricity produced by wind and solar power could exemplary look like. Data about the share of these techniques on the total energy demand are herewith being considered in percentage rates.

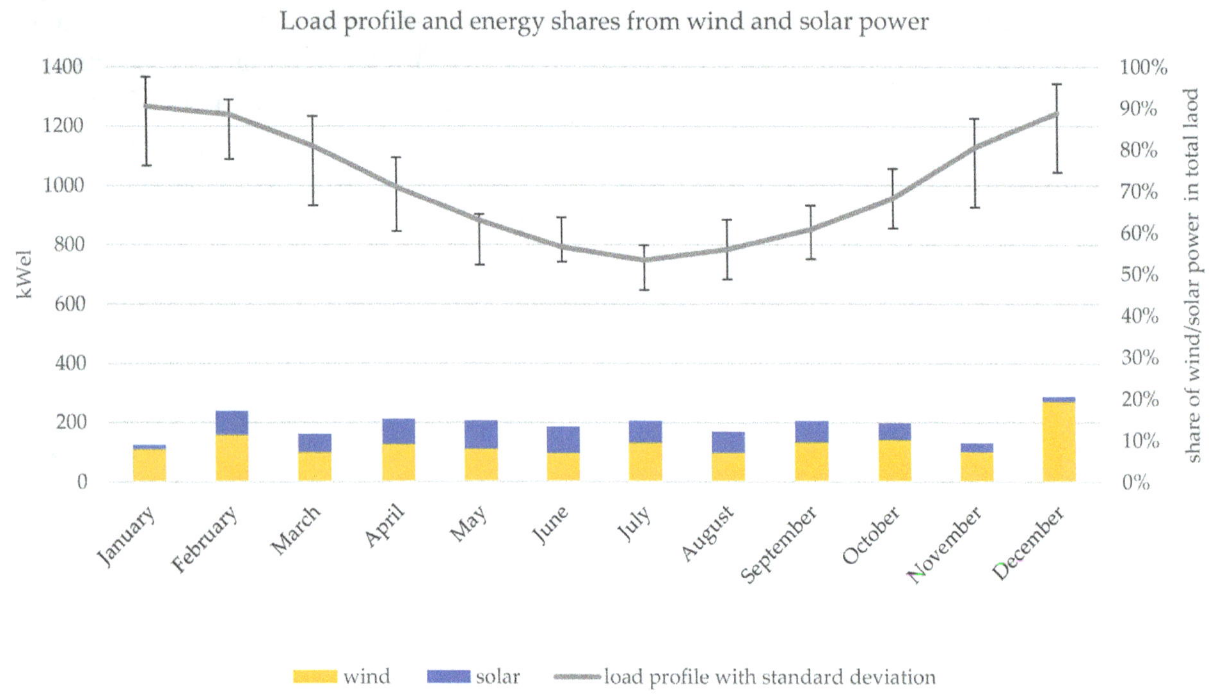

Figure 2. Load profile and energy shares from wind and solar power, data from Refs. [11, 14], adapted.

The annual load profile shows the variations in the electricity demand, which are not consistent. The problem of the electricity from renewable sources becomes obvious as it has been described above. Energy production from wind and solar power cannot be adapted, and it is fluctuating. The operation of biogas plants and production of biogas could be adapted with suitable operational methods, but nowadays the energy production is mostly even at a 24/7 rate. In order to solve the problem of adapted electricity production from renewable sources, there are essentially two possibilities available: (a) to store it or (b) to operate biogas plants adapted to the energy demand. Option (b) means that the production of electricity from biogas provides more features than wind or solar power. The produced biogas or also the produced electricity could be stored for later use. Moreover, a biogas plant provides the option to be operated energy demand oriented. Here the production of electricity depends on the feeding quantity, and therefore, an adapted feeding would lead to a demand-adapted production of electricity.

3. Substrate demand for adapted biogas scenarios

For production of biogas, many different substrates are used [3]. Primarily, manure or sewage sludge was used, but the kind of substrates used for biogas production changed in the last years. Many organic materials, for example, by-products in industrial or agricultural processes or also organic fractions from wastes like residual waste or biowaste from households, can be used. Many data concerning new substrates were collected during European-funded projects REMOWE and ABOWE at Ostfalia University [15]. Because some substrates are arising fluctuating and also with changing biogas yields, a targeted substrate management for an adapted operation of biogas plants is required.

3.1. Optimal operation of biogas plants adapted to the annual load profile

Based on load profiles (graph of the variation in the electricity demand of a region/city/single electricity consumer or other), the operation of a biogas plant can be planned and adapted, so that the electricity will be produced demand-based. Load profiles are provided by the power supplier or can be determined by use of standard load profiles. The substrate amounts, which are necessary for the calculation, will be determined on the basis of regional structural data or known arising substrates. As a conclusion, the exact quantity of demanded energy can be determined. Consequently, the biogas plant can be operated according to a well-structured substrate management.

3.2. Substrate management adapted to the annual load profile

Ostfalia University performed extensive studies concerning the energy demand and the adapted operation of biogas plants within several non-published student thesis works, including the consideration of the substrate demand. Herewith, the load profiles of different energy demands were used for the determination of regional-specific load profiles. The results of an exemplary adapted operation are shown in **Figure 3**.

In **Figure 3**, the exemplary adapted feeding of a biogas plant with biowaste and corn is demonstrated. The biowaste, which arises regularly but not evenly and furthermore with different biogas potentials (dependent on the time of year), has to be fed constantly because it cannot be stored for a longer period.

Therefore, energy crops like ensiled corn, which can be stored over a longer period of time, are being used for the additional feeding. The amount of corn, which has to be fed to reach the amount of the energy demand (based on an exemplary load profile), is being calculated in reference to laboratory substrate assessment results.

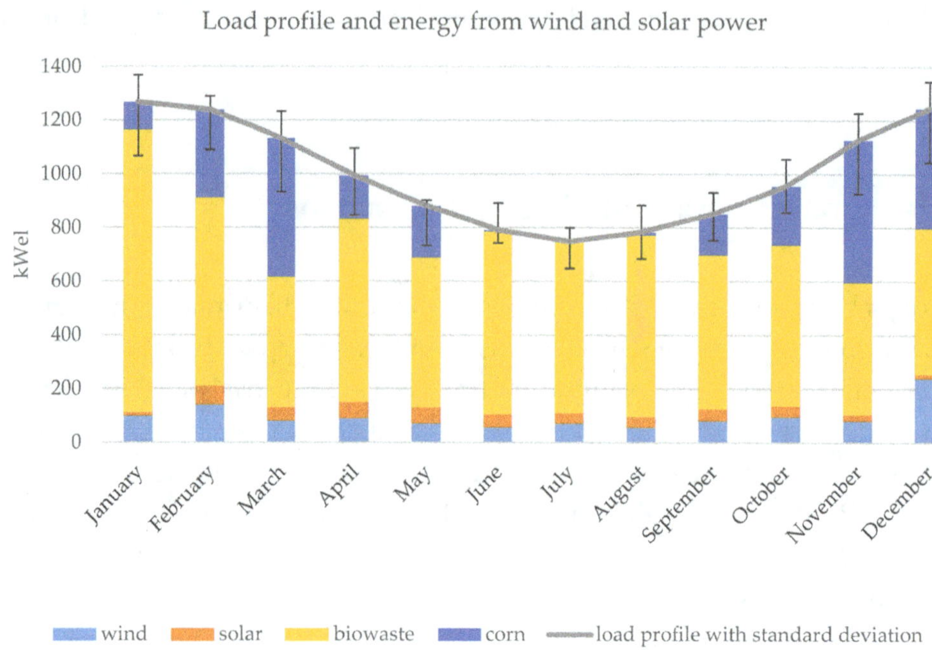

Figure 3. Load profile vs. energy potentials from wind, solar and biomass, data from Refs. [11, 14] and own calculations.

4. Biogas plant operation: from laboratory tests to large-scale plant operation

In order to be able to estimate the gas potentials of pure substrates as well as from substrate mixtures, practical data assessment is required. The following explanations are about samples being exemplary chosen by their availability in the Šilutė region of the country of Lithuania.

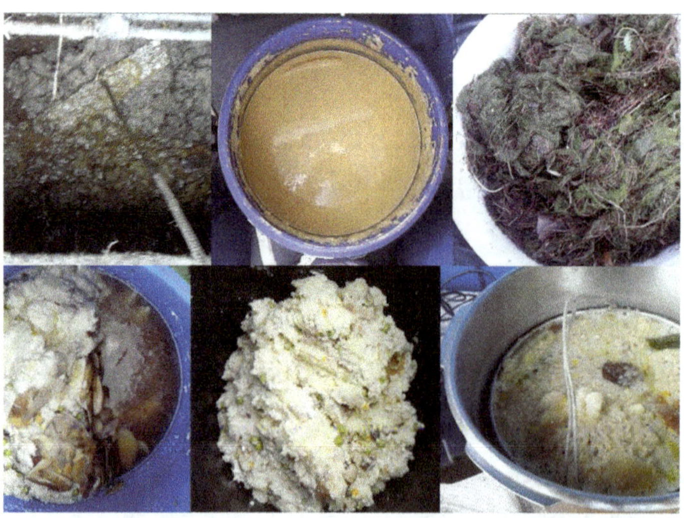

Figure 4. Substrates used in Lithuania. Cow manure (top left), distillery leftovers (top middle), algae (top right), food waste (original, mashed, sanitized) (from bottom left to right) [16].

The Šilutė region is widely settled by farmers mostly cow farms; furthermore, there is a spirit distillery factory which produces considerable amounts of bioethanol process leftovers and residues of food (food waste) from catering business. Other sources for biomass are macro- and microalgae. These have been collected at different locations at the Curonian Lagoon near Klaipeda. The cow manure was delivered by a local farmer, and the distillery leftovers were provided by a local spirit distillery factory. The food waste has been collected in Lithuanian kindergartens.

The substrates are shown in **Figure 4**, and the physical properties are listed in **Table 1**.

Substrate	Methane potential (m_N^3 CH_4/Mg FM)	Methane content (%)	Dry matter content (DM) (%)	Organic dry matter content (oDM) (%)
Food waste	85	57	23.5	22.7
Distillery leftovers	40.5	57	12.3	11.5
Algae, fresh	14.4	55	22.8	14.1
Cow manure	19.3	57	24	10.7

Table 1. Physical properties of the substrates used for the scenario in Lithuania.

4.1. Laboratory tests: batch and continuous

For the determination of anaerobic fermentation data of different substrates, two basic steps have to be done. These are batch fermentation tests to determine the maximum biogas potential and continuous tests to investigate the long-term fermentation behaviour of the tested substrate or substrate mixture. The batch tests are performed in 5-l flasks over a period of 35 days. The vessels for the continuous tests have a volume of 15 l and are being operated constantly over a longer period of time. In **Figures 5** and **6**, batch and continuous fermentation reactors in laboratory scale can be seen.

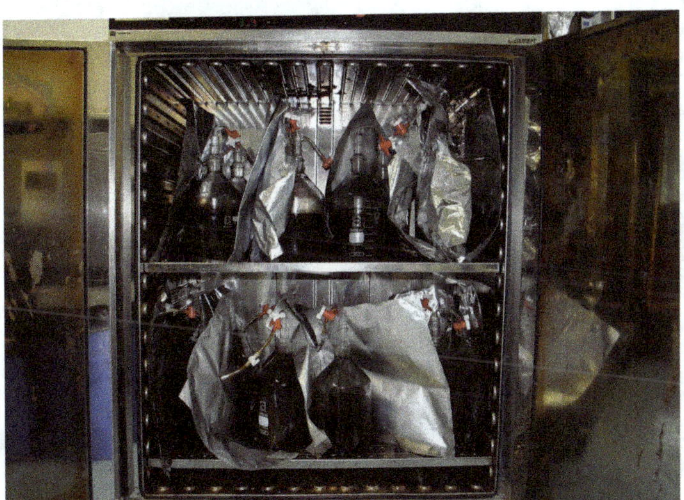

Figure 5. Batch tests in heating cabinet.

Figure 6. Continuous lab fermenter.

4.2. Pilot-scale digester

The pilot plant used during the research is a 550-l plug flow dry digester that was designed for long-term continuous operation to estimate the biogas potential of various substrates. The digester is situated inside a 20-ft container that is equipped with all necessary laboratorial equipment for daily diagnostic of anaerobic digestion conditions, and thus, this pilot plant is totally mobile (**Figures** 7 and **8**).

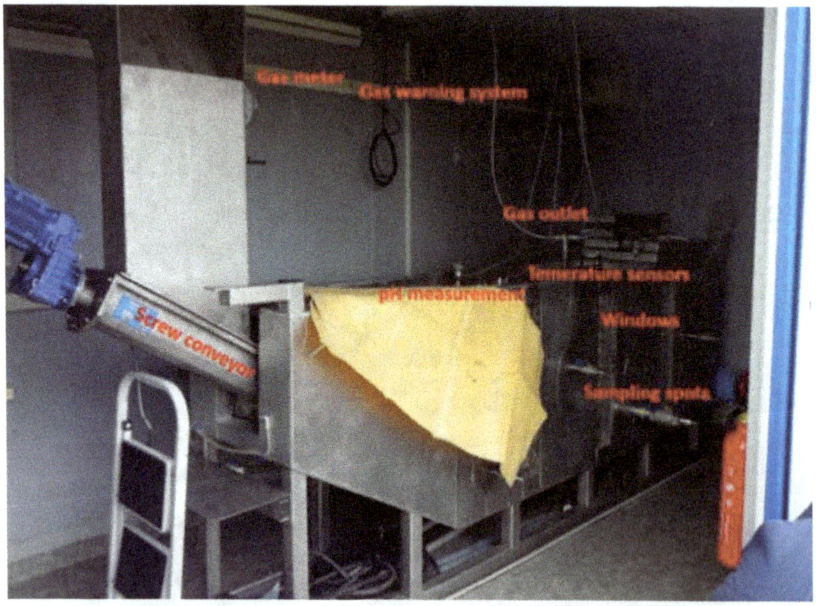

Figure 7. Plug flow dry digestion fermenter with components [16].

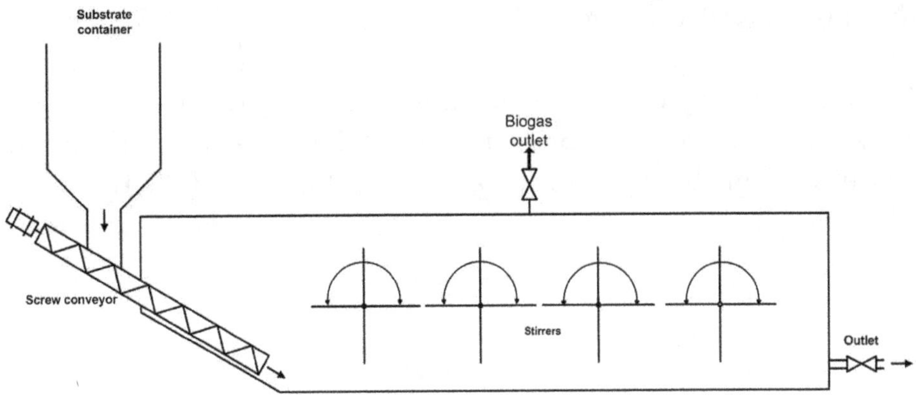

Figure 8. Schematic figure of the pilot plant [17].

Due to its size, real-size material (as for full-scale plants) can be used. This allows the process simulation of full-scale biogas plants.

5. Experimental proof of concept for fermentation strategies

The information obtained from plant operation is compared with the results of a parallel laboratory analysis of the substrates used during the testing period regarding materials and methods. **Figure 9** represents the methane yields of the continuous tests and the expected yields calculated from the batch tests. The results show that these tests correlate to each other. Therefore, the comparability of the continuous tests—relating to the benchmarking of fermentation performances—with the estimated results from previous batch tests is proven [16].

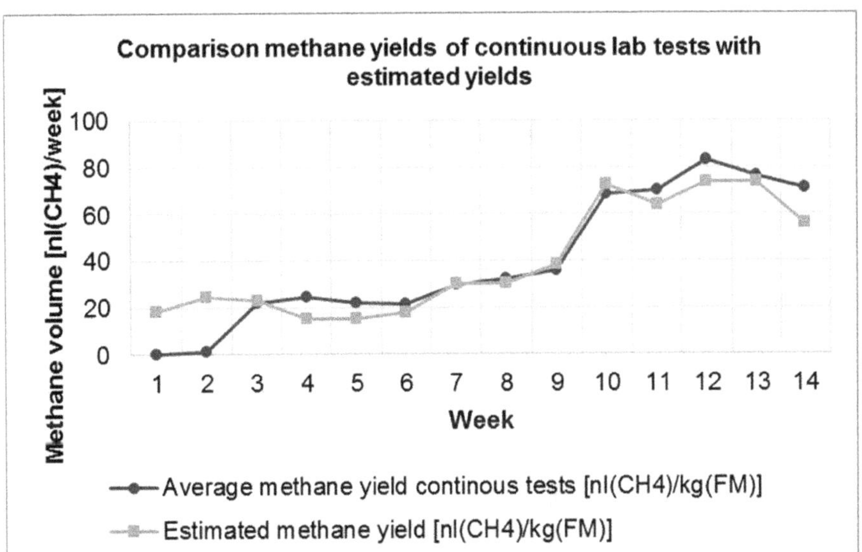

Figure 9. Comparison of methane yields of continuous lab tests with estimated yields calculated from batch tests (substrates: cow manure, food waste, algae, distillery leftovers) [16].

As well as the continuous lab tests, also the measured methane yields from the pilot plant tests correlate to estimated methane yields. The results are shown in **Figure 10**. Accordingly, the comparability of the pilot plant with estimated results from previous performed batch and continuous tests is proven. Because of previous overfeeding, the amount of methane produced in the first 4 weeks is high, but the almost parallel development of the two curves is clearly visible and obvious [16].

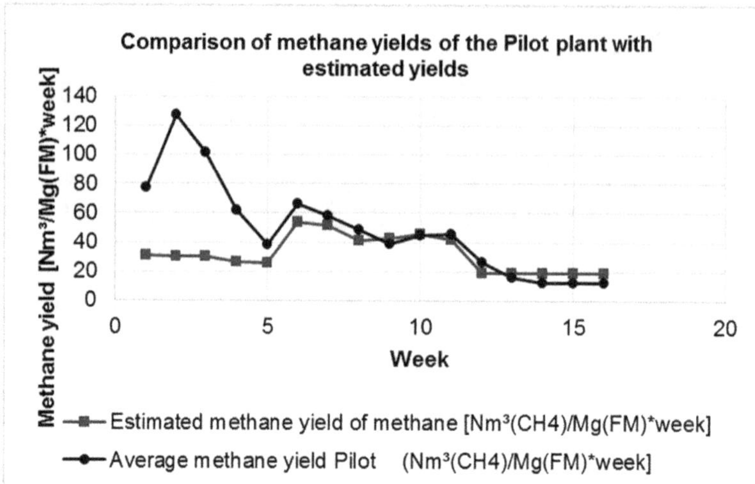

Figure 10. Comparison of methane yields of pilot plant with estimated yields calculated from batch tests (substrates: cow manure, food waste, algae, distillery leftovers).

Finally, the methane yields of the pilot plant were compared with the yields from a German full-scale biogas plant under comparable substrate and operation conditions. As **Figure 11** shows, the comparability of both plants was proven as well. The results of the full-scale plant were thankfully provided by the plant operator (input engineers) and adapted by Bansen [18].

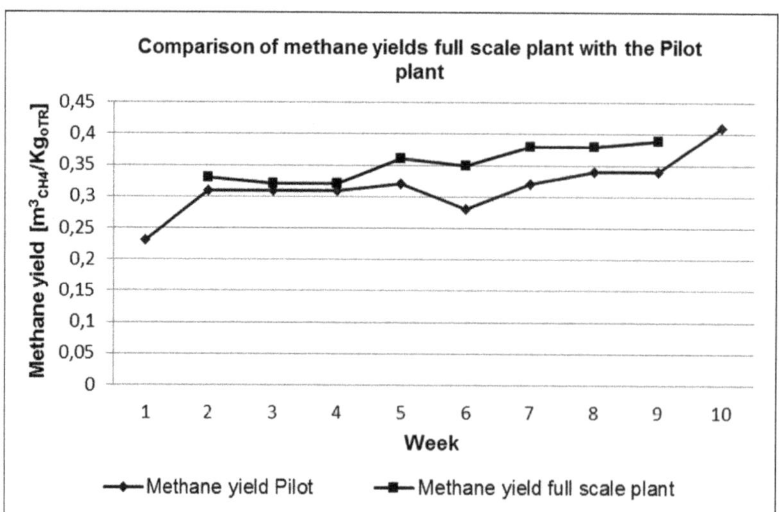

Figure 11. Comparison of methane yields of full-scale plant vs. pilot plant.

The experimental conditions for the lab-scale as well as pilot-scale tests are shown in **Table 2**.

	Temperature (°C)	Retention time (days)
Continuous lab test	42*	56*
Pilot-scale test	42*/55**	50–333*/40**
Full-scale plant	55	40

*For Lithuanian tests;
**for comparison of pilot-scale with full-scale plant.

Table 2. Experimental conditions of lab-scale, pilot-scale and full-scale tests.

A successful process simulation of the fermentation in pilot scale vs. the full scale has been also obtained by Cavinato [19] biogas production tests in 2010 and by Scano et al. [20] where fruit and vegetable wastes were used. Besides applicable experimental results, an economic analysis showed the possible profitability of the planned biogas project. It is important to emphasize that an improvement in the biogas yield was assessed by the operation of the pilot-scale fermenter, which means that the operation of plant engineering in pilot scale is absolutely suitable for technical and economic simulation and development of upscaling scenarios.

6. Financial implementation of biogas technology under market economy conditions

The implementation of biogas technology without any public funding provides the basis for the following considerations concerning the profitability. Implementing biogas technology requires the assessment of many aspects concerning the economic efficiency. The main financial and economic aspects of biogas plant exploitation include investment costs, plant operating cost and possible revenues achieved by the production and use of biogas, utilization of wastes and use of digestate as fertilizer. The resulting operative cash flow (OCF) calculation is based on the difference between annual revenues and annual expenses as well as investments. The discounted cash flow (DCF), that is, the calculation of future cash flows' present value, was a result of the annual OCF multiplied by a discount factor.

The result for the investor is just the amount of money, which could be received after a period of time. Thereby, the time value of money has to be taken into account [21]. The possible cost factors of biogas plants depend on the different sizes, and noticeable biogas plant characteristics concerning the size of the plant and the substrates, which will be used as input materials, are specified.

6.1. Cost factors

For building the biogas plant, investment costs as well as operational plant costs (both in extracts) have to be considered:

- Investment costs: engineering, permission of the authority (application costs), connection to the public grid, land costs, vehicles, offices and functional units such as substrate delivery and pre-treatment, digester, gas storage, biogas treatment, CHP unit, pumps, piping, digestate storing and others.
- Operational expenses: substrate costs, costs for analysis, process energy, consumables, costs for maintenance and repair, labour costs, land costs and others [3].

In particular, the specific investment costs vary, depending on the size of the biogas plant [22].

For the calculation of average investment costs of biogas plants, key values were determined by referring to investment costs of different German agricultural biogas plants and publications during project ABOWE [4, 16].

The operation of biogas plants with substrates, which require special treatment like municipal solid waste or biowaste, causes additional costs for pre-treatment and other necessary modifications. The use of household biowaste especially demands special treatment such as the sanitation of the material [16]. The hygienisation of biowaste used for anaerobic treatment is regulated by EU hygiene regulation (VO 1774/2002/EG) [23] and country-specific by, amongst others, the German Biowaste Ordinance [24]. Accordingly, biowaste could be sanitized by heating it up to 70°C for a time period of 1 h. Consequently, it has to be taken into account that investment costs for biogas plants with plug flow fermenter and the use of biowaste as substrate are about one-third higher than for the use of renewables [25]. For biowaste or MSW, the garage fermenter system makes biogas production with less pre-treatment of the substrate possible, which potentially reduces the operational as well as the investment costs.

In any case, the specific investment costs tend to decrease with larger plant capacities. When assessing the different types of the investment expenses, it becomes obvious that part of the costs for planning and construction of the biogas plant is personnel expenses. These costs should be considered separately, because, as it was pointed out in project ABOWE, there are considerable variations in different countries [4].

Another point is that some parts of the biogas plant have to be replaced regularly, because certain components such as pumps, stirrers and also the CHP unit have a short operational lifespan. The estimated lifetime of pumps is 4 years and that of CHP units about 6 years [26].

One of the most important aspects when implementing biogas technology is to ensure reliable availability of substrates. Biogas plants demand a continuous supply with substrates. Furthermore, the utilization of the produced energy or the conditioned biogas itself as well as the resulting heat or the electric energy generated by the CHP unit need to be guaranteed.

In addition, it has to be taken into consideration that substrates with a high-energy potential should be used, so that costs and expenses for transport are kept low [4, 16].

Considering further economic aspects, the biogas plant (pumps, stirrer and others) itself demands electrical energy amounting to 5–20% of the total of electrical energy produced by CHP technology [25]. The heat demand of the biogas plant (for the heating of the fermenter), which is about 5–25% [25] (according to FNR even at 28% [22]) of the produced heat, can be covered by the heat of the CHP unit. Here it might be economical to sell all of the energy produced and buy the energy needed from the national energy supplier (depending on the feed-in tariff of the electricity generated) [27].

6.2. Country-specific aspects

For the economic and financial implementation, the different cost items vary according to the country in which the biogas plant shall be built.

The calculation done in the project is based on specific data, which vary according to the relevant countries. Therefore, some prior general considerations are necessary:

- Investment costs: it has to be considered which parts of the plant can most cost-effectively be manufactured in the country in which the plant shall be operated.
- Operational costs: most specific costs depending on the relevant countries; the personnel costs in particular are varying strongly.
- Revenues: the feed-in tariffs for electricity and heat are country-specific, also the price for the sale of digestates [4, 16, 17].

General data were collected with respect to the investment and operational costs of biogas plants in EU developing countries and furthermore with data of plant operators as well as plant construction companies.

Based on all above-mentioned aspects, results of the pilot plant operation in certain EU developing and developed countries, lab tests at Ostfalia University and a numerous collection of data (own investigations, literature sources, data from biogas plant operators), many key values were determined and used for the calculation of the operative cash flows over a period of 20 years. The cumulative discounted cash flow then gives an impression of the profitability of a planned biogas project. Spoken fundamentally, a detailed calculation and estimate of the operative cash flows are only possible by defining real system models. Based on these data (referring to commercial offers) and on the investigation of the economic aspects of a planned biogas plant, detailed cash flow calculations are possible.

7. Full simulation of a practical application: implementation of a biogas plant from energy demand to profitability

For illustration of above-constituted considerations concerning the implementation of biogas technology with no public funding, necessary substrate management and load profile adapted operation a scenario were developed. Substrates were used for a complete implementation scenario of biogas technology in Lithuania.

Herewith, food waste, distillery leftovers and algae serve as basic substrates for the production of biogas during the year. Cow manure is available as additional substrate and will be fed as compensation substrate to get an adapted feeding for the production of the needed electricity.

According to the previously described strategy, the following amounts (in **Table 3**) of substrates and methane potentials were used for further calculations. For the scenario, the size of the biogas plant was chosen with 500 kWel.

Substrate	Methane potential (Nm^3 CH_4/Mg FM)	Quantity used (Mg/a)
Food waste	85	8000
Distillery leftovers	40	5000
Algae, fresh	14	2000
Cow manure	19	10,000

Table 3. Substrates used for an implementation scenario (Ostfalia lab).

Table 3 shows the biogas yields, which arose from batch tests of food waste, distillery leftovers and algae. Starting from the assumption that the waste is arising constantly with a constant biogas yield (algae occur from April/May to September/October), the electricity production would be evenly because of a constantly daily feeding amount. The load profile, which constitutes the electricity demand, would therefore not be covered (sum of electricity from food waste, distillery leftovers and algae). Consequently, cow manure will be used as additional feeding material, see **Figure 12**.

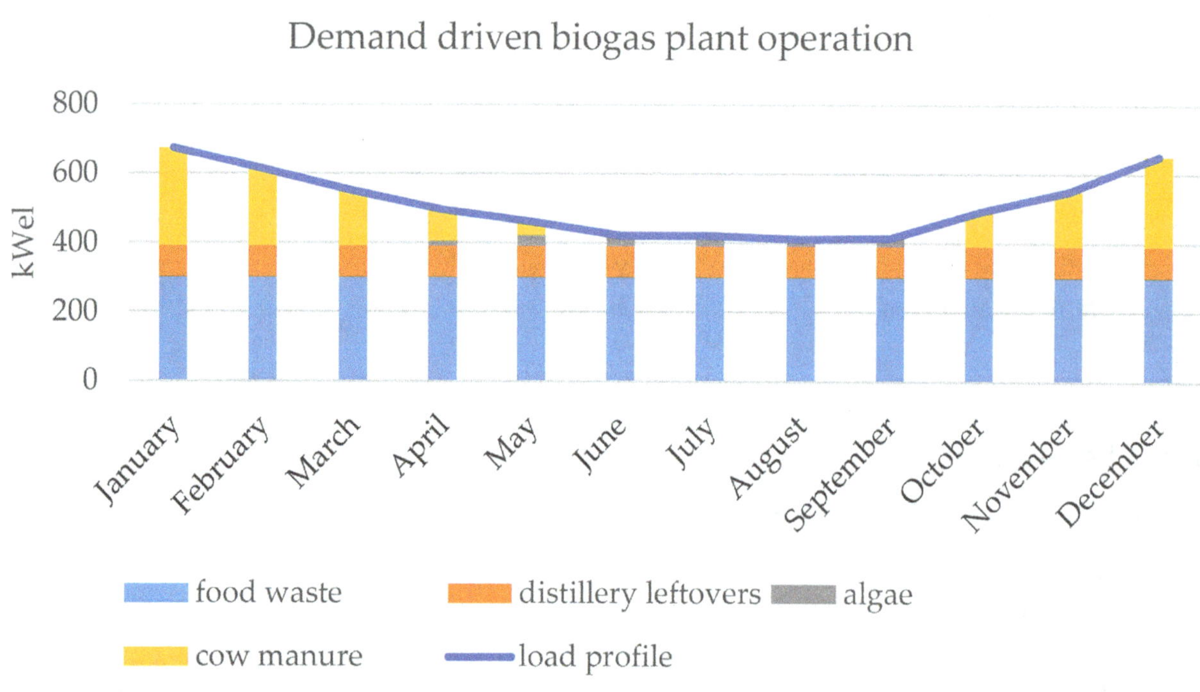

Figure 12. Power from methane production (scenario).

Based on the results of the performed batch tests and the developed scenario, the substrate amounts and mixtures were tested in continuous fermenter tests. The long-term fermentation behaviour is therefore proven, before tests with these substrates will be proceeded in the pilot-scale fermenter. The results of this test procedure lead to the planning of a demand-driven supplied biogas plant by meanings of an adapted production of electricity, no demand for long-time storing capacities and strict avoiding of energy undersupply.

The corresponding calculations were done in reference to basic data, (especially concerning the substrates and feed-in tariffs and other incentives). These data concerning substrate amounts and tariffs are stated by the prospective customer or need to be determined. Therefore, all data are exchangeable and can easily be adapted to other scenarios besides the exemplarily chosen Lithuanian scenario. Assumptions made for the following calculations are summarized in **Table 4**.

	Tariffs (€/Mg)	
	Costs	**Income**
Substrates:	8.66	20
Food waste	–	–
Distillery leftovers		25–50
Algae		
Cow manure		
Digestate		5
Electricity	Public grid: 0.126 (€/kWh_{el}) or less	Feed-in: 0.09 (€/kWh_{el})
Heat		0.04–0.06 (€/kWh_{th}) (by season)
Produced energy	3,929,083 kW_{el}/a and 3,929,083 kW_{th}/a; efficiency rate η_{el} = 40%	
Digestate	22,646.55 Mg/a	
Biogas plant: plug flow fermenter system, investment costs plus 30% for waste hygienisation (assumption based on Ref. [25]).		
CHP unit, 1 × 500 kWel, 1 × 250 kWel: production of electricity and heat, each re-invested after 72-month operating time.		
Substrates: food waste, distillery leftovers, fresh algae, cow manure.		
Sale of electricity and heat (minus own demand).		

Table 4. Tariffs and further background assumptions.

The operational strategy resulting from **Figure 12** requires a flexible operation of the biogas plant. Thus, the biogas plant should be operated with two CHP units, one with an installed power of 500 kWel and another with 250 kWel. The 500 kWel system is in continuous operation and the other is operated in modulation and on-demand. In any case, a suitable dimensioned gas storage facility needs to be provided.

With respect to this substrate scenario, an exemplary financial implementation of the biogas system is developed. The results of the corresponding economic calculations are shown in

Figure 13 as a cumulative discounted cash flow curve progression starting with the initial investment costs and cumulating the operative cash flows over a period of 20 years.

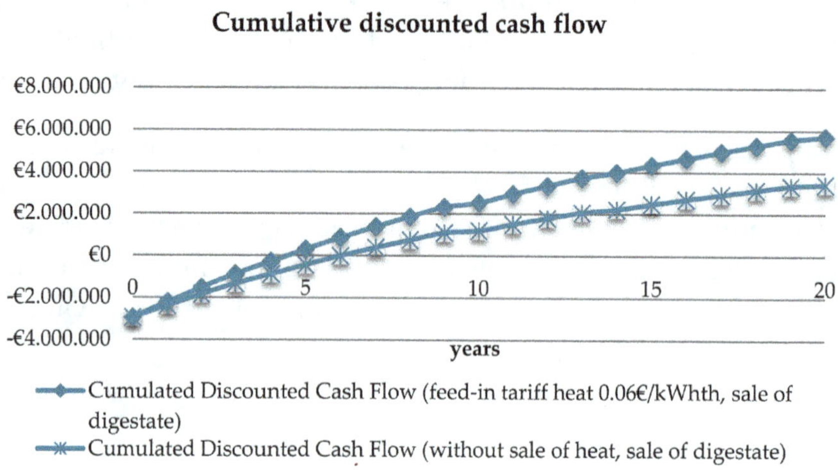

Figure 13. Cumulative discounted cash flow—comparison of two exemplary sale scenarios.

Assuming that the biogas plant operator gets gate fees from the acceptance of waste and all of the produced electricity and heat (minus own heat demand) will be sold, the biogas plant operation reaches the zero line after about 4.5 years. The second curve results in the supposition that the heat could not be sold. Here the zero line would be reached after about 6 years. This example of the variation of just one parameter shows that it is really easy for the user to create scenarios, to vary parameters, to find the most influencing parameters and to get first information about the profitability of a planned project.

The chosen scenario as well as the calculated model biogas plant constitutes a theoretical exemplary situation. The corresponding assumptions are not being meant for a generalized use, because they depend on scenario-specific aspects and were just chosen for the exemplary developed scenario. Nevertheless, all economic influencing variables can easily be changed and adapted to other scenarios and regions, and the resulting impacts are illustrated in the same way as described above. This calculation model provides a good impression of the profitability of a planned biogas implementation project. Therefore, the essential feature of the presented model is a suitable opportunity for easy and rapid profitability calculation for the usual users, respectively, investors such as farmers or rural area authorities.

8. Perspectives towards sustainability and impact to global environmental problems

From the sustainability and circular economy point of view [28], the idea of the biogas production from biowaste is in line with industrial symbiosis mechanism that is essential for achieving social-economic and environmental benefit [29]. Except economic and social aspects that have close relationship and depend on profitability and subsequent improvement in the quality of life, the environmental aspects should be discussed as well. LCA is often used as a tool to evaluate a process, service or product impact to the global environmental problems. The method of LCA is described in ISO standards [30] and starts from the finding of the study system boundaries. The magnitude of the environmental impact of the biomass anaerobic digestion and CHP unit depends on numerous factors, such as fermentation conditions (mesophilic, thermophilic), kinds of substrates (cow manure, bioethanol residue, algae, food wastes), biogas yield, energy demand for the own process of digestion and sanitation, transportation, digestate storage and method of utilization and climate condition. The scientific results related to influence of the renewable energy production and transportation processes to global problem, especially greenhouse effect, sometimes differ a lot [31–33]. Nevertheless, there are some common features, which allow determining the main tendency of the above-mentioned processes and products contribution towards sustained environment.

Global warming potential. The main impact of the GHG emissions is expected from biogas combustion at CHP and methane emissions from the storage of the digestate. Anaerobic biomass digestion itself does not pollute environmental air and does not use oxygen/air for biogas production.

Ozone depletion potential. During biogas burning process at the energy production plant, some amount of CFC/HCFC can get in to the air. In addition, the manufacturing of the fermenter and energy plant could cause CFC/HCFC release in the atmosphere.

Photochemical oxidation potential. This kind of impact relates to the secondary methane emission source—emission from the digestate storage.

Acidification potential. Environmental acidification potential expressed as sulphur dioxide equivalent is the biggest for hydrogen sulphide and ammonia and mostly connected with ammonia emissions due to the digestate storage.

Eutrophication potential. Nitrogen and phosphorus compounds contribute to an increased biomass production in aquatic environment, and these results in additional oxygen consumption for biomass decomposition. Release of ammonia from the digestate storage can increase the eutrophication potential.

Abiotic depletion potential. This kind of depletion can be decreased due to the usage of digestates instead of traditional fertilizers. Anaerobic digestion demand of additional energy consumption increases abiotic depletion, but use of traditional fossil fuel for traditional energy production plant does not reduce abiotic depletion as well. The recycling of the fermenter and CHP plant constructions after usage prevents damage for the environmental and resources.

The environmental impact assessment does not take the digestate spreading into consideration because spreading of fertilizers exists in any case, but the method of the spreading can have significant impact to soil acidification and aquatic body eutrophication. The evaluation of environmental aspects of the study shows that biomass anaerobic digestion and CHP system for energy production have savings in global warming and abiotic depletion potentials. Scientific publications [34–36] regarding biomass anaerobic digestion and energy production support present study environmental results and recommendations.

9. Conclusions

9.1. Technical conclusions

Digestion technology adapted for a Lithuanian scenario with lab-scale examination in batch and continuous tests was realized. Four different kind of substrates were applied: cattle manure, bioethanol distillery waste, food waste and Curonian Lagoon algae.

The similarity of the fermentation process in lab-scale, pilot-scale and full-scale plant was shown.

The technological adequateness of biogas technology for substituting fossil energy resources with respect to fluctuating energy demand and fluctuating supply of renewable energy has been demonstrated, and the possibility of demand-driven operation of a biogas plant has been proven.

9.2. Financial-economic conclusions

The financial-economic aspects as the most important part of the chain of waste-to-energy advanced strategy have been confirmed.

The feasibility study based on certain regional data was detected as the most realistic for biogas development benefit for local society and environment.

A detailed perspective into the strategy of digestion for biogas production is given. Starting with energy demand, load profiles and substrate demand, a detailed substrate management has been performed. Based on this, the profitability and thereby financial implementation without any public funding could be considered. Herewith, a special focus is given in the potential implementers' and investors' demands, which shall be enabled to take reliable decisions on technologies for implementing advanced digestion technology as full-scale plants. With the corresponding developed strategies (experimental as well as economical), new possibilities of producing renewable energy can be implemented and fostered in the Baltic Sea region as well as in other regions of the EU.

Author details

Thorsten Ahrens[1*], Silvia Drescher-Hartung[1] and Olga Anne[2]

*Address all correspondence to: th.ahrens@ostfalia.de

1 Institute for Biotechnology and Environmental Engineering, Ostfalia University of Applied Sciences, Wolfenbüttel, Germany

2 Department of Natural Science, Klaipeda University, Klaipeda, Lithuania

References

[1] Huopana, T. et al: A regional model for sustainable biogas production. Case study:North Savo, Finland, REMOWE project, Savonia University of Applied Sciences (publisher), Kuopio. 2012. ISBN 978-952-203-169-3

[2] Luostarinen, S. (ed.): Energy potential of manure in the Baltic Sea region: biogas potential & incentives and barriers for implementation. Knowledge Report: BALTIC-MANURE WP6 Energy Potentials of Manure. 2013. Available from: http://www.baltic-manure.eu/download/Reports/bm_energy_potentials_web.pdf [accessed: June 2016]

[3] Fachagentur Nachwachsende Rohstoffe e.V. (FNR): Guide to Biogas-from Production to Use. 232 pp. Available from: http://mediathek.fnr.de/media/downloadable/files/samples/g/u/guide_biogas_engl_2012.pdf [accessed: June 2016]

[4] ABOWE project reports, 2014. Available from www.abowe.eu

[5] REMOWE project reports, 2012. Available from www.abowe.eu

[6] Ullrich, H.: economic planning and execution of process plants. Vulkan Verlag, Essen. 1996. 248 pp.

[7] Schacht, U., Fackler, M.: Practical handbook company evaluation. 2. Auflage. Springer Gabler, Wiesbaden. 2009. ISBN 978-3-8349-0633-5

[8] Eurostat, Statistics explained: renewable energy statistics. Available from: http://ec.europa.eu/eurostat/statistics-explained/index.php/Renewable_energy_statistics [accessed: June 2016]

[9] Bundesministerium für Wirtschaft und Energie (BMWi): The Renewable Energy Act, 2014. The key facts on the reform of the EEG. 2014. Available from: https://www.erneuerbare-energien.de/EE/Redaktion/DE/Downloads/Hintergrundinformationen/eeg-2014-infobroschuere-bf.pdf?__blob=publicationFile&v=9 [accessed: June 2016]

[10] Elexon: Load profiles and their use in electricity settlement. 2013. Available from: https://www.elexon.co.uk/wp-content/uploads/2013/11/load_profiles_v2.0_cgi.pdf [accessed: June 2016]

[11] BDEW (Bundesverband der Energie- und Wasserwirtschaft): Standardlastprofile Strom. Available from: https://www.bdew.de/internet.nsf/id/DE_Standartlastprofile [accessed: June 2016]

[12] DVGW/VDE 1. Munich Energy Days: electricity and gas: joint solutions to the pathway to renewable resources in energy supply. Energie, wasser-praxis. Interview. 2013. Available from: [http://www.muenchener-energietage.de/fileadmin/dvgw/angebote/berufsbildung/pdf/1301engelber.pdf [accessed: June 2016]

[13] Ferreira, H. L., Garde, R., Fulli, G., Kling, W., Lopes, J.P.: Characterisation of electrical energy storage technologies. Energy, 2013; 53: 288–297. Available from: http://ac.els-

cdn.com/S0360544213001515/1-s2.0-S0360544213001515-main.pdf?_tid=e34b1ade-2f02-11e6-a511-00000aab0f6c&acdnat=1465560193_bce0f75dddfbb0d2d2572de061785876 [accessed: June 2016]

[14] Burger, B.: Electricity production from solar and wind energy in 2011. Fraunhofer-Institute for Solar Energy Systems ISE. Available from: https://www.ise.fraunhofer.de/de/downloads/pdf-files/aktuelles/vortragsfolien-stromproduktion-aus-solar-und-windenergie-im-jahr-2011.pdf [accessed: June 2016]

[15] Behrendt, A., Drescher-Hartung, S., Ahrens, Th.: Optimization of biogas processes: European experiences. In: E. Dahlquist: Technologies for Converting Biomass to Useful Energy. Sweden: CRC Press; 2013, pp. 293–313. ISBN: 978-0-415-62088-8

[16] ABOWE project reports, 2013. Available from www.abowe.eu

[17] Drescher-Hartung, S., Anne, O., Ahrens, T.: Availability and characteristics of different waste substrates for biogas quality and potential in the Baltic Sea Region. In Book of Proceedings, Bioenergy 2013, Conference and Exhibition, Jyväskylä, Finland. pp. 221–225. Available from: http://www.iat.eu/files/book_of_proceedings_bioenergy2013.pdf [accessed: August 2016]

[18] Bansen, B.: Vergleich der Biogasanlage Hoheneggelsen mit einer Versuchsanlage im Pilotmaßstab zur Entwicklung von prozessspezifischen Kennwerten. Ostfalia University of Applied Sciences Braunschweig/Wolfenbüttel and INPUT engineers. 2010. Sehnde, Germany; not published.

[19] Cavinato, C., Fatone, F., Bolzonella, D., Pavan, P.: Thermophilic anaerobic co-digestion of cattle manure with agro-wastes and energy crops: comparison of pilot and full-scale experiences. Bioresource Technology, 2010; 101: 545–550. ISSN: 0960-8524

[20] Scano, E.A., Asquer, C., Pistis, A., Ortu, L., Demontis, V., Cocco, D.: Biogas from anaerobic digestion of fruit and vegetable wastes; experimental results on pilot-scale and preliminary performance evaluation of a full-scale power plant. Energy Conversion and Management, 2014; 77: 22–30.

[21] Investopedia: Discounted cash flow (DCF). Available from: http://www.investopedia.com/terms/d/dcf.asp [accessed: April 2016]

[22] Fachagentur Nachwachsende Rohstoffe e.V. (FNRa): Faustzahlen Biogas. Available from: http://biogas.fnr.de/daten-und-fakten/ [accessed June 2016]

[23] EU Hygiene Regulation: Regulation (EC) No 1774/2002 of the European Parliament and of the Council of 3 October 2002 laying down health rules concerning animal by-products not intended for human consumption. Available from: http://eur-lex.europa.eu/legal-content/EN/TXT/?uri=celex%3A32002R17742002 [accessed: June 2016]

[24] Federal Ministry of Justice and Consumer Protection: Regulation on the Recycling of biowaste on agricultural, forestry and horticultural used soil (Biowaste Ordinance-

BioAbfV). 1998. Available from: https://www.gesetze-im-internet.de/bioabfv/ [accessed: June 2016]

[25] HEI Hornbachner Energie Innovation. Available from: http://www.biogas-netzeinspeisung.at/technische-planung/ [accessed: May 2016]

[26] Personal information: Armin Weiss, Biokraft, Germany

[27] Al Seadi, T., Rutz, D., Prassl, H., Köttner, M., Finsterwalder, T., Volk, R., Janssen, S.: Biogas handbook. Denmark: University of Southern Denmark Esbjerg [web document]. 2008. Available from: http://www.lemvigbiogas.com/BiogasHandbook.pdf [accessed: June 2016]

[28] European Commission (2015) Closing the loop. an action plan for the European Circular Economy. Available online at: http://ec.europa.eu/environment/circular-economy/index_en.htm

[29] Ghisellini, P., Cialani, C., Ulgiati, S.: A review on circular economy: the expected transition to a balanced interplay of environmental and economic systems. Journal of Cleaner Production. 2015, xxx (online), 1-22. doi: 10.1016/j.jclepro.2015.09.007

[30] Life-Cycle-Based-Sustainability-Standards-Guidelines. Available from https://www.pre-sustainability.com/download/Life-Cycle-Based-Sustainability-Standards-Guidelines.pdf [accessed: June 2016]

[31] Fusi, A., Bacenetti, J., Fiala, M., Azapagic, A.: Life cycle environmental impacts of electricity from biogas produced by anaerobic digestion. Front Bioeng Biotechnol. 2016; 4: 26.

[32] Olsson, L.: Sociotechnical system studies of the reduction of greenhouse gas emissions from energy and transport systems. Dissertations. 2015; 54. doi: 10.3384/diss.diva-116685 (ISSN 0345-7524; 1656, Linköping University Electronic Press)

[33] Holm-Nielsen, J. B., Al Seadi, T., Oleskowicz-Popiel, P.: The future of anaerobic digestion and biogas utilization. Bioresource Technology. 2009; 100: 5478–5484. doi: 10.1016/j.biortech.2008.12.046

[34] Whiting, A., Azapagic, A. Life cycle environmental impacts of generating electricity and heat from biogas produced by anaerobic digestion. Energy. 2014; 70: 181–193. doi: 10.1016/j.energy.2014.03.103 [accessed June 2016]

[35] Poeschl, M., Ward, S., Owende, P.: Environmental impacts of biogas deployment—part I: life cycle inventory for evaluation of production process emissions to air. Journal of Cleaner Production. 2012; 24: 168–183. doi: 10.1016/j.jclepro.2011.10.039

[36] Al Seadi T., Lukehurst C.: Quality management of digestate from biogas plants used as fertiliser. IEA Bioenergy; 2012, May, 1-40. Available from www.iea-biogas.net/files/datenredaktion/download/publi-task37/digestate_quality_web_new.pdf [accessed June 2016] http://www.iea-biogas.net/files/daten-redaktion/download/publi-task37/digestate_quality_web_new.pdf

2

Metal Removal by Seaweed Biomass

Claudia Ortiz-Calderon, Héctor Cid Silva and
Daniel Barros Vásquez

Abstract

Environmental metal pollution is a serious public problem, and it has become an issue leading to research in the effluent remediation area. Techniques involving biosorption processes have been found to be promising due to the low cost of nonliving biomaterials, which have the potential to adsorb metal ions from wastewaters. One of the most promising types of biomasses to be used as biosorbents is the seaweed biomass, particularly from brown algae. The biosorption capability of the seaweed biomass relies on their cell wall chemical composition, mainly composed of alginates and fucoidans, molecules with a high presence of functional groups that interact with metal ions. This book chapter focuses on the use of seaweed biomass for metal biosorption and the chemical basis underlying the process. The current state of the commercial status of biosorption technology based on seaweed biomass is presented. Examples of complementary uses of the algae biomass other than industrial wastewater cleaning processes are presented, and the potential reuse of the biomass after the biosorption focused on biofuel production is discussed.

Keywords: seaweed biomass, metal removal, biosorption, biosorbent, wastewater treatment

1. Introduction

Environmental metal pollution is a serious public problem, and it has become an issue, leading to research in the effluent remediation area. Many techniques have been reported for removing metals from solutions, such as chemical precipitation, adsorption, ion exchange, filtration, chemical oxidation or reduction, electrochemical treatment, membrane processes, and evaporation. It has been found that these methods are limited, because of high operational costs, especially when the initial metal ion concentrations are at the range of 10-100 mg/L [1]. Hence, techniques involving biosorption processes have been found to be promising, due to the low cost of nonliving biomaterials, which have the potential to adsorb metal ions from wastewater.

The biosorption processes occur when metal ions interact with the functional groups present in biopolymers that are part of the biomass. Chemical groups such as amide, hydroxyl, carboxylate, sulfonate, phosphate, and amino are responsible for the quantitative adsorption of metals [2]. Several interaction mechanisms such as complexation, coordination, chelation, ion adsorption, cation exchange, and microprecipitation have been proposed as the participants in the metal biosorption processes [3].

A wide variety of biomasses has been found to be capable of sequestering metal ions from dilute solutions. An interesting approach is the use of the nonliving forms of the biomaterials because they do not need nutrition for the maintenance and avoid metal toxicity problems [4]. One of the most promising types of biomasses suitable for their use as biosorbents is marine algal biomass (seaweeds), which exhibit a high abundance in the oceans [5].

The biosorption capability of algae biomass is mostly related to their cell wall chemical composition, which exhibits a fiber-like structure and an amorphous embedding matrix of polysaccharides such as alginates and fucoidan [6]. In brown algae, alginates have a high affinity for divalent cations and sulfated polysaccharides give account of the uptake of trivalent cations [7]. The physical and chemical nature of the interaction between the metals and the functional groups present in the biomass has been intensively studied, in order to develop technologies for the sequestration of metals to clean, or to recover, valuable metals from industrial effluents [5, 8, 9].

This book chapter focuses on metal biosorption by seaweed biomass and the chemical interactions between the functional groups of this biomass and the cations. To the end, the potential uses of algae biomass in industrial wastewater cleaning processes and its potential reuse are highlighted.

2. Seaweed biomass

Algae are autotrophic organisms that contain chlorophyll and carry out oxygenic photosynthesis; they are widely distributed and have great diversity. Algae do not represent a formal taxonomic group of organisms, but a highly heterogeneous collection of organisms of different evolutionary lineages and high genetic diversity, which is reflected in the huge diversity that algae in morphological terms, ultrastructure, ecological, biochemical, and physiological [10].

Macroalgae, or seaweed, are a group of fast-growing aquatic organisms including about 9000 species. They are commonly classified into three groups according to the color of the thallus, which correspond to the Chlorophyta (green algae), Rhodophyta (red algae), and Heterokontophyta phylum, class Phaeophyceae (brown algae) [11] (**Figure 1**).

The taxonomic classification of these organisms involves much more than this simple designation and is performed considering a combination of features, including the nature of photosynthetic pigments; polymers present in the cell wall and cellular organization. Today,

Figure 1. Brown algae *Durvillaea antarctica*. Left, a specimen freshly collected from the coastal line. Right, a sample of milled *Durvillaea antarctica* biomass with a size of 500–1000 μm.

thanks to molecular systematics, a good progress has been made in the classification of these organisms, solving the problem of underestimation of diversity when considering only morphological characters [12]. There is great interest in the commercial use of the chemical constituents present in the seaweeds, in the field of energy production, agriculture, food, environmental, and pharmaceutical industry. The global harvest seaweed for food and algal products (e.g., Agar, alginates, and carrageenan) exceeds 3 million tons per year, with a potential harvest estimated at 2.6 million tons for red algae and 16 million tons brown algae [13]. Of particular interest is the use of seaweed dead biomass as biosorbent of heavy metals in solution. Multiple studies have shown a high sorption capacity and selectivity for different metal cations attributed to the polysaccharides present in their cell walls [4, 5, 8, 9, 14–18]. The basic organization of their cell walls comprises a fibril skeleton mainly composed of cellulose, and an amorphous matrix of sulfated galactans constituted by carrageenans and agar in red algae and alginates or alginic acid and fucoidan in brown algae (**Table 1**). Studies to assess

Division	Common name	Pigments	Storage product	Cell wall
Chlorophyta	Green algae	Chlorophyll *a*, *b*; α-, β- and γ-carotenes, and several xanthophylls	Starch (amylose and amylopectin) (oil in some)	Cellulose (β-1,4-glucopyroside), hydroxyproline, glucosides, xylans, and mannans
Phaeophyta	Brown algae	Chlorophyll *a*, *c*; β-carotene and fucoxanthin and several other xanthophylls	Laminaran (β-1,3-lucopyranoside, predominantly); mannitol	Cellulose, alginic, acid, and sulfated, mucopolysaccharides (fucoidan)
Rhodophyta	Red algae	Chlorophyll *a* (*d* in some *Florideophyceae*); R- and C-phycocyanin, allophycocyanin; R- and B-phycoerythrin. α- and β-carotene and several xanthophylls	Floridean starch (amylopectin-like)	Cellulose, xylans, several sulfated polysaccharides (galactans) calcification in some; alginate in *Corallinaceae*

Table 1. Physical and chemical characteristics of three algal divisions. Adapted from [2].

the biosorption (mass of metal adsorbed by mass of biosorbent) of different metals (Pb, Cu, Zn, and Cd) by seaweed biomasses have shown that the higher sorption capacity is exhibited by brown algae [5, 8, 9, 18].

Phaeophyceae is a large and diverse class of species ranging from small filamentous algae to highly complex organisms up to 60 m long. This class presents 2046 species organized into 20 orders [19]. It is believed that *Phaeophyceae* emerged between 150 and 200 million years ago in a secondary endosymbiosis event, in which a red algae was captured by an ancestral protist [12, 20, 21]. Its characteristic color is given by large quantities of fucoxanthin present in their chloroplasts. The most important algae orders from the point of view of the biosorption are Fucales and Laminariales, presenting species with a greater structural complexity. The components in their cell walls and reserve polymers have been reported as the main cause of its great capacity for immobilization of metal cations. The cell wall structure varies according to the species, age of the population, climate, and geographical location. Fibril wall structure, which provides structural support, is composed of cellulose-like plants, although displayed in a lower portion, occupying from 1 to 8% of its dry weight [21]. Two groups of anionic polysaccharides that provide strength and flexibility to the cell wall are alginates (between 10 and 40% dry weight) and fucoidan (between 5 and 20% dry weight). Mannitol and laminarans represent energy backup products. Other components such as proteins and phlorotannins also contribute but in a lesser way to the biosorption, representing about 5% of the biomass dry weight (**Figure 2**).

The three major components of the cell wall extracellular matrix of brown algae, cellulose, alginic acids, and polymers like mannuronic and guluronic acids, are complexed with light

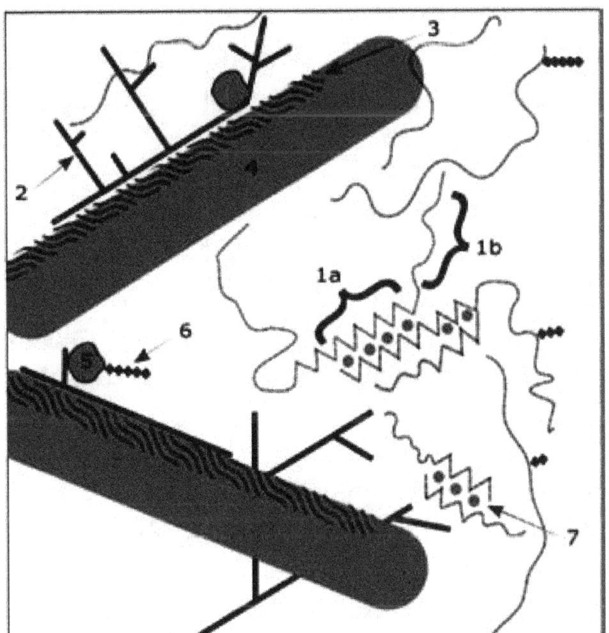

Figure 2. Schematic representation of the main components present in the cell wall of brown algae. Alginates rich in α-1,4-L-guluronate (1a); alginates rich in β-1,4-D-mannuronate (1b); fucoidans (2); hemicellulose (3); cellulose (4); proteins (5); phlorotannins (6); Ca^{2+} (7). Adapted from [2].

metals such as sodium, potassium, magnesium, and calcium, and other polysaccharides [19]. Alginates and sulfated polysaccharides have been reported as the predominant molecular components with reactive groups in brown algae [18]. Biosorption of heavy metals involves several mechanisms that differ qualitatively and quantitatively depending of the chemical species used, the origin of the biomass, and its processing procedure such as reinforcement by crosslinking [20]. Algae biomass possesses several chemical groups that can attract and sequester metals: acetamide, amine, amide, sulfhydryl, sulfate, and carboxyl [2]. This chemical diversity originates a combination of mechanisms for the capture of the metals, including electrostatic attraction, complexation, ion exchange, covalent binding, van der Waals attraction, adsorption, and microprecipitation [3].

Alginates are a family of linear polysaccharides, consisting of two uronic acids units: β-1,4-D-mannuronate (M) and α-1,4-L-guluronate (G). These units are arranged in homopolymer blocks of M, homopolymer blocks of G, and/or heteropolymer blocks of M and G (**Figure 1**). The relative abundance of M and G blocks in the macromolecular structure determines structural properties and affinity of alginates for divalent cations. The affinity of some divalent metal cations varies with M:G ratio [2, 6, 22]. Studies have shown that the affinity of the alginates for cations such as Pb, Cu, Cd, Zn, and Ca increases with a higher content of guluronic acid [23, 24]. The high specificity for divalent cations is explained by the structure of "zig-zag" formed by homopolymers of guluronic acid, which stabilizes the Ca^{2+} and other divalent cations easily (**Figure 1**) [2, 22]. Alginates fibers are able to adopt an ordered conformation in solution through dimerization of homopolymeric regions of guluronic acid, in the presence of calcium or other divalent cations, as they are filled with carboxylic groups and other electronegative oxygen atoms. This description is known as the model of "egg box" [25]. The carboxyl groups are the most abundant functional group in brown algae, determined by the percentage of quantifiable sites by titration, reaching about 70%. Furthermore, most of the metal cations of interest show high sequestering at pH near to the dissociation constant (pK_a) of carboxylic acids [2].

Fucoidans are branched sulfated polysaccharides mainly constituted by α-L-fucose, uronic acids, and a small portion of galactose, xylose, arabinose and/or mannose, glucose, and some-times proteins, presenting an extremely variable molecular weight. They are presented in the form of homopolymers or homofucans called fucans, or heteropolymers called fucoidans. The sulfonate groups in the fucoidans are the second functional group in abundance in brown seaweed, and its role could become prominent, if the binding of the metal occurs at a low pH [2].

Mannitol is a compound derived from monomeric D-mannose present in all brown algae, which can represent up to 30% of the biomass dry weight [2]. The second largest reserve products are laminarans and polysaccharides, which are composed of (1, 3)-β-D-glucans. They consist of residues (1, 3)-β-D-glucopyranose with some 6-O-branches of the main chain. Two types of laminaran chains exist: M, with a mannitol monomer attached to the reducing end, and G, with a glucose monomer attached to the reducing end. All the polysaccharides present hydroxyl groups, but these are less abundant and only are negatively charged at pH above 10, playing a secondary role at low pH [2].

3. Seaweed biomass and metal biosorption

Biosorption of heavy metal ions in wastewater using algae can be ecologically safer, cheap, and efficient. Algae can be used for sorption of toxic and radioactive metal ions [26] and also to recover metal ions like gold and silver [27].

The biosorption of heavy metal ions by seaweed biomass may occur by different mechanisms such as ion exchange, complex formation, and electrostatic interaction [28], being ion exchange the most important [29]. Polysaccharides and proteins present in the algae cell walls provide the metal-binding sites [30]. The sorption capacity of a seaweed cell surface to a specific ion depends on several factors such as the amount of functional groups in the algae matrix, the coordination number of the metal ion to be sorbed, the accessibility of binding groups for metal ions, the complex formation, affinity constants of the metal with the functional group, and the chemical state of these sites [31]. Considering the heterogeneity of the cell wall composition in different seaweed species, the capacity of metal biosorption by the algal strains will vary. For instance, brown algae with alginate in their cell wall composition have a high biosorption affinity for lead ions [5]. Alginate polymers are the primary responsible for heavy metal ions sorption in brown algae, and their capacity to bind the metal directly depends on the number of binding sites on this polymer [32]. In a second place, fucoidans play a key role for heavy metal sequestration.

The functional groups present in the brown and green algae cell wall matrixes, such as carboxyl, hydroxyl, sulfate, phosphate, and amine groups, play a dominant role in the metal binding [30]. The presence of various functional groups and their complexation with heavy metals during biosorption process can be studied by using spectroscopic techniques, such as FT-IR and XPS [33]. The X-ray absorption fine structure spectroscopy and quantum chemistry calculation are also an experimental approach to explain the biosorption mechanisms [34]. An interesting methodology to determine the contribution of different functional groups in the metal adsorption is the derivatization of functional groups, like the pretreatment of the seaweed biomass with methanol in acid media or with propylene oxide, which blocks the action of the carboxyl groups in the biomass [35]. In *Sargassum fluitans*, after 4 h of treatment of the biomass with propylene oxide, about 50% of reduction in the biosorption capacity of Cd and Pb ions was observed [35]. Cid et al. [36] found a 43.3% of decrease in copper biosorption by esterifying biomass of the species *Durvillaea antarctica* using methanol in acid media. The presence of carboxyl, hydroxyl, sulfate, phosphate, and amine functional groups has been found to play a dominant role for the algae-metal interaction by an ion-exchange mechanism that occurs between heavy metals and intrinsic light metals ions such as Ca^{2+} and Mg^{2+}, and Na^{+} and K^{+} [37]. A summary of representative data for efficiency of copper removal for different types of algal biomass is presented (**Table 2**).

Heavy metal ion uptake by algal biomass can be enhanced by physical or chemical treatments that modify the seaweed cell surface structure and provide additional binding sites [32, 41–43]. Physical treatments such as heating/boiling, freezing, crushing, and drying usually lead to an enhanced level of metal ion biosorption. These treatments provide more surface

Brown algae specie	pH	q_m (mmol Cu g^{-1} biomass)	References
Ascophyllum nodosum	4.0	0.91	Romera et al. [5]
Durvillaea antarctica	5.0	1.73	Cid et al. [36]
Fucus serratus	5.5	1.60	Ahmady-Asbchin et al. [37]
Fucus spiralis	4.0	1.10	Romera et al. [5]
Fucus vesiculosus	5.0	1.66	Mata et al. [28]
Padina sp.	5.0	1.14	Sheng et al. [18]
Sargassum sp.	5.0	0.99	Sheng et al. [18]
Sargassum sp.	5.5	1.13	Karthikeyan et al. [38]
Sargassum filipendula	4.5	0.89	Davis et al. [39]
Sargassum filipendula	4.5	1.32	Kleinübing et al. [40]
Sargassum fluitans	4.5	0.80	Davis et al. [39]
Sargassum vulgarie	4.5	0.93	Davis et al. [39]

Table 2. Comparison of Cu(II) biosorption maximum capacities of different types of brown algae. Adapted from [36]

area to increase the biosorption capacity [41] and release cell contents that might bind to metal ions. The most common algal pretreatments are $CaCl_2$, formaldehyde, glutaraldehyde, NaOH, and HCl. Pretreatment with $CaCl_2$ causes calcium binding to alginate that plays an important role in ion exchange [42]. Formaldehyde and glutaraldehyde strengthen the cross-linking between hydroxyl groups and amino groups [43]. NaOH increases the electrostatic interactions of metal ion cations and provides optimum conditions for ion exchange, while HCl replaces light metal ions with a proton and also dissolves polysaccharides of cell wall [32], or denatures proteins, increasing the binding sites for the biosorption process.

4. Industrial uses of seaweed biomass

Over the past four decades, much effort had been devoted to identify readily available non-living seaweed biomass, capable of effectively removing heavy metals, with good hydro-dynamic capacities, physicochemical stability, and with the possibility to enhance their capacities to obtain biosorbents. After years of experimentation on hundreds of raw seaweed biomass for biosorption of heavy metals under different conditions, the optimum conditions for the biosorption process at bench scale have been stated for many seaweed biomasses. This research has conducted the efforts to the development of biosorption technologies for industrial applications, considering the volumes and the complex composition of different wastewaters.

One of the most important issues to consider is the biomass organic leaching phenomenon that is produced by the contact of the biomass with the water to be treated, liberating fractions of the biomass, biopolymers, and another chemical compounds. Organic leaching is an important factor to minimize, because it adds organic pollution to the treated water and generates an important biomass loss, resulting in a decrease on the availability of biomass for the next cycle of biosorption [24]. Also, organic leaching provokes hydraulic problems in column systems, because the biomass tends to accumulate at the exit of the packed columns, generating a clot that impairs the normal flow of the treated water passing by the biosorbent bed and generates elevated levels of TDS [44, 45]. The problems of excessive leaching and swelling can be minimized through proper engineering procedures, but the costs and practicability of these procedures are of concern. To control the swelling in seaweed biomass, Chu and Hashim [46] employed polyvinyl alcohol to immobilize biomass of *Sargassum* to construct packed columns. Omar et al. [47] immobilized *Enteromorpha torta* biomass using calcium alginate and solutions of calcium chloride for the biosorption of Cesium-134 on packed column beads. Other authors used calcium or magnesium salts to improve rheological properties of the biomass by the crosslinking generated by calcium ions and the alginate fibers [2], and used this material to make columns beds [48–52]. Other procedures applied to seaweeds biomass to improve the material properties include protonation treatments with different acids: crosslinking with formaldehyde, glutaraldehyde, bis(ethenyl)sulfone, 1-chloro-2,3-epoxypropane and polyethylene imine [14, 53], thermal treatment [54], and chemical modifications of functional groups, among others [55]. It is important to highlight that the costs associated with the raw biomass process and the ultimate disposal of the derived biomass waste should be considered for a viability of installation of the process. One simple way to prevent swelling of seaweeds in packed columns is to mix seaweed biomass with an inert and stable materials such as sand. This will improve the porosity inside the column and thereby enhance the solute flow pattern [56]. The modification of the surfaces of these biosorbents for further removal of other contaminants, for an integral application on wastewater treatment, is rather challenging.

Another important issue to be considered when using biosorbents to treat metal polluted wastewaters is the complexity of the solution, because it can affect the biosorption process by competition for the exchange sites by cationic chemical species other than metals. Vijayaraghavan et al. [57] studied the nickel biosorption capacities on *Sargassum wightii* in aqueous solutions and residual electroplating solutions, finding that the complex nature of electroplating solutions negatively affected the metal biosorption performance, because of the competition with other ions. Patrón-Prado et al. [58] studied the biosorption of cadmium and copper by *Sargassum sinicola* in aqueous solutions and in saline wastewaters, finding that the salinity of the solution caused a reduction in Cd biosorption from 89 to 5.8%. At the same time, the authors found a clear antagonist effect between both metals in solution. The same antagonist effect was reported on other seaweed species [59, 60].

Also, it is well known that most of the seaweed sorbents have poor affinity for anions such as nitrate, sulfate, and phosphate, due to the predominant anionic sites in their surface. These anions are common in effluents and if not removed may lead to eutrophication and other undesirable effects on the environment [61]. Alginates, one of the major constituents of the seaweed biomass, can be chemically modified to remove anionic contaminants from water solutions and can be used to encapsulate materials such as magnetite, leading to the formation of a multifunctional sorbent that has magnetic properties and can remove both cationic heavy metal ions such as copper ions and anionic contaminants, like arsenic [62].

Alginates can be cross-linked by addition of alkaline metals as Ca or Mg [63], resulting in a encapsulation of raw biomass, improving their biosorption capacities, and generating a solid biosorbent with a better hydrodynamic performance. Mata et al. [15] determined the effect of the immobilization of *Fucus vesiculosus* with alginate xerogels in the biosorption of Cd, Cu, and Pb. The immobilization increased the kinetic uptake and intraparticle diffusion rates for the three metals. Song et al. [64] evaluated the recovery of Li, Sr, and La by Ca-alginate beads at different physicochemical conditions, finding the best performance for Sr and La.

The feasibility of the biosorption process to reduce toxic metals presents some limitations, and a fully understanding of the process in the context of a reactor system is necessary. Engineering considerations are crucial when a seaweed-based biosorption system is designed and developed. In general, biosorption systems using dry seaweed biomass correspond to a solid-liquid contact process. This process ideally implies several cycles of biosorption and desorption stages. The effluents to be treated make contact with the biomass in a batch, semi-continuous or continuous flow system. Banks [65] describes different types of reactors with potential use in biosorption system designs: the conventional stirred tank reactors, packed bed reactors, expanded bed reactors, fluidized bed reactors, and airlift reactors, depending on the final result and the type of effluent to be treated.

Despite the considerable progress in the understanding of seaweed biomass interactions with heavy metals made over four decades of continuous research, most seaweed biosorption processes are still at bench scale. Some proposed processes based on biosorption have been patented for commercial applications, some of them at pilot scale and some at commercial scale, mainly represented by units that were constructed in Canada and USA during the 1990s [66]. Thus, only a few industrial processes or products based on biosorption technology have been implemented, especially if we refer only to seaweed applications. A search in WIPO Patentscope (http://patentscope.wipo.int/search/en/search.jsf) only shows 29 patents related to biosorption and heavy metals removal, and only four consider seaweed biomass in the development. Pohl [67] patented a preparation of a biosorbent based on brown algae, consisting in a method to prepare the raw material for biosorption of heavy metals and hydrocarbons to finally obtain a milled dry seaweed biosorbent material. Other inventions include directly applications of seaweed biomass. For example, Volesky and Kuyucak developed a method for the biosorption of gold using seaweed biomass of *Sargassum genus* [68]. In the market, there is a very limited offer on biosorption technologies. Actually, only five products offer a commercial solution to remove pollutants from water. These commercial technologies include AMT-BIOCLAIM™ from Advanced Mineral Technologies, Inc. (AMT), a sorbent obtained from industrial fermentation process by *Bacillus subtilis* [69, 70]. Bio-Recovery System, Inc. (Las Cruces, USA) developed a microalgae biosorbent named as AlgaSORB™, which consists of *Chlorella vulgaris* biomass immobilized on silica gel polymer matrix, to treat heavy metal ions from diluted solutions (1–100 mg g^{-1}). Metagener and RAHCO Bio-Beads have developed two commercial biosorbents, which could be used as effective materials for removal of heavy metal ions from wastewaters (mainly electroplating and mining industry) [1, 71]. Particularly interesting is the invention BV-SORBEX™, by B.V. SORBEX, Inc. (Montreal, Canada), that offers a family of commercial sorbents based on metal-binding biomass, including different-sized powders and granules made of different types of biomass, including seaweeds [72]. These biosorbents remove a wide range of metal ions from diluted or concentrated solutions with high efficiency (over 99%) in a wide range of pH values and aqueous system conditions [73]. However, these products have not been commercially successful [74, 75]. One reason is related to the lack of a full understanding of the mechanisms, kinetics, and thermodynamics of the biosorption process [76]. Another reason is related to the existence of competing technologies based on physical and chemical treatments such as ion exchange, activated carbon, chemical precipitation, oxidation/reduction methods, electrocoagulation, electrodialysis, ultrafiltration, reverse osmosis, and solvent extraction [77]. However, two major arguments that support the biosorption as a convenient cleanup technology are the low cost of the biosorbents and the constant increase in the demands on environmental regulatory standards [61]. Paradoxically, processes to improve the biosorbent performance not only increase the final price of the product but also raise questions about the toxicity and environmental hazards of the final product. All these issues create a need for other commercially attractive applications of biosorption technology. The main suggestion is the use of biosorption for the recovery of precious metals [61, 75]. Future directions may also include a most integrated solution addressed to clean up industrial effluents containing multiple pollutants and the applications of hybrid technologies between traditional and biosorbent technology. Unfortunately, the increasingly published output on the biosorption field do not reflect an improved knowledge of the process, nor aid to commercial exploitation [78]. Biosorptive processes may contribute to primary or secondary biological water treatments including domestic, municipal and industrial wastes, and in some circumstances, solid wastes. Biosorption technology is still on a developing stage, and commercial success will depend on a better understanding of the process, governed by a pragmatic rationale of its commercial development and potential applications.

The final cost of biosorption treatment certainly involves the harvest, transportation, and processing of the biomass, together with the control of optimal conditions of the process, the regeneration of the biosorbent, and the final disposing of the biomass. Other costs usually not discussed are the capital expenses and plant operation costs, because they depend on the design of the treatment plant and the nature of wastewaters to be treated. Because many species of seaweed are valuable for the production of molecules with nutritional value, cosmetic applications among other uses, the use of residual dead biomass is convenient. Preparation of biosorbents is usually a major cost associated with the biosorption process, and biomass preprocessing is necessary to guarantee a good performance of the biosorbent. Thus, much attention should be taken with the estimation of costs, including the final disposal of the residues. Once a biosorbent life cycle ends, the ultimate disposal should be addressed. Landfilling the biomass, chemical or thermal destruction techniques seems an easy way to manage waste biosorbents, procedures that are not cheap or environmentally friendly. Used biosorbents can also be reused for other applications. Therefore, once heavy metal ions are completely removed from the used seaweeds by a demineralization process, they can be used for other applications.

Because the sorption technology based on biomasses has not been fully developed at industrial level, there are scarce data about the technical-economical evaluation of industrial biosorption applications. Some calculations predict that the prices of a biosorbent system represent about a tenth of the price of resins [79]. A comparison between classic methods for metal sorption and biosorption techniques can be summarized considering advantages and disadvantages of different technological approaches (**Table 3**). From the analysis, it can be concluded that biosorption is a clean technology, reducing the amount of solid wastes generated, gives the possibility to recover various metals, and has low operational cost in terms of energy consumption.

Metal adsorption technology	Advantages	Disadvantages	References
Adsorption	High spectrum of pollutants High capacity and fast kinetics Potentially selective	Performance depends on the type of adsorbent Needs chemical or thermal treatment for better performance	Aderhold et al. [80]
Biosorption	High spectrum of pollutants High capacity and fast kinetics Obtained directly from the nature or from industrial waste (low cost)	Poorly selective without chemical modifications	Davis et al. [2]
Chemical precipitation	High spectrum of metals Low cost of implementation Easy operation	High sludge formation implies cost on maintenance and sludge disposal	Rubio et al. [81]
Ion exchange	Metal selective High regeneration capacity	High initial capital and maintenance cost Limited pH tolerance	Rubio et al. [81]
Coagulation/ flocculation	Capacity to reduce biological viability. Good dewatering performance	High cost on chemicals High costs on sludge management	Rubio et al. [81]
Electrochemical precipitation	Applicable for the treatment of heavily contaminated wastes Easy to operate.	High initial capital, high operation cost, and maintenance Need of power energy	Qin et al. [82]
Membrane filtration	Low solid waste generation Low chemical consumption Small space requirement Metal selective method	Temperature instability Low stability at low pH High initial capital, high cost on operation, and maintenance Easy fouling generation and flow stacks	Madaeni and Mansourpanah [83]

Table 3. Comparison of conventional technologies and biosorption for the treatment of metal contaminated waters.

The reuse of seaweed biomass after the remediation process to obtain other products with commercial value is an approach to make more attractive the biosorption process. The usability of the seaweed biomass after the biosorption has several productive destinies. Seaweed biomass residues traditionally are burned, disposed in landfills, or confined, but also can be used to obtain a final valuable product. A new niche of application of the passive biosorption process of metals is the enrichment of biomass with microelements to be used as biological feed supplements and/or fertilizers. Dietary supplements obtained by this way have already demonstrated good results on animals [84, 85]. Micronutrients, such as Cu(II), Mn(II), Zn(II), and Co(III), are usually targets for biosorption on wastewater remediation processes, and a biomass that incorporates these elements allows obtaining a microelements enriched biological material. Seaweed biomass is rich in many nutritional elements such as carbohydrates, proteins, microelements, and polyphenols, representing a natural fertilizer, but is not a full complement to the all requirements to amend poor soils. Different seaweed biomasses were experimented for microelemental enrichment via biosorption, obtaining material to be used as feed additives, to supplement livestock diet [86–88]. It was observed that for all the studied seaweed, the smaller the content of the microelement in the natural biomass, the higher the enrichment coefficient of the biomass reached. Certainly, the difficulty to integrate a wastewater bioremediation process to recover metals and the use of the biomass as a fertilizer or as a feed supplement is that the microelements present in the wastewater are at minimal quantities. Also, the possibility of contamination of the biomass with undesirable elements has to be avoided. Another interestingly approach is the potential use of biomass for biogas production. Seaweed biomass can be anaerobically digested for the production of methane. Nkemka and Murto [89] experimented with the demineralization of seaweed biomass prior fermentative processes, obtaining an efficient production of biogas and at the same time, a residue to be used as a fertilizer. A good demineralization process can produce a useful biomass for composting soils [90]. The same concept would be applied to obtain bio-oils from residual biomass. Diaz-Vazquez et al. [91] evaluated the demineralization of *Sargassum sp.* biomass by several methods before a hydrothermal liquefaction process (HTL) of the biomass to increase the yield of bio-oil production. Algae biomass has a higher energy content than most of the lignocellulosic biomass used for biofuel production [92]. In contrast to the high lignocellulosic contents of terrestrial plants, seaweeds are primarily composed of elastic polysaccharides such as alginic acid, laminarin, carrageenan, and agarose that make them a more suitable feedstock for thermochemical conversion processes. The HTL process converts complex polymers present in the biomass into simpler molecules that can be converted to bio-oils. One major limitation on the HTL process is the high ash content present on seaweed biomass, which reduces the yields an quality of the generated bio-oils, restricting their use in direct combustion or gasification processes [93, 94], so the biomass has to be pretreated to reduce the ash content. Usually, the pretreatment process is carried out with mineral acids like sulfuric acid, nitric acid, or organic acids as acetic acid and citric acid usually in a rate of 10% (w/w) between biomass and acid [56]. The selection of a specific acid for demineralization relies on the mechanisms of removal, the mechanical stability of the biomass, and the final use of the recycled biomass [56]. A proper demineralization process allows the incorporation of seaweed biomass previously used in biosorption processes into a new productive cycle (**Figure 3**). Once the seaweed biomass has been used for the treatment of metal polluted effluents, the treated effluent is recovered, and the seaweed biomass loaded with the metal can be demineralized in situ. By this procedure, a metal-enriched solution can be recovered, and the biomass can be harvested from the container and be submitted to a secondary treatment for further productive processes (**Figure 3**).

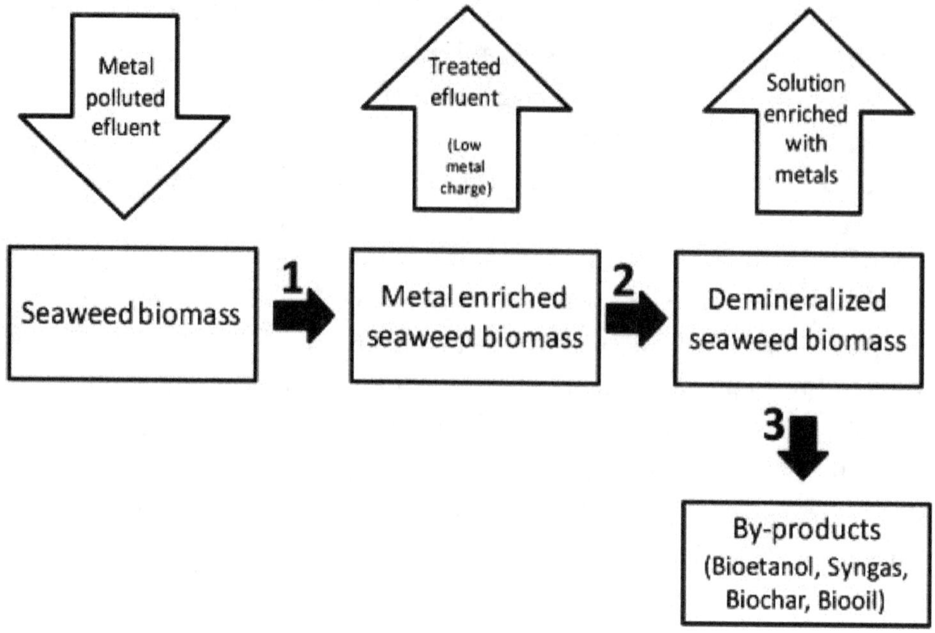

Figure 3. Simplified scheme for industrial use of seaweed biomass. (1) Biosorption process; (2) demineralization process; (3) secondary treatment such as pyrolysis, fermentation, HTL, gasification, among others.

5. Conclusions

A wide variety of biomasses has been evaluated for the sequestration of metal ions from solutions. An interesting approach is the use of the nonliving forms of the biomasses because they do not need nutrition for the maintenance and do not present the problem generated with the toxicity of the metals on living organisms. The biosorption capability of algae biomass is mostly related to the cell wall chemical composition, that is, a fiber-like structure, and an amorphous matrix embedding with polysaccharides such as alginates and fucoidans. In brown algae, alginates have a high affinity for divalent cations and sulfated polysaccharides give account of the capture of trivalent cations. Besides, the reuse of seaweed biomass after the remediation process in order to obtain products with commercial value is an approach that makes attractive the biosorption process at an industrial scale. The chemical composition of brown algae biomass makes it suitable for the production of different by-products such as biofuels, after the biomass has been demineralized. Nevertheless, much more efforts must be done in order to generate quantitative data regarding the performance and the operational costs for biosorption processes using dead seaweed biomass at an industrial level.

Author details

Claudia Ortiz-Calderon*, Héctor Cid Silva and Daniel Barros Vásquez

*Address all correspondence to: claudia.ortiz@usach.cl

Faculty of Chemistry and Biology, Department of Biology, University of Santiago of Chile, Santiago, Chile

References

[1] Atkinson BW, Bux F, Kasan HC. Considerations for application of biosorption technology to remediate metal-contaminated industrial effluents. Water SA. 1998;24(2):129–135.

[2] Davis TA, Volesky B, Mucci A. A review of the biochemistry of heavy metal biosorption by brown algae. Water Res. 2003;37(18):4311–4330.

[3] Raize O, Argaman Y, Yannai S. Mechanisms of biosorption of different heavy metals by brown marine macroalgae. Biotechnol Bioeng. 2004;87(4):451–458.

[4] Jalali R, Ghafourian H, Asef Y, Davarpanah SJ, Sepehr S. Removal and recovery of lead using nonliving biomass of marine algae. J Hazard Mater. 2002;92(3):253–262.

[5] Romera E, González F, Ballester A, Blázquez ML, Muñoz JA. Comparative study of biosorption of heavy metals using different types of algae. Bioresour Technol. 2007;98(17):3344–3353.

[6] Deniaud-Bouët E, Kervarec N, Michel G, Tonon T, Kloareg B, Hervé C. Chemical and enzymatic fractionation of cell walls from Fucales: insights into the structure of the extracellular matrix of brown algae. Ann Bot. 2014;114(6):1203–1216.

[7] Vieira RHSF, Volesky B. Biosorption: a solution to pollution? Int Microbiol. 2000;3(1):17–24.

[8] Murphy V, Hughes H, McLoughlin P. Cu(II) binding by dried biomass of red, green and brown macroalgae. Water Res. 2007;41(4):731–740.

[9] Murphy V, Hughes H, McLoughlin P. Comparative study of chromium biosorption by red, green and brown seaweed biomass. Chemosphere. 2008;70(6):1128–1134.

[10] Rindi F, Soler-Vila A, Guiry MD. Taxonomy of marine macroalgae used as sources of bioactive compounds. In: Hayes M, editor. Marine bioactive compounds: sources, characterization and applications [Internet]. Boston, MA: Springer US; 2012. p. 1–53. doi:10.1007/978-1-4614-1247-2_1

[11] Balboa EM, Conde E, Moure A, Falqué E, Domínguez H. *In vitro* antioxidant properties of crude extracts and compounds from brown algae. Food Chem [Internet]. 2013;138(2–3):1764–1785. doi:10.1016/j.foodchem.2012.11.026

[12] Cock JM, Sterck L, Rouzé P, Scornet D, Allen AE, Amoutzias G, et al. The Ectocarpus genome and the independent evolution of multicellularity in brown algae. Nature. 2010;465(7298):617–621.

[13] Chen JP. Decontamination of heavy metals: processes, mechanisms, and applications. Vol. 97, CRC Press/Taylor and Francis Group; 2012. III–IV.

[14] Holan ZR, Volesky B. Biosorption of lead and nickel by biomass of marine algae. Biotechnol Bioeng. 1994;43(11):1001–1009.

[15] Mata YN, Blázquez ML, Ballester A, González F, Muñoz JA. Biosorption of cadmium, lead and copper with calcium alginate xerogels and immobilized *Fucus vesiculosus*. J Hazard Mater. 2009;163(2–3):555–562.

[16] Rahman MS, Sathasivam KV. Heavy metal adsorption onto *kappaphycus* sp. from aqueous solutions: the use of error functions for validation of isotherm and kinetics models. Biomed Res Int [Internet]. 2015;2015:1–13. Available from: http://www.hindawi.com/journals/bmri/2015/126298/.

[17] Pavasant P, Apiratikul R, Sungkhum V. Biosorption of Cu^{2+}, Cd^{2+}, Pb^{2+} and Zn^{2+} using dried marine green macroalga *Caulerpa lentillifera*. Bioresour Technol. 2006;97:2321–2329.

[18] Sheng PX, Ting YP, Chen JP, Hong L. Sorption of lead, copper, cadmium, zinc, and nickel by marine algal biomass: characterization of biosorptive capacity and investigation of mechanisms. J Colloid Interface Sci. 2004;275(1):131–141.

[19] Guiry MD, Guiry G. AlgaeBase. World-wide electronic publication, National University of Ireland, Galway. [Internet]. 2016 [cited 2016 Jul 8]. Available from: http://www.algaebase.org.

[20] Hamdaoui O, Naffrechoux E. Modeling of adsorption isotherms of phenol and chlorophenols onto granular activated carbon. Part I. Two-parameter models and equations allowing determination of thermodynamic parameters. J Hazard Mater. 2007;147(1–2):381–394.

[21] Michel G, Tonon T, Scornet D, Cock JM, Kloareg B. The cell wall polysaccharide metabolism of the brown alga *Ectocarpus siliculosus*. Insights into the evolution of extracellular matrix polysaccharides in Eukaryotes. New Phytol. 2010;188(1):82–97.

[22] Holan ZR, Volesky B. Biosorption of heavy metals: review. Biotechnol Prog. 1995;11:235–250.

[23] Vierira R, Volesky B. Biosorption: a solution to pollution? Int Microbiol. 2000;3:17–24.

[24] Figueira MM, Volesky B, Ciminelli VST, Roddick FA. Biosorption of metals in brown seaweed biomass. Water Res. 2000;34(1):196–204.

[25] Rowbotham JS, Dyer PW, Greenwell HC, Selby D, Theodorou MK. Copper(II)-mediated thermolysis of alginates: a model kinetic study on the influence of metal ions in the thermochemical processing of macroalgae. Interface Focus. 2013;3:1–16.

[26] Pohl P, Schimmack W. Adsorption of radionuclides (134Cs, 85Sr, 226Ra, 241Am) by extracted biomasses of cyanobacteria (*Nostoc carneum*, *N. insulare*, *Oscillatoria geminata* and *Spirulina laxis-sima*) and Phaeophyceae (*Laminaria digitata* and *L. japonica*; waste products from alginat. J Appl Phycol. 2006;18(2):135–143.

[27] Mata YN, Torres E, Blázquez ML, Ballester A, González F, Muñoz JA. Gold(III) biosorption and bioreduction with the brown alga *Fucus vesiculosus*. J Hazard Mater. 2009;166(2–3):612–618.

[28] Mata YN, Blázquez ML, Ballester A, González F, Muñoz JA. Characterization of the biosorption of cadmium, lead and copper with the brown alga *Fucus vesiculosus*. J Hazard Mater. 2008;158(2–3):316–323.

[29] Michalak I, Chojnacka K. Interactions of metal cations with anionic groups on the cell wall of the macroalga *Vaucheria* sp. Eng Life Sci. 2010;10(3):209–217.

[30] Gupta VK, Rastogi A. Biosorption of lead from aqueous solutions by green algae *Spirogyra* species: kinetics and equilibrium studies. J Hazard Mater. 2008;152(1):407–414.

[31] Mehta SK, Gaur JP. Use of algae for removing heavy metal ions from wastewater: progress and prospects. Crit Rev Biotechnol. 2005;25(3):113–152.

[32] Romera E, González F, Ballester A, Blázquez ML, Muñoz JA. Biosorption with algae: a statistical review. Crit Rev Biotechnol. 2006;26(4):223–235.

[33] Yang F, Liu H, Qu J, Paul Chen J. Preparation and characterization of chitosan encapsulated *Sargassum* sp. biosorbent for nickel ions sorption. Bioresour Technol [Internet]. 2011;102(3):2821–2828. doi: 10.1016/j.biortech.2010.10.038

[34] Zheng Y-M, Liu T, Jiang J, Yang L, Fan Y, Wee ATS, et al. Characterization of hexavalent chromium interaction with Sargassum by X-ray absorption fine structure spectroscopy, X-ray photoelectron spectroscopy, and quantum chemistry calculation. J Colloid Interface Sci [Internet]. 2011;356(2):741–748. doi:10.1016/j.jcis.2010.12.070

[35] Fourest E, Volesky B. Contribution of sulfonate groups and alginate to heavy metal biosorption by the dry biomass of *Sargassum fluitans*. Environ Sci Technol. 1996;30(1):277–282.

[36] Cid H, Ortiz C, Pizarro J, Barros D, Castillo X, Giraldo L, et al. Characterization of copper (II) biosorption by brown algae *Durvillaea antarctica* dead biomass. Adsorption. 2015;21(8):645–658.

[37] Ahmady-Asbchin S, Andrès Y, Gérente C, Cloirec PL. Biosorption of Cu(II) from aqueous solution by *Fucus serratus*: surface characterization and sorption mechanisms. Bioresour Technol. 2008;99(14):6150–6155.

[38] Karthikeyan S, Balasubramanian R, Iyer CSP. Evaluation of the marine algae *Ulva fasciata* and *Sargassum* sp. for the biosorption of Cu(II) from aqueous solutions. Bioresour Technol. 2007;98(2):452–455.

[39] Davis TA, Volesky B, Vieira RHSF. Sargassum seaweed as biosorbent for heavy metals. Water Res. 2000;34(17):4270–7278.

[40] Kleinübing SJ, da Silva EA, da Silva MGC, Guibal E. Equilibrium of Cu(II) and Ni(II) biosorption by marine alga *Sargassum filipendula* in a dynamic system: competitiveness and selectivity. Bioresour Technol. 2011;102(7):4610–4617.

[41] López Errasquín E, Vázquez C. Tolerance and uptake of heavy metals by Trichoderma atroviride isolated from sludge. Chemosphere. 2003;50(1):137–143.

[42] Bishnoi NR, Kumar R, Kumar S, Rani S. Biosorption of Cr(III) from aqueous solution using algal biomass Spirogyra spp. J Hazard Mater. 2007;145(1–2):142–147.

[43] Ebrahimi B, Shojaosadati SA, Ranaie SO, Mousavi SM. Optimization and evaluation of acetylcholine esterase immobilization on ceramic packing using response surface methodology. Process Biochem. 2010;45(1):81–87.

[44] Fosso-Kankeu E, Mulaba-Bafubiandi AF. Review of challenges in the escalation of metal-biosorbing processes for wastewater treatment: applied and commercialized technologies. Afr J Biotechnol. 2014;13(17):1756–1771.

[45] Schiewer S, Volesky B. Ionic strength and electrostatic effects in biosorption of protons. Environ Sci Technol. 1997;31(7):1863–1871.

[46] Chu KH, Hashim MA. Copper biosorption on immobilized seaweed biomass: column breakthrough characteristics. J Environ Sci. 2007;19(8):928–932.

[47] Omar HA, Abdel-Razek AS, Sayed MS. Biosorption of Cesium-134 from aqueous solutions using immobilized marine algae: equilibrium and kinetics. Nat Sci. 2010;8(11):140–147.

[48] Holan ZR, Volesky B, Prasetyo I. Biosorption of cadmium by biomass of marine algae. Biotechnol Bioeng. 1993;41(8):819–825.

[49] Aravindhan R, Madhan B, Rao JR, Nair BU. Recovery and reuse of chromium from tannery wastewaters using *Turbinaria ornata* seaweed. J Chem Technol Biotechnol. 2004;79(11):1251–1258.

[50] Cossich ES, Da Silva EA, Tavares CRG, Filho LC, Ravagnani TMK. Biosorption of chromium(III) by biomass of seaweed *Sargassum* sp. in a fixed-bed column. Adsorption. 2004;10(2):129–138.

[51] Borba CE, Guirardello R, Silva EA, Veit MT, Tavares CRG. Removal of nickel(II) ions from aqueous solution by biosorption in a fixed bed column: experimental and theoretical breakthrough curves. Biochem Eng J. 2006;30(2):184–191.

[52] Tsui MTK, Cheung KC, Tam NFY, Wong MH. A comparative study on metal sorption by brown seaweed. Chemosphere. 2006;65(1):51–57.

[53] Leusch A, Holan ZR, Volesky B. Biosorption of heavy metals (Cd, Cu, Ni, Pb, Zn) by chemically-reinforced biomass of marine algae. J Chem Technol Biotechnol. 1995;62:279–288.

[54] Park D, Yun Y-S, Young Cho H, Park JM. Chromium biosorption by thermally treated biomass of the brown seaweed, *Ecklonia* sp. Ind Eng Chem Res. 2004;43(26):8226–8232.

[55] Park D, Yun YS, Jong MP. Studies on hexavalent chromium biosorption by chemically-treated biomass of *Ecklonia* sp. Chemosphere. 2005;60(10):1356–1364.

[56] Vijayaraghavan K, Balasubramanian R. Is biosorption suitable for decontamination of metal-bearing wastewaters? A critical review on the state-of-the-art of biosorption processes and future directions. J Environ Manage. 2015;160:283–296.

[57] Vijayaraghavan K, Palanivelu K, Velan M. Treatment of nickel containing electroplating effluents with *Sargassum wightii* biomass. Process Biochem. 2006;41(4):853–859.

[58] Patrón-Prado M, Acosta-Vargas B, Serviere-Zaragoza E, Méndez-Rodríguez LC. Copper and cadmium biosorption by dried seaweed *Sargassum sinicola* in saline wastewater. Water Air Soil Pollut. 2010;210(1–4):197–202.

[59] Zhou JL, Huang PL, Lin RG. Sorption and desorption of Cu and Cd by macroalgae and microalgae. Environ Pollut. 1998;101(1):67–75.

[60] Sheng PX, Ting YP, Chen JP. Biosorption of heavy metal ions (Pb, Cu, and Cd) from aqueous solutions by the Marine Alga *Sargassum* sp. in single- and multiple-metal systems. Industrial and Engineering Chemistry Research. 2007;46(8):2438–2444.

[61] Volesky B. Biosorption and me. Water Res. 2007;41(18):4017–4029.

[62] Lim SF, Zheng YM, Zou SW, Chen JP. Characterization of copper adsorption onto an alginate encapsulated magnetic sorbent by a combined FT-IR, XPS, and mathematical modeling study. Environ Sci Technol. 2008;42(7):2551–2556.

[63] He J, Chen JP. A comprehensive review on biosorption of heavy metals by algal biomass: materials, performances, chemistry, and modeling simulation tools. Bioresour Technol. 2014;160:67–78.

[64] Song D, Park S, Kang HW, Park S Bin, Han J. Recovery of Lithium (I), Strontium (II), and Lanthanum (III) using Ca-alginate beads. J Chem Eng. 2013;58:2455–2464.

[65] Banks C. Scavenging trace concentrations of metals. In: Wase J, Forster C, editors. Biosorbents for metal ions. London, United Kingdom: Taylor & Francis Ltd; 1997. p. 115–140.

[66] Tsezos M. Biosorption of metals. The experience accumulated and the outlook for technology development. Hydrometallurgy. 2001;59(2–3):241–243.

[67] Pohl P. Method of production of adsorption material. US patent number 5648313. 1997.

[68] Volesky B, Kuyucak N. Biosorbent for gold. US patent number 4769223. 1988.

[69] Brierley JA, Brieley CL, Goyak GM. AMT-BIOCLAIM: A new wastewater treatment and metal recovery technology. In: Lawrence RW, Branion RMR, Ebner HG, editors. Fundamental and applied biohydrometallurgy. Boca Raton: CRC Press; 1986. p. 291–308.

[70] Brielly J. Production and application of a Bacillus-based product for use in metals biosorption. In: Volesky B, editor. Biosorption of heavy metals. Boca Raton: CRC Press; 1990. p. 305–311.

[71] Chojnacka K. Biosorption and bioaccumulation—the prospects for practical applications. Environ Int [Internet]. 2010;36(3):299–307. doi: 10.1016/j.envint.2009.12.001

[72] Volesky B. Biosorption of heavy metals. In: Volesky B, editor. CRC Press; 1990. 3–5 p.

[73] Bv-sorbex. BV Sorbex Technology Description www.bvsorbex.net/sxProcess.pdf (revised on July 2016).

[74] Vijayaraghavan K, Won SW, Mao J, Yun YS. Chemical modification of *Corynebacterium glutamicum* to improve methylene blue biosorption. Chem Eng J. 2008;145(1):1–6.

[75] Park D, Yun Y-S, Park JM. The past, present, and future trends of biosorption. Biotechnol Bioprocess Eng. 2010;15(1):86–102.

[76] BV SORBEX Inc. BV SORBEX Inc. The company. http://www.bvsorbex.net/sx.htm (revised on July 2016).

[77] Fomina M, Gadd GM. Biosorption: Current perspectives on concept, definition and application. Bioresour Technol [Internet]. 2014;160:3–14. doi:10.1016/j.biortech.2013.12.102

[78] Gadd GM. Biosorption: Critical review of scientific rationale, environmental importance and significance for pollution treatment. J Chem Technol Biotechnol. 2009;84(1):13–28.

[79] Volesky B. Detoxification of metal-bearing effluents: biosorption for the next century. Hydrometallurgy. 2001;59(2–3):203–216.

[80] Aderhold D, Williams CJ, Edyvean RGJ. The removal of heavy-metal ions by seaweeds and their derivatives. Bioresour Technol. 1996;58(1):1–6.

[81] Rubio J, Souza ML, Smith RW. Overview of flotation as a wastewater treatment technique. Miner Eng. 2002;15(3):139–155.

[82] Qin JJ, Wai MN, Oo MH, Wong FS. A feasibility study on the treatment and recycling of a wastewater from metal plating. J Memb Sci. 2002;208(1–2):213–221.

[83] Madaeni SS, Mansourpanah Y. COD removal from concentrated wastewater using membranes. Filtr Sep [Internet]. 2003;40(6):40–46.

[84] Michalak I, Chojnacka K, Witek-Krowiak A. State of the art for the biosorption process—a review. Appl Biochem Biotechnol. 2013;170(6):1389–1416.

[85] Saeid A, Chojnacka K, Korczyński M, Korniewicz D, Dobrzański Z. Biomass of Spirulina maxima enriched by biosorption process as a new feed supplement for swine. J Appl Phycol. 2013;25(2):667–675.

[86] Michalak I, Chojnacka K. Edible macroalga Ulva prolifera as microelemental feed supplement for livestock: the fundamental assumptions of the production method. World J Microbiol Biotechnol. 2009;25(6):997–1005.

[87] Michalak I, Chojnacka K. The application of macroalga *Pithophora varia* Wille enriched with microelements by biosorption as biological feed supplement for livestock. J Sci Food Agric. 2008;88(7):1178–1186.

[88] Michalak I, Chojnacka K. The new application of biosorption properties of *Enteromorpha prolifera*. Appl Biochem Biotechnol. 2010;160(5):1540–1556.

[89] Nkemka VN, Murto M. Evaluation of biogas production from seaweed in batch tests and in UASB reactors combined with the removal of heavy metals. J Environ Manage. 2010;91(7):1573–1579.

[90] Eyras MC, Defossé GE, Dellatorre FG. Seaweed compost as an amendment for horticultural soils in patagonia, Argentina. Compost Sci Util. 2008;16(2):119–124.

[91] Diaz-Vazquez LM, Rojas-Perez A, Fuentes-Caraballo M, Robles IV, Jena U, Das KC. Demineralization of *Sargassum* spp. macroalgae biomass: selective hydrothermal liquefaction process for bio-oil production. Front Energy Res. 2015;3(February):1–11.

[92] Ross AB, Jones JM, Kubacki ML, Bridgeman T. Classification of macroalgae as fuel and its thermochemical behaviour. Bioresour Technol. 2008;99(14):6494–6504.

[93] Bach QV, Sillero MV, Tran KQ, Skjermo J. Fast hydrothermal liquefaction of a Norwegian macro-alga: screening tests. Algal Res. 2014;6(PB):271–276.

[94] Neveux N, Yuen AKL, Jazrawi C, Magnusson M, Haynes BS, Masters AF, et al. Biocrude yield and productivity from the hydrothermal liquefaction of marine and freshwater green macroalgae. Bioresour Technol. 2014;155:334–341.

3

Methods of Estimating Forest Biomass

Lei Shi and Shirong Liu

Abstract

Forest plays a special role in carbon sequestration and thus mitigating climate change. However, the large uncertainty in biomass estimation is unable to meet the requirement of the accurate carbon accounting. The use of a suitable and rigor method to accurately estimate forest biomass is significant. Moreover, the world is increasingly facing the conflicting pressures of economic growth and environmental protection. Improving energy structure and vigorously developing biomass energy has become the development trend of energy utilization in the future. As energy plant is characterized by a large net accumulation of biomass. Therefore, the scientific evaluation of the size and potential of energy from plant also requires a suitable method for estimating biomass. Here, we reviewed the estimate methods, including allometric equation, mean biomass density, biomass expansion factor, geostatistics, etc. For each method, we will present background, rational, applicability, as well as estimation procedure by exemplifying a case. In this chapter, we argued that the new developed technique such as geo-statistics and remote sensing technique (e.g. LIDAR) would be the key tools to improve forest biomass estimation accuracy. However, prior to this, spatial variation of forest biomass at various levels should be explored using multi-source data and multi-approaches.

Keywords: carbon accounting, climate change, field survey, geostatistics, remote sensing technique, scale, uncertainty

1. Introduction

Currently, CO_2 and other greenhouse gas are inducing global warming, and vegetation is the only natural ecosystems to fix atmospheric CO_2. Forest is the main component of vegetation.

Accordingly, forest ecosystem is destined to be paid more attention by governments, academics, and the general public [1]. According to the Global Forest Resources Assessment 2010 [2], the global forest biomass (including above- and belowground) is 600 Pg, with a mean biomass density of 148.8 t/ha (**Table 1**). It is estimated that carbon sequestered in forest can account for about 77% of terrestrial ecosystem [3].

Region	Biomass ($\times 10^6$ t)	Biomass density (t/ha)
Eastern and Southern Africa	33,385	124.8
Northern Africa	3711	47.1
Western and Central Africa	816.3	248.7
Total Africa	**118,700**	**176.0**
East Asia	18,429	72.4
South and Southeast Asia	51,933	176.4
Western and Central Asia	3502	80.5
Total Asia	**73,864**	**124.7**
Europe excl. Russian Federation	25,602	130.7
Total Europe	**90,602**	**90.2**
Caribbean	1092	157.5
Central America	3715	190.5
North America	76,929	113.3
Total North and Central America	**81,736**	**115.9**
Total Oceania	**21,302**	**111.3**
Total South America	**213,863**	**247.4**
World	**600,066**	**148.8**

Table 1. Forest biomass and its density by region, 2010 [2].

Global deforestation is undergoing seriously [2], which contributes to a quarter of carbon released into the atmosphere each year [4]. Land use/land use change (mainly deforestation) is considered to be an important approach to the release of CO_2, which affects the carbon cycle on various spatial and temporal scales, and then global climate change [5, 6]. Therefore, the scientific and real-time monitoring of forest cover change and more accurate estimates of forest biomass and its magnitude is of significance to clarify the contribution of forests in global climate change.

In addition, the current world is facing the dual pressures of economic growth and environmental protection. Adjusting and optimizing energy structure, vigorously developing biomass energy has become the main developing trend of energy in the future. As energy plant, the most important characteristic is to possess a large net accumulation of biomass. Therefore, the scientific evaluation of the size and potential of biomass needs a suitable method used to estimate its biomass potential.

However, a large uncertainty exists in biomass estimation, which is unable to meet the requirement of the accurate carbon accounting required by Kyoto Protocol. The use of a suitable and rigor method to accurately estimate the size and distribution of forest biomass is of significance and also urgently needed. And also, the method used to estimate forest biomass is more likely to vary frequently with scale.

Given this, we reviewed the commonly used methods to estimate forest biomass across the scale in this chapter, for the purpose of operation and guidance, which includes allometric equation, mean biomass density, biomass expansion factor, forest identity, remote sensing- and geostatistics-based estimation methods, etc. For each method, we will present background, rational, applicability, as well as estimation procedure by exemplifying a case. At the end of this chapter, we argued that the new developed techniques such as geostatistics and remote sensing technique (e.g., LIDAR) would be the key tools to improve forest biomass estimation with a high accuracy. However, prior to this, spatial variation in forest biomass at various levels should be first explored using multi-source data and multi-approaches.

2. Allometry and allometric equation

If one organ is correlated to another of a plant, or a certain attribute is to plant size, we can call it allometry [7, 8], which is frequently expressed with a power relationship below [9, 10]:

$$y = ax^b \tag{1}$$

or

$$lgy = lga + blgx \tag{2}$$

where y often represents an attribute of plant (such as metabolic rate, biomass, etc.), x shows the size of the plant body (such as diameter at breast height and/or height), and a and b are coefficients.

In botany, the allometric relationship is able to be used to calculate biomass and other ecological factors by measuring the easily measured diameter at breast height (and/or height). Theoretically, tree D (diameter at breast height) and H (height) can both affect tree biomass. Thus, tree biomass can be estimated by allometric equation (Eqs. (3) and (4)), which includes both D and H [11–13]:

$$w = aD^b H^c \; or \lg w = a' + b\lg D + c\lg H \tag{3}$$

$$w = a(D^2H)^b \; or \lg w = a' + b\lg(D^2H) \tag{4}$$

where D and H represent tree diameter at breast height (cm) and height (m) and a, b, a', and c are regression coefficients.

Tree H is not easy to measure in field survey, so many researchers have used H-D model to estimate the H through easily measured D and then to estimate vegetation biomass [14–16]. In contrast to H, it is easier to measure D, and the measurement error is relatively small while measuring D [17, 18]. Furthermore, it is very common and efficient to use allometric relationship in scientific literatures, which includes D only, while estimating biomass [19, 20]. The allometric equation including D only can be reused by others and also make comparable among regions. The study performed by Wang [12] has also indicated that the including of H variable (i.e., including both D and H) in allometric relationship was unable to improve biomass estimate significantly (increased determination coefficient less than 4%). Therefore, many scholars contain D only as independent variable while fitting biomass allometric relationship (i.e., allometric equations (1) and (2) above) [12, 21, 22].

Such allometric relationship is based on the measured sample tree and aims to estimate vegetation biomass as the mathematical model (hereafter also referred to as a biomass allometric equation). Apparently, plant allometry is the theoretical basis of vegetation biomass estimation, which makes biomass estimate possible. Recently, remote sensing technique has been increasingly applied to estimate the biomass [23, 24]. However, data derived from allometric equation must be verified with field data in the method [25]. Generally, the use of allometric equation is indispensable to estimate biomass for both tree and forest.

3. Procedure of estimating multi-scale forest biomass

Multi-scale aboveground biomass estimation is demonstrated as an example to show the procedure (**Figure 1**). First, a number of plots are set, where field survey is performed (step 1); then several sample trees are cut to fit individual-level allometric equation (step 2); the use of developed allometric equation, together with filed survey data (mainly D), estimates biomass for each tree in plot and sums as stand-level biomass (step 3); finally, such upscaling methods as the mean biomass density, geostatistical technique, and others are used to upscale the regional forest biomass (step 4).

While estimating forest carbon stock, most scholars assumed that carbon content in plant biomass is constant (approximately 50%) [27–29]. Therefore, we estimate forest biomass first, multiplied by 50%, and can calculate the corresponding forest biomass carbon stock. It is not difficult to conclude that the method used to estimate forest carbon stock is almost entirely consistent with one used for biomass estimation; thus, the method of estimating forest biomass was addressed below, which can also be used to estimate forest carbon stock.

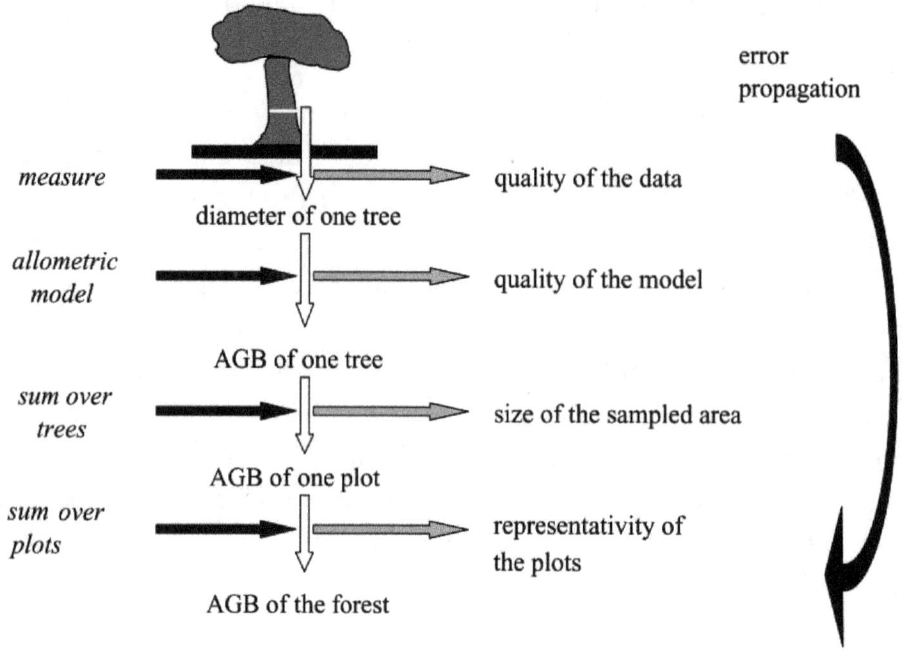

Figure 1. Procedure for biomass estimate and its error propagation [26].

4. Current methods of estimating forest biomass

4.1. Biomass estimate at individual level

Destructive sampling method (or harvesting method) and developed allometric equation can both be used to estimate individual-level biomass. For tree biomass estimate, destructive

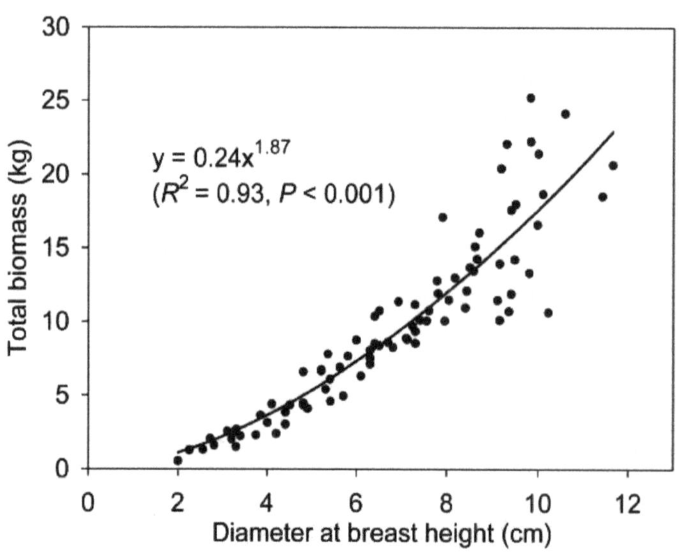

Figure 2. The relationship between the total biomass (y, kg) and D (x, cm) for the moso bamboo (*Phyllostachys edulis*) in South Anhui Province, China. The two variables exhibited a strong power function [30].

sampling method is more accurate than the use of developed allometric equation, because all the developed allometric equations are fitted (derived) from the biomass data based on the destructive sampling method. However, destructive method needs to cut down several sample trees and is thus expensive and time-consuming; moreover, it is not practical to weigh all the biomass for each tree in a stand or forest.

The general procedure for estimating biomass using destructive sampling method is to cut down several sample trees and weigh its different components (e.g., foliage, branch, stem, and root), respectively. After field survey, the components of the sample trees are collected and immediately taken to the laboratory to determine the water content. Subsequently, the (total) biomass can be determined by multiplying the fresh weight by the dry/fresh weight ratio. Then allometric equation can be fitted between the sampling biomass and *D* (and/or H) (e.g., **Figure 2**), and the developed equation can be employed to estimate individual-level biomass for each standing tree.

4.2. Biomass estimate at stand level

4.2.1. Mixed stand: simple allometric equation

The choice of stand-level biomass estimation is varied with the proportion of stand species composition (i.e., mixed or pure forests). Mixed stand-level biomass estimates may be estimated using allometric equation and then obtained by the addition of entire stands.

4.2.2. Pure stand: diameter-distribution model

This stand-level method is similar to the large-scale mean biomass density method described above, which does not take variations in biomass within a stand into account. In addition, the aforementioned simple allometric equation method is unable to fully reflect the developments and changes in stand structures. The corporation of the commonly used simple allometric equation and diameter-distribution functions (e.g., normal, lognormal, gamma, logistic, exponential, Richards, or Weibull functions) into a model (hereafter referred to as a diameter-distribution model) would likely improve the biomass estimation accuracy and strengthen the power of forest dynamics analyses.

The paper reported by Qi et al. [30] has exemplified the diameter-distribution model (Eq. (5)), which combined a three-parameter diameter-distribution function with an allometric equation to estimate the biomass of pure moso bamboo forests in China. The study found that a three-parameter Weibull distribution best characterized the diameter distribution of the moso bamboo stands. The biomass derived using the allometric equation was estimated 52.39 t/ha, smaller than 53.25 t/ha estimated using the Weibull distribution model (**Table 2**); this implied that the use of the common allometric equation alone to estimate forest biomass and carbon stocks may lead to an underestimate. It is concluded that using the diameter-distribution model to estimate forest biomass and carbon stock is expected to improve the accuracy.

Plot	BD_{WD} (t/ha)	BD_{AE} (t/ha)	RE (%)	Plot	BD_{WD} (t/ha)	BD_{AE} (t/ha)	RE (%)
PG1	45.54	46.29	−1.64	PH15	47.48	46.16	2.79
PG2	35.88	34.79	3.04	PH16	31.43	32.17	−2.34
PG3	43.00	42.19	1.87	PH17	34.99	34.59	1.14
PG4	41.46	38.87	6.25	PH18	24.87	23.78	4.37
PH5	18.96	18.02	4.94	PH19	61.27	59.91	2.23
PH6	51.25	50.72	1.03	PH20	13.07	12.68	3.00
PH7	40.39	39.02	3.38	PHS21	87.39	87.36	0.03
PH8	53.54	52.05	2.78	PHS22	63.389	63.28	0.16
PH9	64.17	62.37	2.81	PHS23	49.339	48.96	0.76
PH10	44.38	43.79	1.32	PHS24	61.22	60.13	1.78
PH11	52.57	50.57	3.80	PHS25	100.93	100.79	0.14
PH12	43.42	41.94	3.40	PHS26	102.58	102.10	0.47
PH13	62.56	60.81	2.79	PHS27	102.44	101.54	0.87
PH14	60.40	59.69	1.18	All	53.25	52.39	1.91

Note: The abbreviations PG and PH correspond to the plots at the Guangde Forest Station and the Huangshan Forest Station, respectively. BD_{WD}, BD_{AE}, and RE are estimated biomass density from Weibull-distribution model and allometric equation alone and relative error calculated using the equation RE (%) = (BD_{WD} − BD_{AE})/BD_{WD} × 100%.

Table 2. Comparison of biomass density (BD, t/ha) based on both the Weibull-distribution model and allometric equation alone for the 27 moso bamboo stands.

$$y = N\int_{D_{min}}^{D_{max}} g(x)f(x)dx \tag{5}$$

where x and y denote D and stand total biomass; $g(x)$ is the allometric relationship of stand biomass versus D; $f(x)$ is the probability density function of D for the given stand; D_{min} and D_{max} represent the minimum and maximum D for the stand, respectively; and N is the total culm number within the bamboo stand.

4.3. Large-scale biomass estimate

4.3.1. *Mean biomass density method*

Early in the International Biosphere Plan (IBP) period, Whittaker et al. [31, 32] have assessed forest biomass and carbon stock on the regional and global scales, via mean biomass density method, where one can estimate biomass for a stand or forest by the mean biomass density multiplied by the area.

Shi [33] selected 36 plots of moso bamboo forests to first calculate the mean D and biomass at the stand level using filed survey data and the developed allometric equation (**Figure 2**) and then estimate forest biomass and carbon stock via mean biomass density method (**Table 3**).

According to the sixth National Forest Inventory (NFI) data, bamboo forest area in Anhui Province is about 152,700 ha [34], so bamboo forest biomass of 1999–2003 period in the southern Anhui Province was estimated about 5.70 Tg (=37.33 t ha × 15.27 × 104 ha) (1 Tg = 10^{12} g), approximately 0.05% of the national forest biomass of the same period [35].

Plot	D_{mean} (cm)	Biomass density (t/ha)	Carbon density (t/ha)	Plot	D_{mean} (cm)	Biomass density (t/ha)	Carbon density (t/ha)
P1	9.56	45.94	22.97	P20	4.76	16.56	8.28
P2	7.82	33.03	16.52	P21	4.62	8.81	4.4
P3	9.31	37.99	18.99	P22	6.68	14.66	7.33
P4	6.4	34.26	17.13	P23	6.71	16.64	8.32
P5	6.7	23.33	11.66	P24	8.12	34.5	17.25
P6	8.79	47.79	23.89	P25	10.57	52.75	26.37
P7	7.73	60.46	30.23	P26	8.11	21.46	10.73
P8	8.54	64.65	32.33	P27	6.12	37.49	18.75
P9	8.89	72.69	36.35	P28	5.56	22.51	11.26
P10	7.82	38.89	19.45	P29	5.98	25.26	12.63
P11	8.09	46.84	23.42	P30	7.14	24.35	12.18
P12	7.18	43.36	21.68	P31	5.91	18.44	9.22
P13	7.81	30.61	15.31	P32	7.86	31.12	15.56
P14	9.81	52.73	26.37	P33	9.00	60.95	30.48
P15	7.73	45.80	22.9	P34	10.51	49.96	24.98
P16	8.02	27.09	13.54	P35	10.77	57.13	28.57
P17	8.72	29.62	14.81	P36	8.17	39.05	19.53
P18	6.33	22.09	11.05				
P19	9.07	54.91	27.46	Mean		37.33	18.66

Note: The biomass of *Cunninghamia lanceolata* in each bamboo stand was estimated with the one-parameter power equation developed by Li et al. [36] ($y = 0.1606D^{2.1203}$, $R^2 = 0.99$, $P < 0.001$).

Table 3. Mean biomass method-derived biomass and carbon stock at the stand level in South Anhui Province, China.

However, the measured biomass density (i.e., data or plot) is more limited in this method, and plot location is frequently in the well-growing stand, easily leading to an overestimation [37, 38].

4.3.2. Geostatistics in biomass estimation

As a spatial statistics, geostatistics has become an indispensable method used to study the nature with the dual characteristics of randomness and regularity over the past 50 years [39]. Forest is affected by physical, climate, and other natural disturbances with a high degree of heterogeneity and relevance; therefore geostatistics is also gradually used in forest ecology, including forest biomass estimate [40, 41].

Zhang et al. [42] confirmed the feasibility of geostatistical methods used in estimating bamboo forest biomass. They chose Huangkeng Town, southern Wuyishan Mountain, as the study area and cut 103 sample bamboo culms to develop an allometric equation for moso bamboo of the region, combined with data of field survey and leaf area index data to estimate the biomass of a total of 209 plots at stand level. By means of ARCGIS software, statistical technique was used to estimate bamboo forest biomass of the whole town, and spatial distribution of biomass was also visualized.

4.3.3. National forest inventory and biomass expansion factor

Many countries have implemented national forest inventory (NFI) regularly or irregularly in order to grasp the forest resources and their dynamics, as well as more scientific development of forest policies [43]. Subsequently, many scholars estimated forest biomass using NFI data at the region level, and then a biomass expansion factor (BEF) method came into being [28, 44]. This method assumes that there is a certain relationship between the forest growing stock and biomass; thus, biomass can be estimated based on the growing stock (derived from NFI) multiplied by the BEF conversion factor. Some researchers hold that BEF is a constant; we can call it mean BEF method. Actually, BEF varied frequently with forest age, site class, and stand density [45–47]. Hence, an improved method (called continuous BEF) [38, 45, 48, 49] is gradually being accepted by many scientists.

Current researches regarding forest biomass change are mainly based on NFI data [46, 47, 50–54]. But for the nation with a larger land area, the NFI points are limited, and remote areas are also difficult to reach, thereby often creating a bias in estimates. Moreover, forest resource assessment is incomparable; for example, the error of tropical deforestation rate estimated by FAO may be up to 50%, which is mainly due to the differences in national inventory methods and the definition of forest in tropical region [3]. Additionally, since NFI data have no information recording the spatial distribution of plots; therefore, spatial variation cannot be analyzed while using NFI data.

4.3.4. Remote sensing technique and its application in biomass estimation

Remote sensing technique developed rapidly in the late twentieth century, and remote sensing data with high spatiotemporal resolution, wide coverage, and timely updates has been widely used in the assessment of forest biomass and carbon stock on various scales [55–57]. Currently, remote sensing-derived biomass estimation has become the leading method of large-scale forest biomass and carbon stock estimation. The use of remote sensing technique to assess forest biomass is mainly based on the normalized difference vegetation index (NDVI) datasets. While using NDVI datasets, most researches frequently overlay the vegetation maps of a region and NDVI datasets to explore the spatiotemporal changes. But this method only pays attention to changes in productivity, without considering the change in area caused by land-use change [58, 59].

The method can be exemplified with the study conducted by Shi [24]. In this paper, National Forest Inventory and normalized difference vegetation index (NDVI, comes from Global Inventory Monitoring and Modeling Studies) datasets were integrated via matching forest type for Heilongjiang, Liaoning, and Jilin Provinces while fitting the inverse model (**Figure 3**). The developed inverse models were used to estimate forest growing stock and carbon stock, respectively. Consequently, changes in growing stock and biomass between 1982 and 2006 were analyzed.

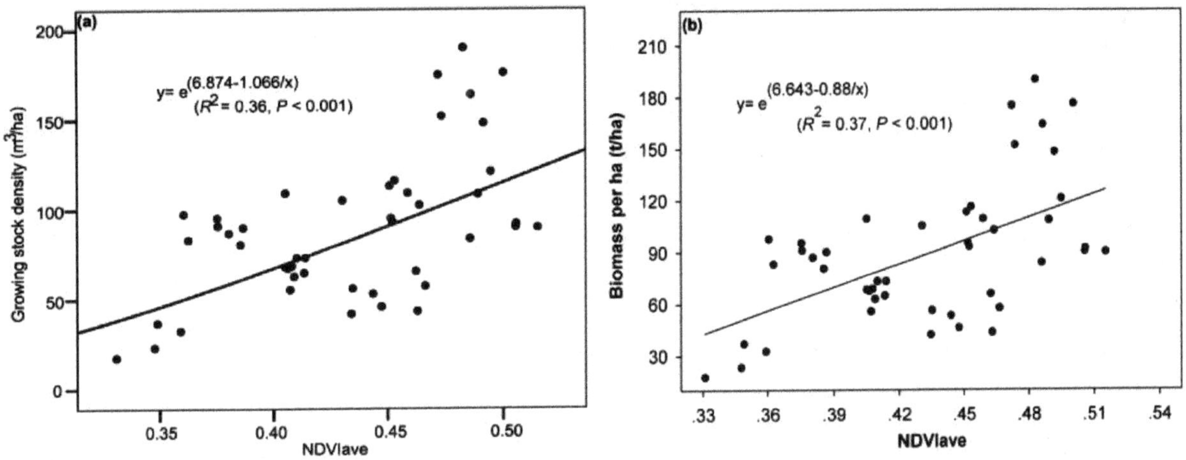

Figure 3. Relationship between forest timber volume density (m^3/ha) (a), biomass density (t/ha) (b), and annual mean NDVI (NDVIave) of northeast China (including Heilongjiang, Liaoning, and Jilin provinces), respectively; four forest types for each inventory period (four inventory periods of and a total of 48 data points) are used in the regression.

4.3.5. Forest identity

For any forest, area, growing stock, biomass, and carbon can be linked using Eqs. (6)–(8). The forest identity method defines these four valued attributes by using measurable variables, and it quantitatively and logically integrates their changes into a causal relationship (i.e., Eqs. (9)–(11)) [60, 61].

$$V\left(\mathrm{m}^3\right) = A\left(\mathrm{ha}\right) \times D\left(\mathrm{m}^3/\mathrm{ha}\right) \tag{6}$$

$$M\left(\mathrm{Mg}\right) = A\left(\mathrm{ha}\right) \times D\left(\mathrm{m}^3/\mathrm{ha}\right) \times B\left(\mathrm{Mg}/\mathrm{m}^3\right) = V\left(\mathrm{m}^3\right) \times B\left(\mathrm{Mg}/\mathrm{m}^3\right) \tag{7}$$

$$Q\left(\mathrm{Mg\ C}\right) = A\left(\mathrm{ha}\right) \times D\left(\mathrm{m}^3/\mathrm{ha}\right) \times B\left(\mathrm{Mg}/\mathrm{m}^3\right) \times C\left(\mathrm{Mg\ C}/\mathrm{Mg}\right) \tag{8}$$

Then, $d\ln(Q)/dt = d\ln(A)/dt + d\ln(D)/dt + d\ln(B)/dt + d\ln(C)/dt$.

Let $q \approx d\ln(Q)/dt$, $a \approx d\ln(A)/dt$,.d $b \approx d\ln(B)/dt$, $d \approx d\ln(B)/dt$, and $c \approx d\ln(C)/dt$.

Then,

$$q = a + d + b + c \tag{9}$$

Similarly,

$$v = a + d \tag{10}$$

And also,

$$m = a + d + b \tag{11}$$

where *V, M, Q, A, D, B,* and *C* represent growing stock (m^3), biomass (Mg), forest carbon stock (Mg C), area (ha), growing stock density (m^3/ha), the conversion ratio of biomass to growing stock (Mg/m^3), and carbon concentration in biomass (Mg C/Mg) at the provincial or national level, respectively, and *v, m, q, a, d, b,* and *c* represent the corresponding derivatives of these attributes with respect to time.

Shi et al. [35] applied both forest identity and regression approaches to explore the temporal changes in China's forest. This paper showed that China's forest area and growing stock density increased by 0.51 and 0.44% annually over the past three decades, while the conversion ratio of forest biomass to growing stock declined by 0.10% annually. These developments resulted in a net annual increase in 0.85% in forest carbon sequestration, which is equivalent to a net biomass carbon uptake of 43.8 Tg per year (1 Tg = 10^{12} g).

In the paper, two regression equations below (i.e., Eqs. (12) and (13)) were used to obtain the derivatives of the forest attributes with respect to time:

$$y = \text{slope} \times x + \text{intercept} \tag{12}$$

$$RR(\text{yr} / \%) = \{\text{slope}/[(y_1 + y_2 + y_3 + y_4 + y_5)/5]\} \times 100 \tag{13}$$

where *y* represents the forest attributes (i.e., area, growing stock density, the conversion ratio, or carbon content) at the provincial or national level; y_1, y_2, y_3, y_4, and y_5 denote the corresponding forest attributes for the inventories of 1977–1981, 1984–1988, 1989–1993, 1994–1998, and 1999–2003, respectively; *RR* (%/yr) is the relative annual change rate of the forest attributes; slope denotes the amplitude and direction of annual absolute change for each forest attribute; and *x* represents the corresponding periods of NFI. In other words, the relative annual change rate (*RR*, %) defined here is equivalent to *q, a, d,* or *b* mentioned above.

5. Method comparison

Each estimation method has its advantages and disadvantages. None of these methods mentioned in this chapter is the best from individual to large scale (**Table 4**). Fitting allometric equation requires some down sampling trees, but the number is small relative to all the standing trees. The equation (coefficients) varies frequently with species, terrain, temperature, and rainfall. To improve its prediction power, combining the field survey with LIDAR and incorporating the variation into allometric coefficients are the two key elements. It is fully convinced that allometric equation, together with LIDAR, is increasing widely used in estimate forest biomass at individual and stand levels in the future.

Method	Scale	Major limitation	Improvement practices
Allometric equation	Individual or stand	Varying frequently with species, terrain, temperature, and rainfall; less sampling trees	Incorporating these factors into allometric coefficients; combine with LIDAR
Mean biomass density	Stand or region	Easily leading to an overestimation	Randomly set more plots
Biomass expansion factor	Stand	Varying frequently with species, terrain, temperature, and rainfall	Incorporating these into conversion factor
Forest identity	Region	Comprehensive analysis	–
Remote sensing	Stand or region	Saturation and bidirectional reflectance of surface features	Higher spatiotemporal resolution, advanced algorithm and technology
Geostatistics	Region	More field data	Constructing the biomass database

Table 4. Comparison of the forest biomass estimation methods.

Mean biomass density easily leads to an overestimation. However, the uncertainty will be reduced if we randomly select some plots across the study region as more as possible.

Like allometric equation, biomass expansion factor also varies frequently with species, terrain, temperature, and rainfall. It is of significance to explore the variation of conversion factor with species and environment; then the conversion factor can be estimated as possible as we can.

Forest identity is a tool used to comprehensively analyze the forest change, which is more likely be employed only by a small number of scientists.

Remote sensing technique has been booming since the 1980s, but saturation of vegetation index (e.g., widely used NDVI) and bidirectional reflectance of surface features still occur. Exploring the higher spatiotemporal resolution, advanced algorithm and technology are needed in the future.

Geostatistics is an important technique to upscale forest biomass but needs more field data. More field data implies more cost and time-consuming. It is, therefore, important to construct the biomass database.

6. Uncertainty

Many scholars have conducted researches to estimate forest biomass and carbon stock on various scales, but estimates have a large uncertainty [62–65]. Dixon et al. [37] used mean biomass density method to estimate China's forest, whose carbon storage and carbon density were 7.0 Pg C and 114 t/ha, while biomass was estimated 6.2 Pg C and 57.07 t/ha based on the 720 plots conducted by Zhou et al. [66]. Estimation error of tropical deforestation rates may be as high as 50% due to the difference by national forest inventory method in tropical regions and different definitions of forest [3]. With regard to carbon stock for the Amazon rain forest, the estimation results varied from 58 [67] to 134 Pg C [68], as well as 89 Pg C estimated by FAO [2]. Given the large uncertainty, Houghton et al. [69] conducted a study regarding the comparison of different methods by different scholars and concluded that carbon storage estimates are rarely considerable, even using similar data and the same estimation method.

As addressed above, the procedure of estimating forest biomass includes four steps (**Figure 1**). Each step can produce uncertainty: step 1 induces measure error, step 2 causes model error, step 3 involves sampling error, and step 4 often ignores biomass spatial variability and increases uncertainty. Uncertainty resulting from the previous steps will affect (propagate) the following biomass estimation process (result). Compared to measure and sampling errors, model and spatial variability errors are much larger but easier to solve. Taking the inverse model (see **Figure 3**), for example, the densities of growing stock and biomass both have significant relationships with NDVI, but determination coefficients (R^2) are very low, implying that the current remote sensing data is unable to fully retrieve forest biomass. This is mainly caused by the complicated soil background, bidirectional reflectance of surface features, and saturation of NDVI. Higher spatiotemporal remote sensing data in the future can improve the prediction accuracy while using inverse model.

In ecological researches, a number of parameters (such as biomass) are not easy to be measured directly; thus, the indirect measurement is available, namely, measuring certain indirect variables first and calculating direct one based on the relationship between indirect and direct variables. Direct measurement error will be transformed and propagated to indirect measurement error by a certain function, namely, error propagation [70]. Overall, the uncertainty is mainly due to the spatial variability of forest carbon stock and error propagated from the estimation process [26, 71, 72]. In general, the total uncertainty can be obtained by summing the error produced during the procedure [73].

Currently, little research has been conducted on the uncertainty of biomass estimation [73], and only a few studies have focused on one or several errors. For example, Ketterings et al. [74] have focused on the error induced by step 2 and presented a simple method to reduce error based on the tree *H* and stand density. Phillips et al. [73] analyzed the uncertainty using NFI data of five states of the USA and found that the uncertainty of large-scale forest volume estimation was mainly from the sampling error. Chave et al. [26] pointed out that sampling, identifying allometric equation, measuring tree *D*, and other stand variables can propagate error to the aboveground biomass estimates and the developed allometric model accounts for the largest source of uncertainty; thus, fitting allometric model should be given adequate attention. Holdaway et al. [75] used Monte Carlo method to analyze the uncertainty while estimating carbon stock in temperate rain forest, New Zealand. Su et al. [76] pointed out that non-geo-referenced plot location would bring biomass estimation error and also estimated the aboveground biomass of China's forest via incorporating sampling error model.

With the increasing value of climate change by world dignitaries and growing global carbon market, the accuracy problems in estimating forest carbon stock on various scales have become increasingly prominent. The uncertainty in forest carbon reserve estimation can involve the climate negotiations and the development of carbon trading market to some extent, thereby affecting the development of government policies and international negotiations regarding the forest response to climate change. Therefore, the scientific community has become increasingly concerned about the uncertainty in the estimation of forest carbon stock. Changes in publication papers from 1990 to date regarding biomass estimation uncertainty can be evident (**Figure 4**); prior to 1997, the number of papers is relatively stable, but article number increases exponentially since 1997 (when Kyoto Protocol effects).

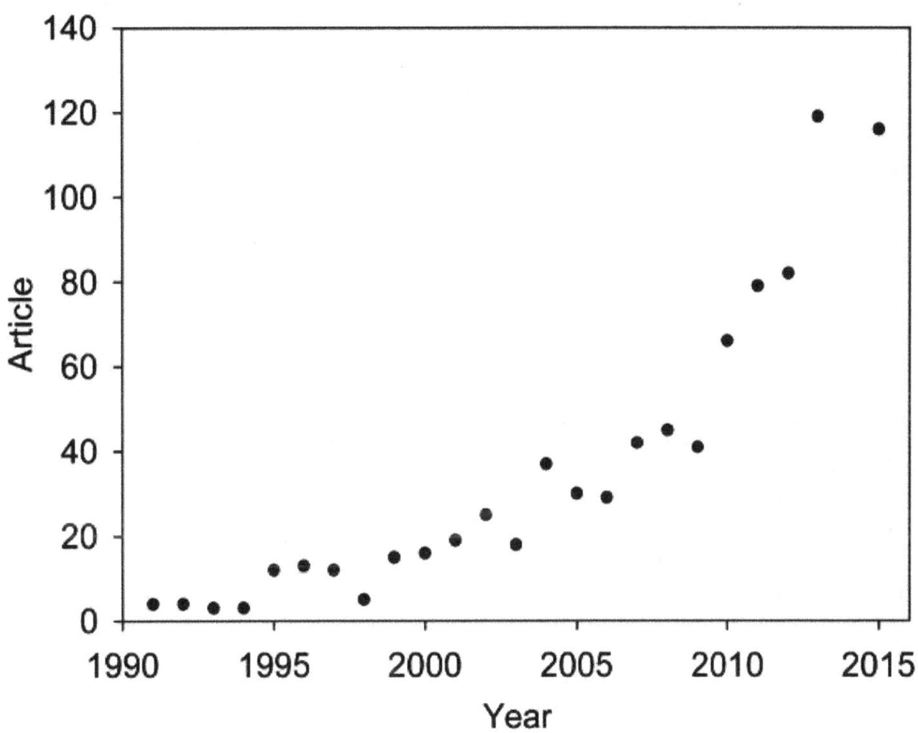

Figure 4. Articles regarding the uncertainty analysis of forest carbon stock versus time. The number is obtained via searching forest, biomass, and uncertainty in core journal of ISI Web of Knowledge.

7. Challenges and prospects

7.1. Global field biomass dataset

Although remote sensing, NFI, geographic information system, and other advanced techniques have been applied to estimate forest biomass, but the accuracy of estimate of forest biomass needs to be improved. In recent years, with the advent of LIDAR, multispectral data, and geostatistical technique, most scholars paid more attention to the application of such advanced technology in biomass estimates but ignored the acquisition and compiling field biomass data. Although acquisition of field biomass data using harvesting method is time-consuming and laborious, the fitted inverse model based on remote sensing data must be verified using the measured data on biomass. Matching remote sensing data and ground-truth data is often a difficult problem during biomass estimation. In various biomass estimation methods, most researchers assume that the ground-truth biomass data are the most accurate. It is thus important to carry out field biomass measurement globally using a unified investigation specification, as well as the establishment of a unified biomass database.

7.2. Upscaling in biomass estimation

Scale has been one of the core focuses of ecology. Scale and upscale affect the accuracy of forest biomass estimation and are also an important issue in biomass estimation. Biomass estimation method varied with scale. The mean biomass density method, BEF method, remote sensing, and geostatistic method are used to estimate large-scale forest biomass and carbon stock, which involves scale biomass (upscaling); allometric equation is used for small-scale biomass estimation. However, large-scale biomass estimate must be based on the allometric equation. If forest biomass estimation method you use is different from another, estimate may be different even if the same data is used. The rapid development of remote sensing data, geostatistical technique, and others make scaling possible while estimating biomass. Therefore, upscaling forest biomass from stand scale to regional and even global scale should be one of the key elements for future biomass researches. Only by addressing the scaling issue and clarifying the spatial variation of forest biomass can a unified framework or system for estimating forest biomass be presented. Then national and international standards of estimating forest biomass can be developed to improve biomass estimation with a high accuracy.

7.3. High-accuracy biomass estimate needs remote sensing

Remote sensing data with a high spatiotemporal resolution, wide coverage, and timely updates have been widely used in the assessment of forest biomass and carbon stock on various scales, which play an important role in improving estimation accuracy.

In addition, scaling biomass cannot do without application of remote sensing data. However, remote sensing data have no continuity of the time series, the resolution is not high enough, and the fitted inverse model is often saturated. Moreover, the quality of the current multispectral data is not high, and LIDAR data and UAV technology applied in forest biomass estimate are not yet mature. Improving the quality of multispectral data, LIDAR and UAV technology are two of the most active frontiers in the future.

7.4. Geographical variation of allometric equation

As mentioned above, allometric equation is indispensable in biomass estimation. However, the coefficients of the equation varied frequently with species, terrain, temperature, and rainfall [77–79]. Small variation in the allometric coefficients is likely to result in a larger biomass estimation error [80]. Therefore, the clarification of the variation in the allometric coefficients along different gradients is also the element to improve the accuracy of biomass estimation.

Biomass estimate may be different using different allometric equations while estimating vegetation biomass. Thus, it is likely to come to inconsistent results, even when different scholars estimate the same forest. In other words, the choice of the appropriate predictors and optimization of allometric equation and its parameters can contribute to reduce biomass estimation uncertainty and improve the ability to predict the biomass and carbon stock. It is worth noting that recently developed metabolic rate theory (fractal networks) is contrary to conventional wisdom. The theory inference indicated that the forest biomass (*M*) and its *D* on the species and biota scale both showed a constant relationship, which can be written as $M \propto D^{8/3}$ [81–83]. Whether the coefficients of biomass allometric equation varied with the species, site, and climatic factors or not is a hot debate in recent years [84]. If the relationship between forest biomass and the corresponding *D* remains to be constantly exponential on various scales, a unified allometric biomass estimation model is more likely to be developed, which helps to save costs in in situ measurements of biomass, but also can contribute to promote the in-depth study of biomass estimation methods.

8. Conclusions

Plant allometry is the theoretical basis of vegetation biomass estimation. Generally, the use of allometric equation is indispensable to estimate biomass for both tree and forest. By reviewing the methods used for estimating forest biomass, we can conclude that each estimation method has its advantages and disadvantages, and none of these methods mentioned is always the best from individual to large scales. A large uncertainty exists in biomass estimation, which is unable to meet the requirement of the accurate carbon accounting required by Kyoto Protocol. It is of significance and urgently needed to develop the more suitable and rigor method to accurately estimate the size and distribution of forest biomass. To achieve this goal, we argued that the new developed technique such as geostatistics and remote sensing technique (e.g., LIDAR) would be the key tools to improve forest biomass estimation with a high accuracy. However, prior to this, spatial variation in forest biomass at various levels should be explored using multi-source data and multi-approaches.

Acknowledgements

This study was supported by the Special Fund for Basic Scientific Research of International Center for Bamboo and Rattan (1632015005) and the National Natural Science Foundation of China (31300177).

Author details

Lei Shi[1*] and Shirong Liu[1,2]

*Address all correspondence to: leishi@icbr.ac.cn

1 State Forestry Administration Key Laboratory for Science & Technology of Bamboo & Rattan, International Center for Bamboo and Rattan, Beijing, China

2 Institute of Forest Ecology, Environment and Protection, Chinese Academy of Forestry, Beijing, China

References

[1] FAO (Food and Agriculture Organization, the United Nations). State of the world's forests 2011. Rome: UN Food and Agriculture Organization; 2011.

[2] FAO (Food and Agriculture Organization, the United Nations). Global Forests Resources Assessment 2010: Main Report. Rome: UN Food and Agriculture Organization; 2010.

[3] Watson RT, Noble A, Bolin B, Ravindranath NH, Verado D. Land use, land-use change, and forestry. Intergovernmental Panel on Climate Change Special Report. Cambridge: Cambridge University Press; 2000.

[4] IPCC (Intergovernmental Panel on Climate Change) Climate Change 2007: Synthesis Report. Core Writing Team P, Pachauri RK, Reisinger A, Editor. Geneva, Switzerland: Contribution of Working Groups I, II and III to the Fourth Assessment Report of the Intergovernmental Panel on Climate Change; 2007.

[5] Jackson RB, Jobbagy EG, Avissar R, Roy SB, Barrett DJ, Cook CW, Farley KA, le Maitre DC, McCarl BA, Murray BC. Trading water for carbon with biological sequestration. Science. 2005; 310: 1944–1947. DOI: 10.1126/science.1119282.

[6] Piao SL, Friedlingstein P, Ciais P, de Noblet-Ducoudre N, Labat D, Zaehle S. Changes in climate and land use have a larger direct impact than rising CO_2 on global river runoff trends. Proceedings of the National Academy of Sciences of the United States of America. 2007; 104: 15242–15247. DOI: 10.1073/pnas.0707213104.

[7] Niklas KJ, Owens T, Reich PB, Cobb ED. Nitrogen/phosphorus leaf stoichiometry and the scaling of plant growth. Ecology Letters. 2005; 8: 636–642. DOI: 10.1111/j.1461-0248.2005.0075 9.x.

[8] Huxley JS. Problems of relative growth. London: Methuen. 1932. 276.

[9] Enquist BJ, Niklas KJ. Invariant scaling relations across tree-dominated communities. Nature. 2001; 410: 655–660. DOI: 10.1038/35070500.

[10] Cheng DL. Study on the correlation between plant biomass allocation model and growth rate [Thesis]. Lanzhou: Lanzhou University; 2007.

[11] Shanmughavel P and Francis K. Above ground biomass production and nutrient distribution in growing bamboo (*Bambusa bambos* (L) Voss). Biomass & Bioenergy. 1996; 10: 383–391.

[12] Wang CK. Biomass allometric equations for 10 co-occurring tree species in Chinese temperate forests. Forest Ecology and Management. 2006; 222: 9–16. DOI: 10.1016 / j.foreco.2005.10.074.

[13] Gao X, Jiang ZH, Guo QR, Zhang Y, Yin H, He F, Qi LH, Shi L. Allometry and biomass production of *Phyllostachys edulis* across the whole lifespan. Polish Journal of Environmental Studies. 2015; 24: 511–517.

[14] Feldpausch TR, Banin L, Phillips OL, Baker TR, Lewis SL, Quesada CA, Affum-Baffoe K, Arets E, Berry NJ, Bird M, Brondizio ES, de Camargo P, Chave J, Djagbletey G, Domingues TF, Drescher M, Fearnside PM, Franca MB, Fyllas NM, Lopez-Gonzalez G, Hladik A, Higuchi N, Hunter MO, Iida Y, Salim KA, Kassim AR, Keller M, Kemp J, King DA, Lovett JC, Marimon BS, Marimon BH, Lenza E, Marshall AR, Metcalfe DJ, Mitchard ETA, Moran EF, Nelson BW, Nilus R, Nogueira EM, Palace M, Patino S, Peh KSH, Raventos MT, Reitsma JM, Saiz G, Schrodt F, Sonke B, Taedoumg HE, Tan S, White L, Woll H, Lloyd J. Height-diameter allometry of tropical forest trees. Biogeosciences. 2011; 8: 1081–1106. DOI: 10.5194/bg-8-1081-2011.

[15] Peng C, Zhang L, Liu J. Developing and validating nonlinear height-diameter models for major tree species of Ontario's boreal forests. Northern Journal of Applied Forestry. 2001; 18: 87–94.

[16] Gao X, Li ZD, Yu HM, Jiang ZH, Wang C, Zhang Y, Qi LH, Shi L. Modeling of the height-diameter relationship using an allometric equation model: a case study of stands of *Phyllostachys edulis*. Journal of Forestry Research. 2016; 27: 339–347. DOI: 10.1007/s11676-015-0145-6.

[17] Gower ST, Kucharik CJ, Norman JM. Direct and indirect estimation of leaf area index, f(APAR), and net primary production of terrestrial ecosystems. Remote Sensing of Environment. 1999; 70: 29–51. DOI: 10.1016/s0034-4257(99)00056-5.

[18] Mowrer HT and Frayer WE. Variance propagation in growth and yield projections. Canadian Journal of Forest Research-Revue Canadienne De Recherche Forestiere. 1986; 16: 1196–1200. DOI: 10.1139/x86-213.

[19] Pilli R, Anfodillo T, Carrer M. Towards a functional and simplified allometry for estimating forest biomass. Forest Ecology and Management. 2006; 237: 583–593. DOI: 10.1016/j.foreco.2006.10.004.

[20] Andrade HA, Campos RO. Allometry coefficient variations of the length-weight relationship of skipjack tuna (*Katsuwonus pelamis*) caught in the southwest South Atlantic. Fisheries Research. 2002; 55: 307–312. DOI: 10.1016/s0165-7836(01)00305-8.

[21] Singh AN, Singh JS. Biomass, net primary production and impact of bamboo plantation on soil redevelopment in a dry tropical region. Forest Ecology and Management. 1999; 119: 195–207.

[22] Isagi Y, Kawahara T, Kamo K. Biomass and net production in a bamboo *Phyllostachys-Bambusoides* stand. Ecological Research. 1993; 8: 123–133.

[23] Shi L. Changes in China's forests over the past 25 years [Thesis]. Beijing: Peking University; 2011.

[24] Shi L. Changes of forest in Northeast China over the past 25 years: an analysis based on remote sensing technique. Proceedings of SPIE. 2011; 8203: The Society of Photo-Optical Instrumentation Engineers (SPIE); International Society for Digital Earth; UNDESA GAID e-SDDC; Asian Association on Remote Sensing; IEEE Beijing Section.

[25] Mitchard ETA, Saatchi SS, Lewis SL, Feldpausch TR, Gerard FF, Woodhouse IH, Meir P. Comment on 'A first map of tropical Africa's above-ground biomass derived from satellite imagery'. Environmental Research Letters. 2011; 6: 6. DOI: 10.1088/1748-9326/6/4/04 9 001.

[26] Chave J, Condit R, Aguilar S, Hernandez A, Lao S, Perez R. Error propagation and scaling for tropical forest biomass estimates. Philosophical Transactions of the Royal Society of London Series B-Biological Sciences. 2004; 359: 409–420. DOI: 10.1098/rstb.2003.1425.

[27] Matthews G. The carbon content of trees. Edinburgh, Scotland: Forestry Commission; 1993.

[28] Birdsey RA. Carbon storage and accumulation in United States forest ecosystems. Washington D.C.: U.S. Department of Agriculture, Forest Service, Washington Office; 1992.

[29] Brown S. Estimating biomass and biomass change of tropical forests: a primer. Rome: Food and Agriculture Organization of the United Nations; 1997.

[30] Qi LH, Liu XJ, Jiang ZH, Yue XH, Li ZD, Fu JH, Liu GL, Guo BH, Shi L. Combining diameter-distribution function with allometric equation in biomass estimates: a case study of *Phyllostachys edulis* forests in South Anhui, China. Agroforestry Systems. 2015. DOI: 10.1007/s 10457-015-9887-6.

[31] Whittaker RH. Net production of heath balds and forest heaths in the Great Smoky Mountains. Ecology. 1963; 44: 176–182.

[32] Woodwell GM and Whittaker RH. Primary production in terrestrial ecosystems. American Zoologist. 1968; 8: 19–30.

[33] Shi L. Carbon stock estimation of *Phyllostachys edulis* forest: a case study in South Anhui Province, China. Beijing: International Center for Bamboo and Rattan (ICBR). 2014; p. 55.

[34] SFA (State Forestry Administration). National Forest Resources Statistics (1999–2003). Beijing: China Forestry Publishing House; 2004.

[35] Shi L, Zhao SQ, Tang ZY, Fang JY. The changes in China's forests: an analysis using the forest identity. PLoS One. 2011; 6: e20778. DOI: 10.1371/journal.pone.0020778.

[36] Li Y, Zhang JG, Duan AJ, Xiang CW. Selection of biomass estimation models for Chinese fir plantation. Chinese Journal of Applied Ecology. 2010; 21: 3036–3046.

[37] Dixon RK, Brown S, Houghton RA, Solomon AM, Trexler MC, Wisniewski J. Carbon pools and flux of global forest ecosystems. Science. 1994; 263: 185–190.

[38] Fang JY, Oikawa T, Kato T, Mo WH, Wang ZH. Biomass carbon accumulation by Japan's forests from 1947 to 1995. Global Biogeochemical Cycles. 2005; 19: Gb2004. DOI: 10.1029/2004gb002253.

[39] Feng YM, Tang SZ, Li ZY. Application of spatial statistic analysis in forestry. Scientia Silvae Sinicae. 2004; 40: 149–155.

[40] Yadav BKV, Nandy S. Mapping aboveground woody biomass using forest inventory, remote sensing and geostatistical techniques. Environmental Monitoring and Assessment. 2015; 187: DOI: 10.1007/s10661-015-4551-1.

[41] Barra M, Petitgas P, Bonanno A, Somarakis S, Woillez M, Machias A, Mazzola S, Basilone G, Giannoulaki M. Interannual changes in biomass affect the spatial aggregations of anchovy and sardine as evidenced by geostatistical and spatial indicators. Plos One. 2015; 10: e0135808. DOI: 10.1371/journal.pone.0135808.

[42] Zhang Y, Yue XH, Qi LH, Jiang ZH, Shi L. Estimation of *Phyllostachys edulis* forest biomass in southern Wuyishan Mountain using allometric equation and geostatistical technique. Chinese Journal of Ecology. 2016; 35: 1957–1962.

[43] Kangas A, Maltamo M. Forest inventory: methodology and applications. Dordrecht, the Netherlands: Springer Verlag. 2006.

[44] Turner DP, Koerper GJ, Harmon ME, Lee JJ. A carbon budget for forests of the conterminous United-States. Ecological Applications. 1995; 5: 421–436.

[45] Brown S, Lugo AE. Aboveground biomass estimates for tropical moist forests of the Brazilian Amazon. Interciencia. 1992; 17: 8–18.

[46] Fang JY, Chen AP, Peng CH, Zhao SQ, Ci L. Changes in forest biomass carbon storage in China between 1949 and 1998. Science. 2001; 292: 2320–2322.

[47] Fang JY, Wang GG, Liu GH, Xu SL. Forest biomass of China: an estimate based on the biomass-volume relationship. Ecological Applications. 1998; 8: 1084–1091.

[48] Fang JY, Guo ZD, Piao SL, Chen AP. Terrestrial vegetation carbon sinks in China, 1981–2000. Science in China Series D-Earth Sciences. 2007; 50: 1341–1350. DOI: 10.1007/s11430-007-0049-1.

[49] Schroeder P, Brown S, Mo JM, Birdsey R, Cieszewski C. Biomass estimation for temperate broadleaf forests of the United States using inventory data. Forest Science. 1997; 43: 424–434.

[50] Corona P, Fattorini L, Franceschi S. Estimating the volume of forest growing stock using auxiliary information derived from relascope or ocular assessments. Forest Ecology and Management. 2009; 257: 2108–2114. DOI: 10.1016/j.foreco.2009.02.017.

[51] Fazakas Z, Nilsson M, Olsson H. Regional forest biomass and wood volume estimation using satellite data and ancillary data. Agricultural and Forest Meteorology. 1999; 98-9: 417–425.

[52] Franco-Lopez H, Ek AR, Bauer ME. Estimation and mapping of forest stand density, volume, and cover type using the k-nearest neighbors method. Remote Sensing of Environment. 2001; 77: 251–274.

[53] Al-Bakri JT and Taylor JC. Application of NOAA AVHRR for monitoring vegetation conditions and biomass in Jordan. Journal of Arid Environments. 2003; 54: 579–593. DOI: 10.1006/jare.2002.1081.

[54] Peng CH, Zhou XL, Zhao SQ, Wang XP, Zhu B, Piao SL, Fang JY. Quantifying the response of forest carbon balance to future climate change in Northeastern China: model validation and prediction. Global and Planetary Change. 2009; 66: 179–194. DOI: 10.1016/ j.gloplacha.2008.12.001.

[55] Lu DS. The potential and challenge of remote sensing-based biomass estimation. International Journal of Remote Sensing. 2006; 27: 1297–1328. DOI: 10.1080/01431160500486732.

[56] Nelson R, Krabill W, Tonelli J. Estimating forest biomass and volume using airborne laser data. Remote Sensing of Environment. 1988; 24: 247–267.

[57] Timothy D, Onisimo M, Cletah S, Adelabu S, Tsitsi B. Remote sensing of aboveground forest biomass: a review. Tropical Ecology. 2016; 57: 125–132.

[58] Piao SL, Fang JY, Zhou LM, Ciais P, Zhu B. Variations in satellite-derived phenology in China's temperate vegetation. Global Change Biology. 2006; 12: 672–685. DOI: 10.1111/j.1365-2486.2006.01123.x.

[59] Myneni RB, Dong J, Tucker CJ, Kaufmann RK, Kauppi PE, Liski J, Zhou L, Alexeyev V, Hughes MK. A large carbon sink in the woody biomass of northern forests. Proceedings of the National Academy of Sciences of the United States of America. 2001; 98: 14784–14789.

[60] Waggoner PE. Using the Forest Identity to grasp and comprehend the swelling mass of forest statistics. International Forestry Review. 2008; 10: 689–694.

[61] Kauppi PE, Ausubel JH, Fang JY, Mather AS, Sedjo RA, Waggoner PE. Returning forests analyzed with the forest identity. Proceedings of the National Academy of Sciences of the United States of America. 2006; 103: 17574–17579. DOI: 10.1073/pnas.0608343103.

[62] Mitchard ETA, Feldpausch TR, Brienen RJW, Lopez-Gonzalez G, Monteagudo A, Baker TR, Lewis SL, Lloyd J, Quesada CA, Gloor M, ter Steege H, Meir P, Alvarez E, Araujo-Murakami A, Aragao L, Arroyo L, Aymard G, Banki O, Bonal D, Brown S, Brown FI, Ceron CE, Moscoso VC, Chave J, Comiskey JA, Cornejo F, Medina MC, Da Costa L, Costa FRC, Di Fiore A, Domingues TF, Erwin TL, Frederickson T, Higuchi N, Coronado ENH, Killeen TJ, Laurance WF, Levis C, Magnusson WE, Marimon BS, Marimon BH, Polo IM, Mishra P, Nascimento MT, Neill D, Vargas MPN, Palacios WA, Parada A, Molina GP, Pena-Claros M, Pitman N, Peres CA, Poorter L, Prieto A, Ramirez-Angulo H, Correa ZR, Roopsind A, Roucoux KH, Rudas A, Salomao RP, Schietti J, Silveira M, de Souza PF, Steininger MK, Stropp J, Terborgh J, Thomas R, Toledo M, Torres-Lezama A, van Andel TR, van der Heijden GMF, Vieira ICG, Vieira S, Vilanova-Torre E, Vos VA, Wang O, Zartman CE, Malhi Y, Phillips OL. Markedly divergent estimates of Amazon forest carbon density from ground plots and satellites. Global Ecology and Biogeography. 2014; 23: 935–946. DOI: 10.1111/ geb.12168.

[63] Yuen JQ, Fung T, Ziegler AD. Review of allometric equations for major land covers in SE Asia: uncertainty and implications for above- and below-ground carbon estimates. Forest Ecology and Management. 2016; 360: 323–340. DOI: 10.1016/j.foreco.2015.09.016.

[64] Chen Q, Laurin GV, Valentini R. Uncertainty of remotely sensed aboveground biomass over an African tropical forest: propagating errors from trees to plots to pixels. Remote Sensing of Environment. 2015; 160: 134–143. DOI: 10.1016/j.rse.2015.01.009.

[65] Ahmed R, Siqueira P, Hensley S, Bergen K. Uncertainty of forest biomass estimates in north temperate forests due to allometry: implications for remote sensing. Remote Sensing. 2013; 5: 3007–3036. DOI: 10.3390/rs5063007.

[66] Zhou YR, Yv ZL, Zhao SD. Carbon storage and budget of major Chinese forest types. Chinese Journal of Plant Ecology. 2000; 24: 518–522.

[67] Olson JS, Watts JA, Allison LJ. Carbon in live vegetation of major world ecosystems. Oak Ridge National Lab., Oak Ridge, TN (USA). 1983.

[68] Fearnside PM. Greenhouse gases from deforestation in Brazilian Amazonia: net committed emissions. Climatic Change. 1997; 35: 321–360. DOI: 10.1023/a:1005336724350.

[69] Houghton RA, Lawrence KT, Hackler JL, Brown S. The spatial distribution of forest biomass in the Brazilian Amazon: a comparison of estimates. Global Change Biology. 2001; 7: 731–746. DOI: 10.1046/j.1365-2486.2001.00426.x.

[70] Yang ZC. Error Theory. Changsha: Central South University of Technology Press. 1987.

[71] Rejou-Mechain M, Muller-Landau HC, Detto M, Thomas SC, Le Toan T, Saatchi SS, Barreto-Silva JS, Bourg NA, Bunyavejchewin S, Butt N, Brockelman WY, Cao M,

Cardenas D, Chiang JM, Chuyong GB, Clay K, Condit R, Dattaraja HS, Davies SJ, Duque A, Esufali S, Ewango C, Fernando RHS, Fletcher CD, Gunatilleke I, Hao Z, Harms KE, Hart TB, Herault B, Howe RW, Hubbell SP, Johnson DJ, Kenfack D, Larson AJ, Lin L, Lin Y, Lutz JA, Makana JR, Malhi Y, Marthews TR, McEwan RW, McMahon SM, McShea WJ, Muscarella R, Nathalang A, Noor NSM, Nytch CJ, Oliveira AA, Phillips RP, Pongpattananurak N, Punchi-Manage R, Salim R, Schurman J, Sukumar R, Suresh HS, Suwanvecho U, Thomas DW, Thompson J, Uriarte M, Valencia R, Vicentini A, Wolf AT, Yap S, Yuan Z, Zartman CE, Zimmerman JK, Chave J. Local spatial structure of forest biomass and its consequences for remote sensing of carbon stocks. Biogeosciences. 2014; 11: 6827–6840. DOI: 10.5194/bg-11-6827-2014.

[72] Gregoire TG, Naesset E, McRoberts RE, Stahl G, Andersen HE, Gobakken T, Ene L, Nelson R. Statistical rigor in LiDAR-assisted estimation of aboveground forest biomass. Remote Sensing of Environment. 2016; 173: 98–108. DOI: 10.1016/j.rse.2015.11.012.

[73] Phillips DL, Brown SL, Schroeder PE, Birdsey RA. Toward error analysis of large-scale forest carbon budgets. Global Ecology and Biogeography. 2000; 9: 305–313. DOI: 10.1046/j.1365-2699.2000.00197.x.

[74] Ketterings QM, Coe R, van Noordwijk M, Ambagau Y, Palm CA. Reducing uncertainty in the use of allometric biomass equations for predicting above-ground tree biomass in mixed secondary forests. Forest Ecology and Management. 2001; 146: 199–209. DOI: 10.1016/s0378-1127(00)00460-6.

[75] Holdaway RJ, McNeill SJ, Mason NWH, Carswell FE. Propagating uncertainty in plot-based estimates of forest carbon stock and carbon stock change. Ecosystems. 2014; 17: 627–640. DOI: 10.1007/s10021-014-9749-5.

[76] Su YJ, Guo QH, Xue BL, Hu TY, Alvarez O, Tao SL, Fang JY. Spatial distribution of forest aboveground biomass in China: estimation through combination of spaceborne lidar, optical imagery, and forest inventory data. Remote Sensing of Environment. 2016; 173: 187–199. DOI: 10.1016/j.rse.2015.12.002.

[77] Wang XP, Fang JY, Tang ZY, Zhu B. Climatic control of primary forest structure and DBH-height allometry in Northeast China. Forest Ecology and Management. 2006; 234: 264–274. DOI: 10.1016/j.foreco.2006.07.007.

[78] Lopez-Serrano FR, Garcia-Morote A, Andres-Abellan M, Tendero A, del Cerro A. Site and weather effects in allometries: a simple approach to climate change effect on pines. Forest Ecology and Management. 2005; 215: 251–270. DOI: 10.1016/j.foreco.2005.05.014.

[79] Berninger F and Nikinmaa E. Implications of varying pipe model relationships on Scots Pine growth in different climates. Functional Ecology. 1997; 11: 146–156. DOI: 10.1046/j.1365-2435.1997.00067.x.

[80] Geron CD, Ruark GA. Comparison of constant and variable allometric ratios for predicting foliar biomass of various tree Genera. Canadian Journal of Forest Research-Revue Canadienne De Recherche Forestiere. 1988; 18: 1298–1304. DOI: 10.1139/x88-200.

[81] West GB, Brown JH, Enquist BJ. A general model for the structure and allometry of plant vascular systems. Nature. 1999; 400: 664–667.

[82] Enquist BJ, Brown JH, West GB. Allometric scaling of plant energetics and population density. Nature. 1998; 395: 163–165. DOI: 10.1038/25977.

[83] West GB, Brown JH, Enquist BJ. A general model for the origin of allometric scaling laws in biology. Science. 1997; 276: 122–126. DOI: 10.1126/science.276.5309.122.

[84] Marshall AR, Willcock S, Platts PJ, Lovett JC, Balmford A, Burgess ND, Latham JE, Munishi PKT, Salter R, Shirima DD, Lewis SL. Measuring and modelling above-ground carbon and tree allometry along a tropical elevation gradient. Biological Conservation. 2012; 154: 20–33. DOI: 10.1016/j.biocon.2012.03.017.

4

Review of Biomass Thermal Gasification

Mohammed Abed Fattah Hamad,
Aly Moustafa Radwan and Ashraf Amin

Abstract

Gasification of biomass is one of the most attractive methods for producing hydrogen rich gas. Syngas production from biomass is an attractive solution for energy crisis. The production of energy from biomass reduces the dependence of developing countries on fossil fuels, as ample biomass is available in the developing countries and is renewable. Downdraft gasifiers are fixed bed gasifiers where the gasifying agent and biomass are flowing downwards, developed for high-volatile fuels such as wood or biomass gasification. Cocurrent flow regime throughout the oxidation and reduction zones reduces the tars and particulates in syngas, which will reduce the necessity of complicated cleaning methods compared to updraft gasifiers especially if the gas is used as a burnable gas in a small community. It is important to ensure homogenous distribution of gasifying agent at the downdraft gasifier throat. This chapter presents latest trends in gasification of biomass using downdraft gasification.

Keywords: gasification, hydrogen, agriculture waste, catalysts, downdraft gasifier

1. Introduction

Gasification of biomass is one of the most attractive methods for producing hydrogen rich gas. Syngas production from biomass is an attractive solution for energy crisis. The production of energy from biomass reduces the dependence of developing countries on fossil fuels; as ample biomass is available in the developing countries and is renewable. Downdraft gasifiers are fixed bed gasifiers where the gasifying agent and biomass are flowing downwards, developed for high volatile fuels like wood or biomass gasification. Cocurrent flow regime throughout the oxidation and reduction zones reduces the tars and particulates in syngas, which will reduce the necessity of complicated cleaning methods compared to updraft gasifiers especially if the gas is used as a burnable gas in a small community. It is important to ensure homogenous distribution of gasifying agent at the downdraft gasifier throat [1–3].

Gasification is a process under development to utilize the energy conserved in biomass. Gasification can be used as a source of energy in rural and off-grid areas to fill the power gaps. The limited supply and the increasing demands of fossil fuels have led the world to investigate alternative energy sources. Renewable energy sources have been studied widely, and biomass appears as the most promising renewable energy source. Biomass can be used to overcome the depletion of fossil fuels and to reduce the environmental impact of the conventional fuels such as greenhouse gas emissions using one of these four technologies: direct combustion, thermochemical processes, biochemical processes, and agrochemical processes. Biomass is the third energy source after coal and oil. Biomass covers 35% of the energy demand of the developing countries corresponding to 13% of the world energy demand. Biomass is widely available in quantities enough to meet the world energy demand [1–7].

The oldest way to generate energy is to burn biomass. Due to environmental and technical difficulties associated with burning biomass, innovative processes should be developed to utilize biomass [5, 8–10]. Developing more effective techniques to utilize biomass will reduce the disposal problem and create profits. Hydrolysis, pyrolysis, gasification, and hydrogenation are the principal processes for biomass conversion in the literature [7, 11]. Gasification represents efficient and environmentally friendly method for producing the syngas as a biofuel from different sources of biomass [12–14], and to produce second-generation biofuels such as methanol, ethanol, and hydrogen [8, 10, 12, 15]. Gasification can be defined as the partial (incomplete) combustion of biomass, and gasification could extract up to 60–90% of the energy stored in biomass [16, 17]. To develop second-generation biofuels, economical and Feasible clean technologies of syngas are required. [15]. However, economical gasification of biomass may produce burnable gases, which can be used to provide heat requirements instead of LPG [12]. Gasifiers were developed to replace biomass burners. Gasifiers will prevent the necessity of on-site power generation [18, 19]. Gasification is the conversion of biomass into a combustible gas and charcoal by partial oxidation of biomass at temperature range of 800–900°C [6, 19, 20].

The charcoal is finally reduced to H_2, CO, CO_2, O_2, N_2, and CH_4 [6, 8–10, 21]. Char gasification starts at temperatures above 350°C [7]. The products of gasification consist of the following components: ash, volatile alkali metals, tars, and syngas. Tars represent a challenge for the commercialization of gasification product as an alternative fuel [22]. Frequently using tar may result in complete shutdown and repair of the industrial unit [18, 22]. Tars set and amount vary considerably based on reaction conditions and gasifier type [18]. Gas produced from gasifier can be cleaned by removing tars either physically or chemically [18]. Physical removal can be achieved using bag filters or wet scrubbers. Chemical removal methods depend on converting tars to lighter hydrocarbons either using thermal conversion or catalytic conversion processes [18, 22].

Gasification of such material may help in reducing the gap between electricity requirements and available energy sources. Decentralized power regeneration units will help to fill power gap in rural and off-grid locations [4]. Yet, it is still difficult to develop a decentralized power generation unit based on biomass energy which can be used to fill the gap in energy needs in rural areas and farms [4]. Technical difficulties prevent further commercialization of gasification units in accordance to lower conversion efficiency [23, 24]. Leung et al. [25] proposed a governmental support to accomplish faster steps toward gasification units commercialization. However, all over the world, biomass energy has been widely incorporated in the power generation system; U.S. started partial and full conversion of conventional power plants to biomass [24]. Throughout this chapter, we will discuss the latest trends in agricultural waste gasification. Our goal is to provide a full description of the process starting from basic understanding and ending by design of a gasification unit.

2. Chemistry of gasification

The reactions taking place in the gasifier can be summarized as indicated below [3, 4, 21]:

Partial oxidation:

$$\mathrm{C} + \tfrac{1}{2}\mathrm{O_2} \leftrightarrow \mathrm{CO} \quad \Delta H = -268\ \mathrm{kJ/mol} \tag{1}$$

Complete oxidation:

$$\mathrm{C} + \mathrm{O_2} \leftrightarrow \mathrm{CO_2} \quad \Delta\mathrm{H} = -406\ \mathrm{kJ/mol} \tag{2}$$

Water gas phase reaction:

$$\mathrm{C} + \mathrm{H_2O} \leftrightarrow \mathrm{CO} + \mathrm{H_2} \quad \Delta H = 118\ \mathrm{kJ/mol} \tag{3}$$

Boudouard reaction:

$$\mathrm{C} + \mathrm{CO_2} \leftrightarrow 2\mathrm{CO} \quad \Delta H = 170.7\ \mathrm{kJ/mol} \tag{4}$$

The heat required for water gas phase and Boudouard reactions is provided by complete and partial oxidation reactions, and complete oxidation provides around 60% of the heat requirements during gasification [3, 17]. In addition to the previous reactions that are common in combustion and gasification, hydrogen, steam, and carbon monoxide undergo further reactions as shown below [3, 24]:

Water gas shift reaction:

$$\mathrm{CO} + \mathrm{H_2O} \leftrightarrow \mathrm{CO_2} + \mathrm{H_2} \quad \Delta H = -42\ \mathrm{kJ/mol} \tag{5}$$

Methane formation:

$$\mathrm{CO} + 3\,\mathrm{H_2} \leftrightarrow \mathrm{CH_4} + \mathrm{H_2O} \quad \Delta H = -88\ \mathrm{kJ/mol} \tag{6}$$

The water gas shift and methane formation reactions are in equilibrium and the governing parameters are: pressure, temperature, and concentration of reaction species.

3. Gasifier design

The unit design is a very important factor in determining the syngas quality and heating value [15]. The gasifier will hold two processes: conversion of biomass to charcoal and then conversion of charcoal to hydrogen and carbon monoxide. The mixture of hydrogen and carbon monoxide can be used for direct heating in rural areas [16]. Leung et al. [25] identified four types of gasifiers: updraft, open core, downdraft, and circulating fluidized bed (CFB) gasifiers. The maximum efficiency of the updraft, downdraft, and CFB gasifiers may reach to 75%, the maximum energy output is 10E6, 4E6, and 40E6 kJ/h, respectively. According to Chopra and Jain [13], the fixed bed gasifiers can be further divided into: updraft, Imbert downdraft, throatless downdraft, cross draft, and two-stage gasifiers. The fixed-bed gasifier is suitable for producing low heating value gas for small and medium applications [13, 26]. The downdraft gasifier is distinguished by a simple design, high carbon conversion, long residence time, low cost, low pressure, relatively clean gas, and low gas velocity. The downdraft gasifier is suitable for producing low heating value burnable gas or for generating electricity of small-scale systems in the range of 10 kW up to 1 MW [12, 26–28].

3.1. Design of downdraft gasifiers

Downdraft gasifiers are fixed bed gasifiers where the gasifying agent and biomass are flowing downwards, developed for high volatile fuels like wood or biomass gasification. Cocurrent flow regime throughout the oxidation and reduction zones reduces the tars and particulates in syngas, which will reduce the necessity of complicated cleaning methods compared to updraft gasifiers especially if the gas is used as a burnable gas in a small community [12, 17]. It is important to ensure homogenous distribution of gasifying agent at the downdraft gasifier throat.

Bhavanam and Sastry [24] provided design procedures for different types of downdraft gasifiers. The gasification reaction in a downdraft gasifier undergoes several steps, starting with drying step at 100°C, followed by pyrolysis step between 200 and 300°C resulting in release of around 70% of biomass weight as volatile matter and tars [16, 24]. After pyrolysis, the remaining biomass and volatile matter react with the incoming oxygen in the combustion step. Finally, various reactions take place in the reduction zone including carbon and steam reaction to produce CO and hydrogen, water-gas shift reaction, and CO and steam to form methane and carbon dioxide [24]. The four gasification reaction steps are illustrated in **Figure 1**. However, a limited experience has been gained in the field of biomass gasification while it represents an attractive renewable energy route [16]. **Table 1** illustrates the design specifications for two types of downdraft gasifiers: Imbert and stratified downdraft gasifiers. **Table 1** is developed based on extensive discussion in Bhavanam and Sastry [24].

Imbert downdraft gasifier is a cylindrical chamber of varying inner diameter across chamber length. The upper part of the cylindrical chamber is loaded with biomass according to

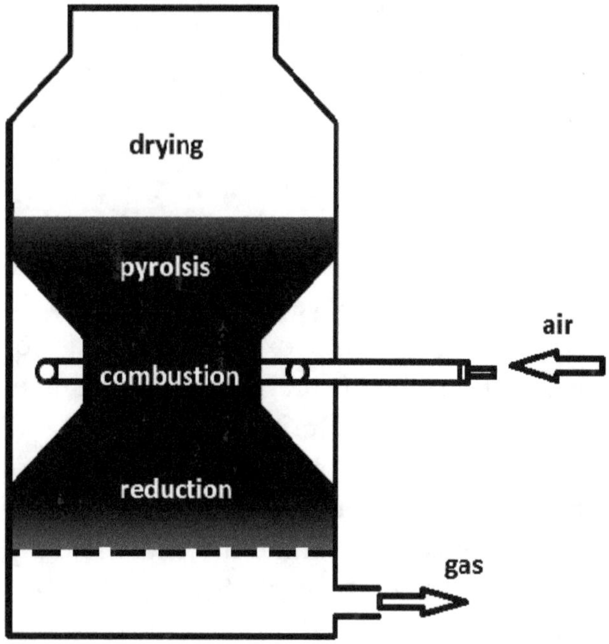

Figure 1. Different reaction zones in downdraft gasifier.

Design considerations		Imbert	Stratified
Biomass	Material	Uniform woody	Small size
	Moisture content	<20%	<20%
	Ash content	<5%	–
Reactor type		Packed bed supported on a throat	No-throat cylindrical packed bed with open top
Biomass feeding		Hopper	Open top
Gas feeding		Nozzle in the combustion zone	Enters from top mixed with biomass
Produced gas		Tar oils <1% $T = 700°C$	Less tar
Maximum capacity		500 kW	Easy to scale up

Table 1. Design consideration for Imbert and stratified downdraft gasifiers.

requirement. Air nozzles, attached to distribution manifold, permit air to be drawn into biomass to improve mixing of gasifying agent and biomass. A charcoal balance is established around the nozzles. Below the air nozzles, a classical Imbert hearth forms the reduction part. Insulating the reduction hearth reduces the amount of tars in the produced syngas and increases gasification efficiency. The hot gases are forced to go through the hot zone due to hearth constriction. The char bed on the grate removes the dust, which should be cleaned eventually to prevent clogging, and dropping in airflow or channeling [17].

Stratified or open-top downdraft gasifier is a uniform diameter gasifier, usually made of a cylindrical vessel with a hearth near the bottom. The stratified gasifier is an improved, easy to design gasifier compared to Imbert downdraft gasifier. The open-top helps in maintaining uniform access of gasifying agent to the pyrolysis zone, which prevents localized heating. Biomass is added through the open-top to the top layer of gasifier. The length of the gasifier can be divided into four reaction zones: unreacted biomass zone at which air enters, the flaming pyrolysis zone at which air reacts with biomass, adiabatic char gasification zone at which gases from flaming pyrolysis zone reacts with charcoal, and finally the unreacted charcoal zone that is located just above the grate which acts as a buffer for ash and charcoal. The stratified downdraft gasifier can be mathematically modeled easily as a plug flow reactor at which air and biomass are uniformly mixed. With such simple design, it is expected that stratified downdraft gasifier will replace the Imbert downdraft gasifier in commercial applications [17].

Wander et al. [29] illustrated the design of 12 kg/h downdraft stratified gasifier for sawmill dust gasification. The reactor is a cylindrical body of 270 mm internal diameter and 1100 mm of height made of SAE 1020 steel. Internal rods are used to mix the sawdust in the reduction zone. Ash box is used to reduce the ash content of the produced syngas. The reactor is insulated using 50 mm of rock wall. Air was introduced from the open top as a gasifying agent and a secondary air was used to provide air required for internal burner. A gas chromatography was used to analyze the gas samples and three water condensers in an ice bath were used to measure tars and humidity content.

Zainal et al. [30] developed a downdraft gasifier for the gasification using around 50 kg/h of wood chips. The temperature in the combustion zone may reach 1000°C, which reduces the tar content of produced syngas. The gasifier is made of mild steel pipe with a diameter of 0.6 m and a height of 2.5 m. A cone structure is used inside the gasifier above the combustion zone with an inclination of 60° to facilitate the wood material movement. The air supply was accomplished using a 40 mm diameter stainless-steel pipe with eight 10 mm diameter nozzles. The air is preheated by positioning the supply tube inside the gasifier. The heating value of the produced syngas was in the range of 4.65–5.62 MJ/Nm^3 depending on operating conditions.

Panwar et al. [31] developed an open-top downdraft gasifier for wood gasification to provide the heat requirement for the food processing industry. The downdraft gasifier was lined with ceramic and designed for a wood input of 60 kg/h equivalent to 180 kW. The gasifier body is made of mild steel. The air distribution system consists of six air tuyeres of 20 mm in diameter. A cyclone was used to remove solid particulates from produced syngas. The complete combustion of the syngas is achieved in a premixed burner to provide heat needed for the food processing industry unit. Note that 30 kg of charcoal and 10 kg of wood were used to provide hear required for gasifier start up.

Sheth and Babu [1] showed a design of an Imbert downdraft gasifier for wood waste gasification with a total height of 1.1 m. The diameter of pyrolysis zone and reduction zone are 0.31 and 0.15 m, respectively. The gasifier has throated combustion zone, a bed of char supported by a grate follows the combustion zone. The air is supplied through two nozzles in the oxidation zone. The high temperature in the combustion zone ensures cracking of tars into volatiles and water. The diameter of pyrolysis, reduction, and oxidation zones is 310, 150, and 53 mm, respectively. The grate is movable to unclog it for removing ash.

Vervaeke et al. [32] illustrated the design of a small-scale pilot plant downdraft gasifier equivalent to 100 kW of electricity generation. The downdraft unit used in this study is a pilot scale of the Xylowatt gasifier. The downdraft gasifier is a batch gasifier with a capacity of 90–105 kg. The gasification system consists of downdraft gasifier and inside it are ash collection container, cyclone, filter, and a scrubber.

Lv et al. [33] developed a downdraft gasifier to produce hydrogen from biomass using air and oxygen/steam as gasifying agents. Total 5 kg of char were supported on the grate to reduce tar content and to act as a catalyst to upgrade syngas. Biomass is pine wood blocks used in cubes cut into 3 cm × 3 cm × 3 cm. The gasifier height is 1.3 m and the diameter is 35 cm. The gasifying agent is preheated in a chamber inside the gasifier. Gas is cleaned using triple-stage spray shower filled with steel wire rings. The internal diameter of the gasifier is calculated according to the power output. The height is calculated based on batch operation time. The internal diameter is calculated in meters using the following equation [26]:

$$D = \left(\frac{1.27 * \mathrm{FCR}}{\mathrm{SGR}}\right)^{0.5} \tag{7}$$

where FCR is the fuel consumption rate (kg/h) and SGR is the specific gasification rate (kg/h/m^2).

The height can be determined in meter using the following equation [26]:

$$H = \frac{\mathrm{SGR} \times t}{\rho} \tag{8}$$

where t is the operation time (h) and ρ is the feedstock bulk density (kg/m^3).

The power output P_0 can be calculated in kW from the following equation [26]:

$$P_0 = \frac{\mathrm{FCR} \times \mathrm{HHV} \times \eta}{3.6} \tag{9}$$

where HHV is the higher heating value of the feedstock in MJ/kg and η is the efficiency of the gasifier usually around 0.7. The amount of air needed during operation can be calculated in Nm^3/h from the following equation [26]:

$$\mathrm{AFR} = \frac{\varepsilon \times \mathrm{FCR} \times \mathrm{SA}}{\rho_a} \tag{10}$$

where ε is the equivalence ratio, FCR is the fuel consumption rate, SA is the stoichiometric amount of air required for chemical reaction, and ρ_a is the density of air (1.18 kg/m^3). Finally, the size of the air nozzle, which is required for uniform air distribution, can be calculated in mm^2 from the following equation [26]:

$$A = \frac{\mathrm{AFR} \times 10^3}{v \times 3.6} \tag{11}$$

where v is the inlet velocity of air (m/s).

4. Factors affecting on the gasification process

Zhou et al. [34] discussed the ongoing gasification projects taking place in China. The biomass gasification units were divided based on scale: small-, medium-, and large-scale biomass gasification and power generation units. Pretreatment of biomass includes size reduction, size screening, separation of magnetic materials, and storing as wet biomass. Then prior to gasification, drying and storing as dry material are accomplished to reduce the moisture content to 10–15% [35]. Feedstock type and feedstock preparation are important factors affecting the yield and quality of produced syngas. Shredding and drying are two processes conducted to prepare the biomass raw material for gasification process [14]. The main parameters affecting the gasification are clarified below:

Equivalence ratio (ER): The equivalence ratio is the air/biomass ratio divided by the theoretical air/biomass ratio. Increasing ER will decrease the heating value of the produced gas due to decreasing H_2 and CO concentration and increasing CO_2 concentration. Higher ER helps in reducing tars and provides more O_2 to react with volatiles. Typical values of ER ranges between 0.2 and 0.4 [24]. Guo et al. [36] reported that increasing ER decreases the concentration of combustible gases (H_2, CO, CH_4, and C_nH_m). The heating value was higher than 4 MJ/Nm^3 when ER is kept lower than 0.4. Increasing ER improves the reaction temperature and carbon conversion, and reduces the tar yield. For a downdraft stratified gasifier, Wander et al. [29] suggested an equivalence ratio of 0.3:0.35 kg-O_2/kg-wood. A higher ratio is required when higher heat loss is expected, an equivalence ratio of 2:2.4 kg-air/kg-wood is optimum for producing a syngas with low heating value of 4–6 MJ/Nm^3. For woody material in a downdraft gasifier, Zainal et al. [30] suggested an equivalence ratio of 0.268–0.43 with 0.38 showed optimum value (corresponding to a heating value of 5.62 MJ/Nm^3). While Sheth and Babu [1] defined that the optimum equivalence ratio for wood gasification in Imbert downdraft gasifier is 0.205.

Effect of biomass characteristics: Biomass characteristic is a major factor affecting produced syngas quality. The physical properties that may have major effect are: absolute and bulk density, and particulate size. The chemical composition parameters that are of major importance to define the syngas [17] quality including volatile matter, moisture content, fixed carbon, ash content, and gross calorific value and the ultimate analysis comprises the carbon, oxygen, nitrogen, and sulfur of the dry biomass on a weight% [19].

Moisture content: The moisture content can be determined by complete drying of biomass sample. The moisture content is calculated by subtracting the sample weight after drying from fresh sample weight. Maximum allowable moisture content in downdraft gasifier is 40% on dry weight basis. Updraft gasifier can handle biomass with higher moisture content. The higher moisture content in biomass will increase the consumed energy for drying, and will reduce the pyrolysis of biomass. As a general rule, increasing moisture content decreases the conversion [1, 24].

Superficial velocity: The superficial velocity is the ratio of the syngas production rate at normal conditions and the narrowest cross-sectional area of gasifier. Lower superficial velocity is linked with high yield of char, and large quantities of unburned tars, which may deactivate catalyst, plug lines, and destroy compressors. On the other hand, higher superficial velocity results in reduced amount of char and low overall process efficiency [24].

Operating temperature: Operating temperature affects conversion, tar content, gas composition, gas heating value, and char conversion. To select the optimum temperature, gasifier type, and biomass source should be considered. Usually, temperature higher than 800°C should be used to obtain high conversion and low tar content in the produced syngas [24]. Low temperature is associated with low tar content, low H_2 and CO content in the produced syngas [12]. Increasing temperature will increase gas yield, hydrogen, heating value, and ash agglomeration. To overcome the ash agglomeration problem, practical temperature does not exceed 750°C [24].

Gasifying agent: Gasifying agents in use are air, steam, steam/oxygen mixture, and CO_2. Gasifying agent affects the heating value of the produced syngas. The heating value increases with increasing steam content of the gasifying agent, while heating value decreases as air increases in the gasifying agent [24]. The steam/oxygen mixture represents a zero nitrogen-gasifying agent which increase heating value and allow liquefying the produced gas after proper treatment [37]. Using almond shells, the lower heating value was 5.9–6.7, 6.3–8.4, and 10.9–11.7 MJ/Nm^3 using the gasifying agent: 35 wt.% O_2 enriched air, 50 wt.% O_2 enriched air, and steam/oxygen mixture, respectively. Campoy et al. [38] reported a heat value of syngas produced from gasification to have an average value of 4–6 and 9–13 MJ/Nm^3 using air and oxygen/steam mixture, respectively. In addition to lower efficiency compared to air/steam mixture, enriched oxygen-air requires high capital cost for oxygen [38]. The addition of steam will shift toward the reforming reaction and heterogeneous gasification reactions.

Residence time: Residence time has a remarkable impact on the composition and produced tars. Increasing residence time decreases the fraction of oxygen-containing compounds, decreases yield of one and two atomic ring compounds, and increases three and four ring compounds [24].

Pressure: Atmospheric and higher pressures are commonly used in gasification process. Selecting the optimum pressure depends on the application of the produced syngas. If the syngas is used for producing methanol or synthetic auto-fuels, higher pressures are preferred to improve the process yield and to reduce tar content. For generating burnable gases, atmospheric pressure should be used [12]. High pressure applications are recommended for large-scale gasification, while atmospheric pressure is recommended for small-scale gasification [35]. High pressure gasification is still not well developed and further research is needed to further commercialize such process [39].

Catalyst: Catalyst type is a very important factor affecting gasification quality and produced syngas. Catalyst affects the composition of the syngas by manipulating the percentage volume of hydrogen, carbon dioxide, methane, and carbon monoxide. Optimum catalyst should play a role in minimizing the gas content of carbon dioxide and maximizing the gas content of hydrogen, carbon monoxide, and methane [40]. The catalyst type and loading on the gasification of cotton stalks and saw dust were studied. The catalysts selected are USY zeolite, dolomite, CaO, granulated slag, red brick clay, olivine, and cement kiln dust. The results demonstrate that the cement kiln dust and calcium hydroxide are more effective for increasing the gas yield and decreasing the char yield [8, 10].

Effect of biomass/steam ratio: Biomass/steam ratio affects hydrogen content in the produced syngas. Contradictory reports are found in literature, while Lv et al. [41] reported a positive effect on hydrogen content when biomass/steam ratio increases. Lv et al. [33] reported a negative effect of biomass/steam ratio increase on syngas hydrogen content. This variation can be understood by considering that biomass/steam ratio effect is altered according to the entire system configuration.

Lower values of biomass/steam ratio shift the reaction to produce more solid carbon and methane, since number of moles of steam increases in the feed. While at higher values of biomass/steam ratio, Co and H_2 are increased in the syngas as carbon and methane produced are decreased consequently.

5. Cost of biomass gasification process

The cost of any industrial process is governed by the capital cost and the running cost. Selection of best gasifier type depends on cost of fabrication, ease of manufacture, tar content, lower heating value, feedstock elasticity, and application of syngas [26]. The fixed bed gasifiers are more suitable for small- and medium-scale applications, while fluidized bed gasifiers are suitable for large-scale applications (equivalent to >15 MW) [25]. For example in China rice hulls, fluidized bed gasifiers are used in a production scale equivalent to 1–1.2 MW while downdraft gasifiers are used in a production scale equivalent to 60–200 kW [42]. The capital cost of the gasifier is divided into three items: gasifier and gas cleaning system cost, fuel gas utilization equipment cost, and fitting and system construction cost [25]. Cleaning systems and removing tars will add a significant cost to the produced syngas, which reduce the feasibility of using syngas in internal combustion engines [12]. Optimizing tar content can be achieved by varying the operating conditions and feedstock [43].

Upgrading using catalytic treatment represents the most economical and efficient method for syngas upgrading since it provides a way for removing tars and other particulates and converting tars to hydrocarbons [12]. Downdraft gasifier represents a reasonable cost production method for generating syngas with low tar content [29]. Especially small gasifiers that has proven economic feasibility [27].Wu et al. [44] recommended implementing biomass gasification depending on the low biomass price. By comparing different technologies to generate electricity based on 1 MW scale, Wu et al. [44] mentioned that the capital cost of fluidized bed gasifier system for biomass gasification-power generation system is 60–70% of the capital cost of coal power station and much lower compared to the capital cost of conventional power station. For producing combustible gases, Bridgwater et al. [45] reported that for syngas produced from fluidized bed, updraft, and downdraft gasifiers: hydrogen volume percentage is 9, 11, and 17%, respectively; CO volume percentage is 14, 24, and 21%, respectively; and a heating value of 5.4, 5.5, and 5.7 MJ/Nm^3, respectively. The downdraft represents the ideal solution to produce combustible (burnable) gases for household uses.

Biomass gasification economics are very sensitive to the scale of produced MW [44]. Leung et al. [25] mentioned two disadvantages of small- and medium-size gasifiers: capital cost limitation that may prevent incorporating important processes like tar removal, and the environmental demands imposed by new regulations which is difficult to be met by different biomass gasification technologies. Wu et al. [44] identified 160 kW as a critical scale of biomass gasification unit, less than 160 kW biomass gasification units loses the economical attraction. Note that 1–5 MW was recommended as the most competitive size for biomass gasification unit. Lower than a unit capacity of 160 kW, the price of kWh increases sharply from 0.4 to 1.8 Yuan RMB/kWh for very small capacities. For unit with capacities higher than 160 Wh, the price will decrease gradually as the unit size increases. At the 600 kW capacity, the price will be around 0.3 Yuan RMB/kWh; while the price may reach 0.25 for a unit capacity of 1000 kW [44].

It is recommended to conduct gasification at pilot plant scale to mimic large scale to figure out the approximate industrial process scale economics [38]. The steam enhances the reforming and heterogeneous gasification reactions, the temperature inside the gasifier should be kept enough to support such reactions [38]. Combining gasification unit with heat and power generation systems will improve the economics of the process [3]. Gasification units combined with heat and power generation systems are expected to have an overall efficiency of 85% compared to a maximum efficiency of 35–55% for conventional power station, in addition to a substantial saving in carbon emissions. Total 1000 kg/year of carbon are saved for each MW when gasification units hybrid with heat and power generation systems [3].

Downdraft gasifiers are economically competitive even to conventional LPG heating unit. Panwar et al. [31] found that replacing LPG heating system with a downdraft wood gasification system could save $13,850 US for 3000 h of operation. The payback period of the gasification system was only 1100 h. According to the extensive study of literature, the recommended gasification process consists of the following steps [13, 17, 19]:

1. Straw collection and preparation (milling and pelletization of straws).
2. Belt conveyor for feeding of the gasifier.
3. Downdraft gasifier.
4. Blower for suction of air and gas produced.
5. Gas cleaning and separation of tars.
6. Gas holder for storage of gas.
7. Gas distribution net.
8. Gas application devices.
9. Gas metering devices.

6. Preliminary techno-economic studies of downdraft gasifier

The aim of this chapter is to illustrate a detailed design of biomass gasification system to generate syngas for household applications. The stratified gasifier is selected based on the

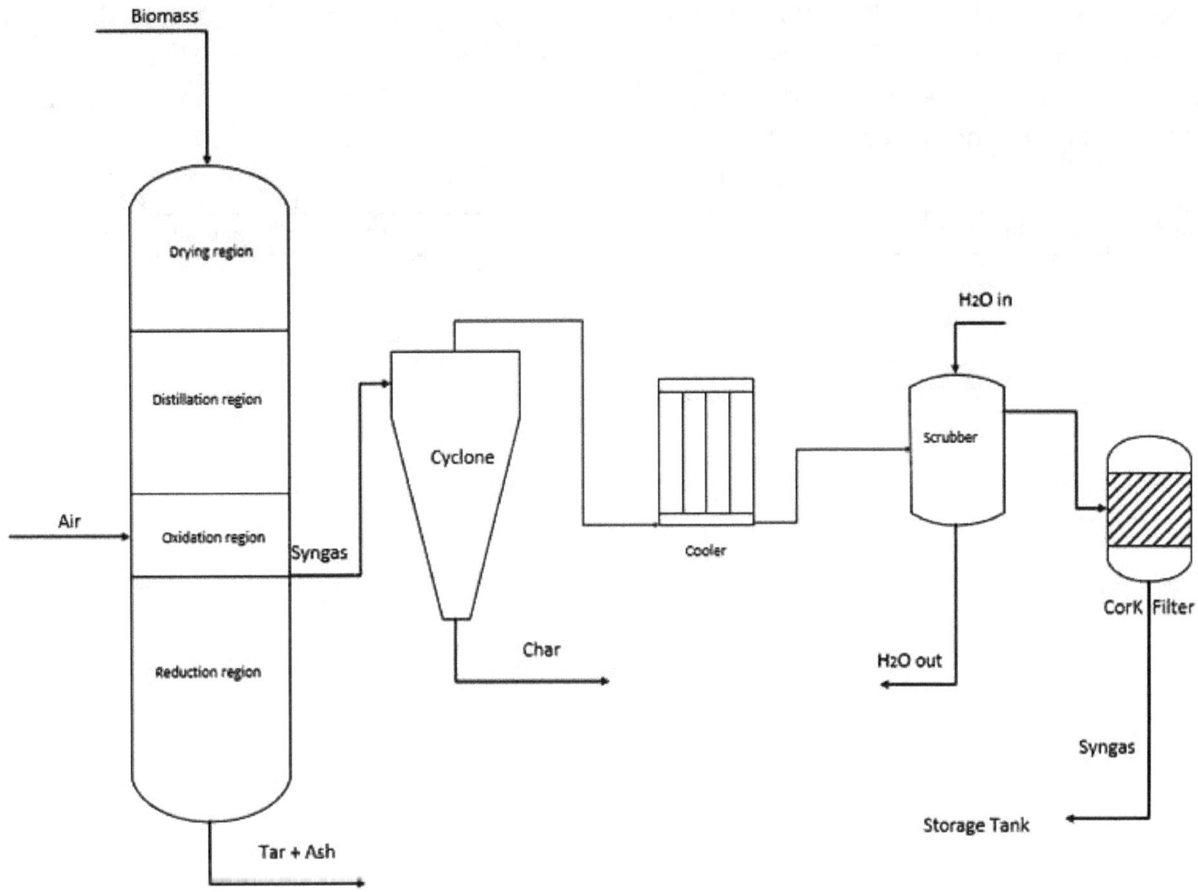

Figure 2. Gasification system for producing burnable gas.

following parameters: easiness in design and scaling up, and production of syngas with tar content lower than that of Imbert gasifier [16–18, 26]. The gasification system comprises of a downdraft burnable syngas gasifier followed by a gas cleaning and distributing system. Throughout this chapter, the design specification of the downdraft gasifier is presented. This system can be used to convert solid agricultural waste to a syngas that is a burnable gas used to provide energy requirements for small communities, as shown in **Figure 2**.

The energy (household) requirements: for 50 families.

The gas demand per day is: 500 m^3/day.

The syngas gas will be produced on two batches: Morning and afternoon (each one will last for 250 m^3/batch).

The first batch will take place from 7 to 10 am.

The second batch will take place from 2 to 5 pm.

The storage unit will hold around 200 m^3 gas and accordingly will provide heating requirements during the period of nonoperation.

The system consists of the following units: biomass shredding, grinding unit, gasification unit, air controlling system, air heating and gas precooling unit, cyclone (acts like cyclone to remove dust), gas cooler, water filter (scrubbing unit), gas distribution system, control system, cork filter, and storage tank.

No.	Object	Cost, Egyptian pounds
1	Gasifier with control system	30,000
2	Gas cleaner	10,000
3	Belt conveyor	15,000
4	Milling and pelletization of biomass	50,000
5	Gas holder	15,000
6	Gas distribution system 100 × 500	60,000
7	Gas stoves 100 ×350	35,000
8	Gas meters 100 ×400	40,000
9	Erection	10,000
10	Contingency (10%)	26,000
	Total	2,90, 000

$ = 8.5 Egyptian pounds.
Assume that:
Cost of material = 300 L.E/ton
Cost of preparation = 100 L.E/ton
Cost of raw materials = 0.25 ×400 × 360 = 36 000 L.E/year
Labors required = 2 ×1500 = 3000 L.E/month = 3000 ×12 = 36000 L.E/year
Income gas production = 500 m^3/day
≈60 kg L.P.G/day
= 60 × 360 = 21,600 Kg/year
= 21,600 × 7 = 151,200 L.E/year
Profit = income − raw material − depreciation
Profit = 151,200 – 36,000 – 36,000 – 29,000 = 49,400
Return on Investment = profit/initial cost
Return on Investment = (49,400/290,000)×100 = 17%
Therefore, the payback period is about 6 years based on the international prices of L.P.G.

Cost of equipment for gasification system.

The cost of land required for erection of the plant is not included in this calculation of the feasibility study.

7. Conclusions

Gasification represents a viable solution to overcome the energy shortage by developing commercial gasification units in rural and off-grid areas. An integrated system comprising a gasification-electrical generation method represents an ideal solution from the technical and economical points of view.

However, due to the wide varieties of available biomass feedstock, it is recommended to manipulate different systems in each location depending on the feedstock, produced syngas, and energy demands. Downdraft gasifier is recommended for small-scale applications in rural areas. The co-current nature of air and biomass flow reduces the tar content and increases CO and H_2 in the produced syngas. The syngas produced from downdraft gasifier can be used after a simple purification process in thermal applications. From a cost study, the payback period of a gasification system is around 5 years.

Acknowledgements

The authors would like to express deepest thanks for the Science and Technology Development Fund (STDF Egypt) for financing and supporting the development of the thermal gasification of biomass.

Author details

Mohammed Abed Fattah Hamad*, Aly Moustafa Radwan and Ashraf Amin

*Address all correspondence to: hamadnrc@hotmail.com

National Research Centre, Cairo, Egypt

References

[1] Sheth PN, Babu B. Experimental studies on producer gas generation from wood waste in a downdraft biomass gasifier. Bioresour Technol. 2009;100:3127–33.

[2] Khater EMH, El-Ibiary NN, Ki-Iattab IA, Hamad MA. Gasification of rice hulls. Biomass Bioenerg. 1992;3:329–33.

[3] Zabaniotou AA, Skoulou VK, Mertzis DP, Koufodimos GS, Samaras ZC. Mobile gasification units for sustainable electricity production in rural areas: The Smart-CHP project. Ind Eng Chem Res. 2011;50:602–8.

[4] Buragohain B, Mahanta P, Moholkar VS. Biomass gasification for decentralized power generation: The Indian perspective. Renew Sustainable Energy Rev. 2010;14:73–92.

[5] Karellas S, Karl J. Analysis of the product gas from biomass gasification by means of laser spectroscopy. Opt Las Eng. 2007;45:935–46.

[6] Ni M, Leung DYC, Leung MKH, Sumathy K. An overview of hydrogen production from biomass. Fuel Process Technol. 2006;87:461–72.

[7] Kirubakaran V, Sivaramakrishnan V, Nalini R, Sekar T, Premalatha M, Subramanian P. A review on gasification of biomass. Renew Sustainable Energy Rev. 2009;13:179–86.

[8] Radwan AM, Hamad MA, Singedy AM. Synthesis gas production from catalytic gasification of saw dust. Life Sci J. 2015;12:104–18.

[9] Yoon SJ, Choi YC, Lee JG. Hydrogen production from biomass tar by catalytic steam reforming. Energy Convers Manage. 2010;51:42–7.

[10] Hamad MA, Radwan AM, Heggo DA, Moustafa T. Hydrogen rich gas production from catalytic gasification of biomass. Renew Energy. 2016;85:1290–300.

[11] Klass DL. Biomass for Renewable Energy, Fuels, and Chemicals. Academic Press. San Diego, CA., 1998.

[12] Asadullah M. Barriers of commercial power generation using biomass gasification gas: A review. Renew Sustainable Energy Rev. 2014;29:201–15.

[13] Jain A, Chopra S. A review of fixed bed gasification systems for biomass. Agricultural Engineering International Journal. 2007; IX (5).

[14] Begum S, Rasul M, Cork D, Akbar D. An experimental investigation of solid waste gasification using a large pilot scale waste to energy plant. Proc Eng. 2014;90:718–24.

[15] Zhang W. Automotive fuels from biomass via gasification. Fuel Process Technol. 2010;91:866–76.

[16] Moharkar SP, Padole P. Design and development of downdraft gasifier for rural area. IJCA Proceedings on International Conference on Emerging Frontiers in Technology for Rural Area (EFITRA-2012): Foundation of Computer Science (FCS); 2012. p. 24.

[17] Reed T, Reed TB, Das A, Das A. Handbook of Biomass Downdraft Gasifier Engine Systems. Biomass Energy Foundation. Golden, Colorado,1988.

[18] Brown D, Gassner M, Fuchino T, Marechal F. Thermo-economic analysis for the optimal conceptual design of biomass gasification energy conversion systems. Appl Therm Eng. 2009;29:2137–52.

[19] Radwan AM. An overview on gasification of biomass for production of hydrogen rich gas. Der Chem Sin. 2012;3:323–35.

[20] Goyal HB, Seal D, Saxena RC. Bio-fuels from thermochemical conversion of renewable resources: A review. Renew Sustainable Energy Rev. 2008;12:504–17.

[21] Tavasoli A, Ahangari MG, Soni C, Dalai AK. Production of hydrogen and syngas via gasification of the corn and wheat dry distiller grains (DDGS) in a fixed-bed micro reactor. Fuel Process Technol. 2009;90:472–82.

[22] Tasaka K, Furusawa T, Tsutsumi A. Biomass gasification in fluidized bed reactor with Co catalyst. Chem Eng Sci. 2007;62:5558–63.

[23] Gil J, Aznar MP, Caballero MA, Frances E, Corella J. Biomass gasification in fluidized bed at pilot scale with steam-oxygen mixtures. Product distribution for very different operating conditions. Energy Fuels. 1997;11:1109–18.

[24] Bhavanam A, Sastry R. Biomass gasification processes in downdraft fixed bed reactors: A review. Int J Chem Eng Appl. 2011;2:425–33.

[25] Leung DY, Yin X, Wu C. A review on the development and commercialization of biomass gasification technologies in China. Renew Sustainable Energy Rev. 2004;8:565–80.

[26] Guangul FM, Sulaiman SA, Ramli A. Gasifier selection, design and gasification of oil palm fronds with preheated and unheated gasifying air. Bioresour Technol. 2012;126:224–32.

[27] Faaij AP. Bio-energy in Europe: Changing technology choices. Energy Policy. 2006;34:322–42.

[28] Schmidt DD, Gunderson JR. Opportunities for hydrogen: An analysis of the application of biomass gasification to farming operations using microturbines and fuel cells. 2000 US DOE Hydrogen Program Review. San Ramon, California, US: DOE; 2000.

[29] Wander PR, Altafini CR, Barreto RM. Assessment of a small sawdust gasification unit. Biomass Bioenerg. 2004;27:467–76.

[30] Zainal Z, Rifau A, Quadir G, Seetharamu K. Experimental investigation of a downdraft biomass gasifier. Biomass Bioenerg. 2002;23:283–9.

[31] Panwar NL, Rathore NS, Kurchania AK. Experimental investigation of open core downdraft biomass gasifier for food processing industry. Mitig Adapt Strategies Glob Chang. 2009;14:547–56.

[32] Vervaeke P, Tack F, Navez F, Martin J, Verloo M, Lust Nl. Fate of heavy metals during fixed bed downdraft gasification of willow wood harvested from contaminated sites. Biomass Bioenerg. 2006;30:58–65.

[33] Lv P, Yuan Z, Ma L, Wu C, Chen Y, Zhu J. Hydrogen-rich gas production from biomass air and oxygen/steam gasification in a downdraft gasifier. Renew Energy. 2007;32:2173–85.

[34] Zhou Z, Yin X, Xu J, Ma L. The development situation of biomass gasification power generation in China. Energy Policy. 2012;51:52–7.

[35] Tijmensen MJ, Faaij AP, Hamelinck CN, van Hardeveld MR. Exploration of the possibilities for production of Fischer Tropsch liquids and power via biomass gasification. Biomass Bioenerg. 2002;23:129–52.

[36] Guo F, Dong Y, Zhang T, Dong L, Guo C, Rao Z. Experimental study on herb residue gasification in an air-blown circulating fluidized bed gasifier. Ind Eng Chem Res. 2014;53:13264–73.

[37] Barisano D, Canneto G, Nanna F, Alvino E, Pinto G, Villone A, et al. Steam/oxygen biomass gasification at pilot scale in an internally circulating bubbling fluidized bed reactor. Fuel Process Technol. 2016;141:74–81.

[38] Campoy M, Gomez-Barea A, Villanueva AL, Ollero P. Air-steam gasification of biomass in a fluidized bed under simulated autothermal and adiabatic conditions. Ind Eng Chem Res. 2008;47:5957–65.

[39] Iaquaniello G, Mangiapane A. Integration of biomass gasification with MCFC. Int J Hydrogen Energy. 2006;31:399–404.

[40] Demirbas A. Pyrolysis and steam gasification processes of black liquor. Energy Convers Manage. 2002;43:877–84.

[41] Lv PM, Xiong ZH, Chang J, Wu CZ, Chen Y, Zhu JX. An experimental study on biomass air-steam gasification in a fluidized bed. Bioresour Technol. 2004;95:95–101.

[42] Lan W, Chen G, Yan B, Li W. Biomass gasification for power generation in China. Adv Biomed Eng. 2012;9:72.

[43] Carpenter DL, Bain RL, Davis RE, Dutta A, Feik CJ, Gaston KR, et al. Pilot-scale gasification of corn stover, switchgrass, wheat straw, and wood: 1. Parametric study and comparison with literature. Ind Eng Chem Res. 2010;49:1859–71.

[44] Wu C, Huang H, Zheng S, Yin X. An economic analysis of biomass gasification and power generation in China. Bioresour Technol. 2002;83:65–70.

[45] Bridgwater A, Toft A, Brammer J. A techno-economic comparison of power production by biomass fast pyrolysis with gasification and combustion. Renew Sustainable Energy Rev. 2002;6:181–246.

5

Effects of Fertilizers on Biomass, Sugar Content and Ethanol Production of Sweet Sorghum

Tran Dang Xuan, Nguyen Thi Phuong and Tran Dang Khanh

Abstract

Sweet sorghum (*Sorghum bicolor*) is a promising alternative crop for bioethanol production in developing countries. However, to extend the cultivative area of this crop, it needs to develop an appropriate growing protocol for farmers. This chapter describes the examination of different doses of fertilizers combined with manure and micronutrients, in various applied times, on biomass, sugar content and ethanol production of sweet sorghum. It was observed that the application of 90 N + 90 P_2O_5 + 60 K_2O provided maximum stem yield and optimum contents of sugar and ethanol yield, however nontreatment of any among P, P_2O_5 and K_2O caused significant reduction of biomass and ethanol production. Higher fertilization >90 N may provide greater productivity of this crop but it may cause lodging and economic deficit for farmers in developing countries. It was also found that the applied times of fertilization should be at 3–4 to 7–8 leaf stage. In contrast, when the fertilization was as close to the flowering stage caused remarkable reduction of stem yield and ethanol production. The supplementation of $(NH4)_2MO_7O_2.4H_2O$ at 5 kg/ha provided an increase of 10–12 tons/ha of stem yield and a remarkable enrichment of ethanol production. Findings of this study are useful for farmers and agricultural extensionists to promote biomass and ethanol productivity of this crop for bioethanol production. This research also highlights a greater possibility of exploiting sweet sorghum cultivation in infertile and hilly, abandoned areas for ethanol production.

Keywords: sweet sorghum, fertilizer, stem yield, biomass, ethanol yield, growing protocol, biofuel crops

1. Introduction

It has been estimated that there are 700 million light duty vehicles, automobiles, light trucks, SUVs and minivans on roadways around the world. These numbers may increase to 1.3 billion by 2030 and to over 2 billion vehicles by 2050, with most come from developing countries [1]. This growth will affect the stability of ecosystems and global climate as well as global oil reserves [2]. The use of biofuels can contribute to the mitigation of greenhouse gas emissions, provide a clean and sustainable energy source and may promote the agricultural income for rural poor in developing countries. Currently, biofuels are predominantly produced from biomass resources, because they are the renewable resource that could be sustainably developed, resulting in no net release of carbon dioxide and very low sulfur content and economic potential as fossil fuel prices can be increased in the future [2]. The principal sources of biofuels are liquid or gaseous fuels made from plant matter and residues, including agricultural crops, municipal wastes and agricultural and forestry by-products.

Bioethanol has a high octane number (108), broader flammability limits, higher flame speeds and higher heats of vaporization. It helps for a higher compression ratio and shorter burn time, which lead to efficiency advantage over gasoline in an IC engine [3]. Bioethanol can be directly used or it can be blended with gasoline in different ratio from 3 to 25%, depending on countries. The most common that has been widely used is E10 (10% bioethanol blended with 90% gasoline), however in Brazil, bioethanol 100% (E100) or bioethanol 25% (E25) are used [4].

Recently, the production of bioethanol has focused on crops, such as maize, sugarcane and soybean. It includes three major groups: sucrose-containing feedstocks (sugarcane, sugar beet, sweet sorghum and fruits), starchy materials (corn, milo, wheat, rice, potatoes, cassava, sweet potatoes and barley) and lignocellulosic biomass (wood, straw and grasses) [5]. Major countries produced bioethanol include Brazil (sugarcane 100%), the USA (corn 98%, sorghum 2%), China (corn 70%, wheat 30%), EU (wheat 48%, sugar beet 29%) and Canada (corn 70%, wheat 30%) [2]. Although the production of bioethanol can provide environmental and economic benefits, but food security in developing countries has been a key factor to the instability, whereas a large amount of bioethanol is produced from crops in developed countries, that has been the crucial social controversy for bioethanol production.

Sweet sorghum (*Sorghum bicolor*) is a C_4 crop and is considered as one of the most important food and fodder crops in arid and semiarid regions of the world [6]. This crop occupies approximately 45 million ha in Africa and India, accounts for >80% of the global cultivated area. Similar to grain sorghum, sweet sorghum has rapid growth and wide adaptability [7], good drought resistance [8], salinity tolerance [9] and high yield of biomass [10]. Sweet sorghum has played an important role in promoting the development of agricultural production, livestock husbandry [11], biofuel [12, 13], refining sugar and paper making [10]. Carbohydrates in sweet sorghum can be nonstructural as sugars and starch, or structural such as cellulose, hemicellulose and pectic substances [14]. The major sugars in sweet sorghum are the monosaccharides, glucose and fructose; the disaccharides, sucrose and maltose; and the trisaccharide, raffinose. Depending on the kind of sugar in the stalk, there are saccharin and syrup types of sweet sorghum [14]. Sugars in sweet sorghum stalk are mostly sucrose and invert sugars are glucose, fructose, maltose and xylose [15]. It was reported that other sugars such as mannose, galactose and arabinose were not found in juice of the crop; therefore, the production of bioethanol from sweet sorghum is appropriate, as the carbohydrates present are easily converted to bioethanol. The bagasse and grain of the crop are also possible to produce bioethanol, but they are not economically efficacy [16].

Although sweet sorghum has been suggested as a potential crop for bioethanol production in developing countries [6, 17, 18], but it is still acknowledged as a new crop in many countries. To earn permission for cultivation in a large area for bioethanol production, the cultivation protocol for this crop should be critically examined. In our previous report, we have selected a cultivar 4a from 66 sweet sorghum cultivars through 3 years of cultivation in Vietnam and evaluated the effects of sowing times, densities and soils to biomass and bioethanol production [19]. In this chapter, we describe the application of different fertilizers, including the combinations, doses and time of fertilization and examined their effects on the productivity of biomass, sugar content and ethanol production of sweet sorghum.

2. Materials and methods

The cultivar 4a (India origin) was selected from a previous experiment as the most potential of high biomass and sugar content [19]. Seeds of this cultivars were tested and their germination rate was >95%.

The experiments were conducted in three different locations in northern Vietnam including Phu Tho, Hoa Binh and Bac Giang provinces, in continuous cropping seasons from 2010 to 2013. In Vietnam, all major ethanol factories are currently using cassava as a material for production, with exception of Phu Tho factory where both cassava and sugarcane can be produced for ethanol production. Therefore, Phu Tho province is the first priority to be chosen as a place of sweet sorghum production. Bac Giang and Hoa Binh are selected because their close distance to the ethanol factory in Phu Tho province (less than 100 km). In addition, the three provinces have a large area of hilly and abandoned areas which can be exploited for cultivation of sweet sorghum.

2.1. Applied doses

Different ratios of N, P_2O_5 and K_2O from 30, 60, 90 kg/ha with 2 tons/ha of manure were combined. The commercial fertilizers were purchased from Ha Bac nitrogenous fertilizer & chemicals company limited (Bac Giang city, Vietnam). The details of fertilizer ratios are presented in **Table 1**. Each treatment was conducted in thrice in a plot of 30 m^2 and repeated in three provinces Hoa Binh, Phu Tho and Bac Giang provinces, Vietnam. Effects of different-treated doses on stem yield, sugar content and ethanol production of the cultivar 4a were measured.

Doses	Fertilizer combinations
1	0 N + 60 P_2O_5 + 60 K_2O + 2 tons/ha manure
2	30 N + 60 P_2O_5 + 60 K_2O + 2 tons/ha manure
3	60 N + 60 P_2O_5 + 60 K_2O + 2 tons/ha manure
4	90 N + 60 P_2O_5 + 60 K_2O + 2 tons/ha manure
5	90 N + 0 P_2O_5 + 60 K_2O + 2 tons/ha manure
6	90 N + 30 P_2O_5 + 60 K_2O + 2 tons/ha manure
7	90 N + 90 P_2O_5 + 60 K_2O + 2 tons/ha manure
8	90 N + 60 P_2O_5 + 0 K_2O + 2 tons/ha manure
9	90 N + 60 P_2O_5 + 30 K_2O + 2 tons/ha manure
10	90 N + 60 P_2O_5 +90 K_2O + 2 tons/ha manure

Table 1. Applied doses of fertilizers.

2.2. Chemical composition of soils in locations selected for the experiment

The criteria for selecting locations in each Hoa Binh, Phu Tho and Bac Giang provinces are where there are abandoned and difficult for cultivating food crops because of their low nutrient soils, to examine the possibility of exploiting the cultivation of sweet sorghum for ethanol production. There were nine villages were selected (total 27 villages) and soils in the selected area in each villages were taken and analyzed for pH, contents of percentages of N, P, K, OC (organic carbon), CEC (cation exchanged capacity) (cmol$^{(+)}$/kg), Zn (mg/kg), Fe (mg/kg) and Al (mg/kg) (**Table 2**). The analysis of these criteria was conducted by conventional methods.

Locations	pH	N (%)	P (%)	K (%)	OC (%)	CEC (cmol$^{(+)}$/kg)	Zn (mg/kg)	Fe (mg/kg)	Al (mg/kg)
Hoa Binh	6.4	0.234	0.059	1.392	0.286	8.06	265.48	58116.93	85879.64
Phu Tho	5.1	0.219	0.062	0.561	0.174	5.48	89.84	93138.45	75458.45
Bac Giang	5.6	0.027	0.224	0.135	0.962	5.46	64.01	66112.50	78512.50

Table 2. Chemical compositions of soils for sweet sorghum cultivation in different locations.

2.3. Applied time of fertilization

This experiment was conducted in a plot of 30 m^2 in thrice, in the three locations in Hoa Binh, Phu Tho and Bac Giang provinces in summer-autumn cropping season 2010 and repeated in 2011. The dose of fertilizers per ha included 90 N +60 P_2O_5 +60 K_2O + 2 tons manure. Applied times of fertilization are presented in **Table 3**.

2.4. Improvement of biomass and ethanol production by micronutrient fertilizers

This experiment was conducted to examine the role of micronutrient fertilizers by an additional application of $(NH4)_2MO_7O_2.4H_2O$ including 54% Mo (5kg/ha) on biomass and ethanol production of the sweet sorghum 4a. The based fertilization (BF) was 90 N + 60 P_2O_5 + 60 K_2O + 2 tons/ha manure. Four trials were conducted: (i) BF, (ii) BF + $(NH_4)_2MO_7O_2.4H_2O$, (iii) BF + 60 kg/ha P_2O_5 and (iv) BF + 60 kg K_2O. This experiment was also conducted in the designated areas in the three provinces in the cropping season of summer-autumn 2010 and repeated in 2011.

Trials	Applied time
1	1st fertilization: plants at 3–4 leaf period: 1/2 total N + 1/2 total % K_2O 2nd fertilization: plants at 7–8 leaf period: 1/2 total N + 1/2 total % K_2O
2	1st fertilization: plants at 3–4 leaf period: 1/3 total N + 1/3 total % K_2O 2nd fertilization: plants at 7–8 leaf period: 1/3 total N + 1/3 total % K_2O, trimming seedlings at 20 cm space per plant 3rd fertilization: after the 2nd fertilization 20 d, provided the remained fertilizers
3	1st fertilization: plants at 3–4 leaf period: 1/3 total N 2nd fertilization: plants at 7–8 leaf period: 1/4 total N + 1/2 total K_2O 3rd fertilization: before flowering period: 1/2 total N + 1/2 total % K_2O

Table 3. Times of fertilization application.

2.5. Determination of biomass, sugar content (Brix %) and ethanol production

Sweet sorghum at flowering stages in different trials was harvested in random. The biomass weight, sugar content (Brix %) and ethanol production were determined by a method described in Ref. [19].

3. Statistical analysis

All trials of the study were conducted in a completely randomized design with three replications. Means and differences between the treatments were determined by using a two-way analysis of variance (ANOVA) with values of least significant difference (LSD) at $P \leq 5\%$ level from IRRISTAT program version 5.0.

4. Results

4.1. Effects of different applied doses of fertilizers on stem yield, sugar content and ethanol production of sweet sorghum

There were 10 different applied doses were used in the designated locations in Hoa Binh, Phu Tho and Bac Giang provinces. The effects of these different fertilization doses on stem yield, sugar content and ethanol production of the cultivar 4a are presented in **Table 4**.

It is observed that the stem yield of sweet sorghum varied between 44.7 and 77.3 tons/ha, depending on applied doses of fertilizers and locations (**Table 1**). The most effective dose was 90 N + 90 P_2O5 + 60 K_2O, followed by 90 N + 60 P_2O5 + 60 K_2O. It was found that nonapplication of any N, K_2O and P_2O_5 caused severe marked reduction of stem yield, up to 20 tons/ha and the amount of N played a more significant role on the yield of this crop, than either K_2O or $P2O_5$ application (**Table 1**). Accordingly, the ethanol yield was the highest in the doses of 90 N + 90 P_2O5 + 60 K_2O and 90 N + 60 P_2O5 + 60 K_2O. Nontreatment of N, K_2O and P_2O_5 also resulted in a critical decrease of ethanol yield and the lowest ethanol yield was the nontreatment of nitrogen fertilizer (2963.6 l/ha). Differently, the applied doses did not strongly influence on sugar contents of the cultivar 4a. The dose 90 N + 90 P_2O5 + 60 K_2O also recorded the highest sugar quantity, but the treatment of 90 N + 60 P_2O5 + 60 K_2O showed a significantly lower sugar content (12.0–13.3%). In addition, the dose 90 N + 60 P_2O5 + 90 K_2O did not provide the highest stem yield (53.2–59.5 tons/ha), but it has a remarkable high sugar content and ethanol yield than other doses. Regarding to the location, in general, Bac Giang province has a significantly higher sugar content and ethanol yield than Hoa Binh and Phu Tho provinces (**Table 4**).

Doses	Evaluated categories								
	Stem yield (tons/ha)			Sugar content (Brix %)			Ethanol yield (l/ha)		
	Hoa Binh	Phu Tho	Bac Giang	Hoa Binh	Phu Tho	Bac Giang	Hoa Binh	Phu Tho	Bac Giang
90 N + 60 P_2O_5 + 60 K_2O	64.6	66.0	66.9	12.0	12.0	13.3	4283.0	4375.8	4916.0
0 N + 60 P_2O_5 + 60 K_2O	44.7	45.0	47.7	12.0	12.0	12.7	2963.6	2983.5	3347.0
30 N + 60 P_2O_5 + 60 K_2O	53.2	53.2	55.6	13.0	13.0	13.0	3821.1	3842.6	3993.5
60 N + 60 P_2O_5 + 60 K_2O	58.0	58.0	61.0	12.0	12.0	13.3	3845.4	3845.4	4482.4
90 N + 0 P_2O_5 + 60 K_2O	46.0	46.0	54.3	12.0	12.0	10.3	3049.8	3049.8	3090.1
90 N + 30 P_2O_5 + 60 K_2O	48.7	48.5	66.4	12.0	13.0	11.7	3228.8	3483.5	4292.3
90 N + 90 P_2O_5 + 60 K_2O	68.6	69.0	77.3	14.0	14.0	13.7	5306.2	5337.2	5851.0
90 N + 60 P_2O_5 + 0 K_2O	49.2	53.0	50.5	12.0	12.0	12.0	3262.0	3513.9	3315.0
90 N + 60 P_2O_5 + 30 K_2O	52.6	52.0	54.4	13.3	13.3	10.0	3778.0	3734.9	3005.6
90 N + 60 P_2O_5 + 90 K_2O	53.2	53.5	59.5	14.0	14.0	13.3	4408.9	4138.2	4372.2
LSD (5%)	8.76			0.15			1.1		
CV (%)	9.7			3.3			5.1		

Table 4. Effects of applied doses of fertilizers on sugar content, sugar yield and ethanol production of the sweet sorghum variety 4a.

4.2. Effects of applied times of fertilizers on stem yield, sugar content and ethanol production of sweet sorghum

There were three trials of fertilizer application in different times and fertilizer doses (**Table 4**), applied in the three locations in Hoa Binh, Phu Tho and Bac Giang provinces (**Table 5**). It indicated that times of fertilizer application at 3–4 to 7–8 leaf stages provided the highest stem yield and ethanol yield in every location and their values were markedly different as compared among the 1st, 2nd and 3rd applied times (**Table 5**). However, when the amount of fertilizers were divided into three times of application, of them, the third fertilization was either after 20 d of the second application (trial 2) and before flowering, the yields of stems and ethanol production were significantly reduced, as compared with other treatments (**Table 5**). The stem yield was markedly decreased >20 tons/ha, whereas the amount of ethanol production was significantly reduced > tons/ha. For location, Phu Tho province obtained higher yields of stems and ethanol production than Hoa Binh and Bac Giang provinces. For the sugar content (Brix %), different times of fertilizer application did not influence on the sugar content of sweet sorghum, as no significant difference among treatments were observed (**Table 5**).

Locations	Trials	Stem yield (tons/ha)	Sugar content (Brix %)	Ethanol yield (l/ha)
Hoa Binh	1	68.6	14.0	5306.2
	2	62.7	14.0	4849.9
	3	47.5	14.0	3674.1
Phu Tho	1	72.5	14.0	5607.9
	2	65.5	14.0	5066.4
	3	49.5	14.0	3828.8
Bac Giang	1	69.5	14.0	5375.8
	2	61.5	14.0	4757.0
	3	46.6	13.0	3347.0
LSD (5%)		0.4	1.4	60.0
CV (%)		0.7	5.8	0.7

Note: Trials 1–3 are different applied times as indicated in **Table 3**.

Table 5. Effects of different times of fertilizer application on stem yield, sugar content and ethanol production of sweet sorghum.

4.3. Effects of micronutrient fertilizers on stem yield, sugar content and ethanol production of sweet sorghum

In this experiment, the control was the based fertilization (BF) that included 90 N + 60 P_2O_5 + 60 K_2O + 2 tons/ha manure. Other treatments consisted of BF + $(NH4)_2MO_7O_2.4H_2O$ (Mo), BF + 60 K_2O and BF + 60 P_2O_5. The effects of these treatments on stem yield, sugar content and ethanol productivity are presented in **Table 6**.

Treatments	Stem yield (tons/ha)			Sugar content (Brix %)			Ethanol yield (l/ha)		
	Hoa Binh	Phu Tho	Bac Giang	Hoa Binh	Phu Tho	Bac Giang	Hoa Binh	Phu Tho	Bac Giang
Based fertilization (BF)	64.4	66.0	62.4	14.8	15.0	14.8	5266.0	5469.8	5012.5
BF + Mo	74.3	79.0	75.9	14.9	15.0	14.9	6116.6	6547.1	6248.3
BF + P_2O_5	70.4	75.4	71.5	15.5	15.0	15.5	6028.9	6248.8	6123.1
BF + K_2O	68.5	69.5	70.5	15.5	15.0	14.9	5866.2	5759.8	5073.7
LSD (5%)	0.9			0.8			138.7		
CV (%)	0.7			3.1			11.3		

Table 6. Effects of micronutrient fertilizers on stem yield, sugar content and ethanol production of sweet sorghum.

It was found that by the application of based fertilization, the stem yield of sweet sorghum was 62.4–66.0 tons/ha. However, the supplemented application of Mo significantly increased the stem yield 10–12 tons/ha and it was higher than either the supplementation of P_2O_5 or K_2O (**Table 6**). For the sugar content, there was no remarkable difference among treatments (14.8–15.5 Brix %). Following the promoted amount of stem yield, the ethanol productivity was the maximum with Mo application, followed by P_2O_5 and K_2O supplementation. They were all significantly higher than the application of unique-based fertilizers (**Table 6**).

5. Discussion

Although sweet sorghum is well adaptable to many soils and climates and has a strong resistance to pests and diseases, an effective fertilization protocol is crucial to farmers and agricultural extensionists to promote the biomass and ethanol productivity of this crop. Trials of this study were replicated in the three locations in three different provinces, where are the most suitable for ethanol production in the North of Vietnam [19]. Among different doses of fertilizers, this study found that the application of 90 N + 90 P_2O_5 + 60 K_2O was the best for maximum biomass and ethanol productivity of the crop. Their values were significantly higher than those of other treatments (**Table 4**). Nontreatment of N, P_2O_5 and K_2O was examined and the lack of fertilization of any among these nutrients caused severe markedly reduction of stem yield, sugar content and ethanol production; however, N was the most important for biomass and ethanol productivity (**Table 4**).

It has been described that sweet sorghum can become a profitable crop for farmers in developing countries, however its agronomic practices should be optimized to improve the productivity. Among them, nitrogen is the most important nutrient element as this crop responds well to N fertilization [20]. Turgut et al. [21] investigated the optimum N input for maximizing sweet sorghum yield in a clay loam soil. Reddy et al. [22] also studied the effects of N input on sweet sorghum growth. Uchino et al. [20] reported that the amount of N from 90 to 120 kg/ha was the optimum for ethanol productivity, but the amount of N >150 kg/ha may cause lodging. In this study, the application of N >90 kg/ha was not studied as it may cause economic troublesome for farmers in developing countries such as Vietnam.

This research also verified the application of K_2O and P_2O_5 in combination with nitrogen and manure (**Table 4**) to reduce the dependency on nitrogen application. The use of manure, K_2O and P_2O_5 with reduced quantity of nitrogen fertilizers are efficacy for soil health and maintain the biomass and ethanol productivity of sweet sorghum. The time of application was also examined to optimize the time of application and doses of fertilizers. The best time for fertilization was found to be on 3–4 to 7–8 leaf stage (**Table 5**). Findings of this study indicate that fertilization should not be conducted after 7–8 leaf stage, as the application of fertilizers at 20 d after the second application (7–8 leaf stage) to flowering stage caused significant reduction of stem yield and ethanol production, although the sugar content (Brix %) was not influenced (**Table 5**). This also helps to save the labor costs as the too much fertilization may attribute to heavy fieldwork and economic deficit in sweet sorghum cultivation.

In this study, for micronutrient fertilization, 5 kg/ha of $(NH4)_2MO_7O_2.4H_2O$ (Mo) was used as supplement to the base fertilizer application of 90 N + 60 P_2O_5 + 60 K_2O + 2 tons/ha manure. The supplements of 60 kg/ha of P_2O_5 and K_2O were also conducted to compare with that of the Mo supplement. Interestingly, the use of Mo was the most effective in enhancing the biomass and ethanol production and it was higher than that of either P_2O_5 or K_2O supplements (**Table 6**). The use of Mo was found to be more effective than the adding of P_2O_5 and K_2O. In addition, the application of Mo at 5 kg/ha provided less heavy fieldwork than the supplementation of 60 kg of either P_2O_5 or K_2O. The application of micronutrients such as $ZnSO_4$ and sodium borate was effective on the stalk yield of the sweet sorghum ICSV 93046 variety, which is widely cultivated in India and Philippines [23]. The supplements of micronutrients of Zn and Fe significantly promote the stalk yield of sweet sorghum [24]. It is therefore proposed that the application of micronutrient fertilizers is effective to improve the productivity and profitability of sweet sorghum than the supplement of unique either P_2O_5 or K_2O.

The influences of fertilization including times of application and applied doses on energy crops have been extensively studied. It has been reported that the application of N fertilizer significantly improves the production of bioenergy crops, such as switchgrass (*Panicum virgatum* L.) and corn (*Zea mays* L.) [25]. Individual applications of N, P, or K fertilizers markedly improved cassava (*Manihot esculenta* Crantz) yield. The best yield performance in this research to increase the yield of cassava (62.8 t/ha) was the use of 400, 200 and 400 kg/ha of N, P, K fertilizers, respectively, although the use of P fertilizer incorporation with high levels of N and K fertilizers was not effective to improve cassava yields [26]. In addition, Olugbemi and Ababyomi [27] examined different treatment of nitrogen fertilizer on the growth and ethanol yield of four sweet sorghum varieties and found that the application of 120 kg N/ha was the best for ethanol production. There are also several research studies on optimizing the nitrogen requirements of sweet sorghum for various environments and soil types [20–21]. The applied doses on sugarcane were important on the yield and ethanol production in Brazil [28]. Appropriate doses of fertilizers are useful to increase the yield and ethanol production of biofuel crops, but the overuse of fertilizers may cause critical environmental pollution [29]; therefore, the examination of applied doses and times of application as conducted in this research is indispensable for the cultivation of biofuel crops.

6. Conclusions

Sweet sorghum is a promising alternative crop for bioethanol production in developing countries to reduce the social criticisms on the use of food crops for biofuel production. Findings of this study, together with a previous work, examined the sowing times, densities and soil chemicals [19] to determine the cultivar 4a which was continuously examined in this research for different doses of fertilizers and times of application. Achievements of the two research studies are useful to establish a protocol of sweet sorghum cultivation for farmers and agricultural extensionists in developing countries, as the fertilization of different doses and times was examined in different locations with low nutrient contents that were difficult to cultivate other crops. This study also highlighted the higher possibility of using sweet sorghum in infertile soil in hilly areas than other bioenergy crops. The combination of manure and micronutrients with a reduced amount of inorganic fertilizers is crucial to establish a sustainable agriculture in sweet sorghum production.

Acknowledgements

The author express sincere thanks to Ministry of Trade, Vietnam, for funding to this research. Thanks are also due to Phung Thi Tuyen, La Hoang Anh, Truong Ngoc Minh and Do Tan Khang for their assistance to the manuscript's preparation.

Author details

Tran Dang Xuan[1,*], Nguyen Thi Phuong[2] and Tran Dang Khanh[3]

*Address all correspondence to: tdxuan@hiroshima-u.ac.jp

1 Graduate School for International Development and Cooperation (IDEC), Hiroshima University, Higashi Hiroshima, Japan

2 Vietnam Academy of Agricultural Science (VASS), Hanoi, Vietnam

3 Agricultural Genetics Institute (AGI), Hanoi, Vietnam

References

[1] Hansen G. Driving technology in the motor vehicle industry. In: Proceedings of the IPCC expert meeting on the industrial technology development, transfer and diffusion. Tokyo, September 21–23. 2004.

[2] Balat M, Balat H. Recent trends in global production and utilization of bio-ethanol fuel. Applied Energy 2009; 86: 2273–2282. DOI: 10.1016/j.apenergy.2009.03.015

[3] Balat M. Global bio-fuel processing and production trends. Energy Exploration & Exploitation. 2007; 25 (3): 195–218. DOI: 10.100/15435070802107322

[4] Oliveria MED, Vaughan BE, Rykiel Jr. EJ. Ethanol as fuel: energy, carbon dioxide balance and ecological footprint. Bioscience 2005; 55(7): 593–602. DOI: 10.1525/bio.2013.63.4.6

[5] Balat M. Global trends on the processing of bio-fuels. International Journal of Green Energy 2008; 5(3): 212–238. DOI: 10.1080/15435070802107322

[6] Basavaraj G, Rao PP, Basu K, Reddy CR, Kumar AA, Rao PS, Reddy BVS. Assessing viability of bio-ethanol production from sweet sorghum in India. Energy Policy 2013; 56: 501–508. DOI: 10.1016/j.enpol.2013.01.012

[7] Reddy BVS, Ramesh S, Reddy PS, Ramaiah B, Salimath PM, Kachapur R. 2005. Sweet sorghum - a potential alternate raw material for bio-ethanol and bio-energy. International Sorghum Millets Newsletter 2005; 46, 79–86.

[8] Tesso TT, Claflin LE, Tuinstra MR. Analysis of stalk rot resistance and genetic diversity among drought tolerant sorghum genotypes. Crop Science 2005; 45(2): 645–652. DOI: 10.2135/cropsci2005.0645

[9] Almodares A, Hadi MR, Ahmadpour H. Sorghum stem yield a soluble carbohydrate under phonological stages and salinity levels. African Journal of Biotechnology 2008; 7(2), 4051–4055.

[10] Almodares A, Hadi M.R. Production of bioethanol from sweet sorghum: a review. African Journal of Agricultural Research 2009; 4(9), 772–780.

[11] Fazaeli H, Golmohhammdi HA, Almodares A, Mosharraf S, Shaei A. Comparing the performance of sorghum silage with maize silage in feedlot calves. Pakistan Journal of Biological Sciences 2006; 9, 2450–2455. DOI: 10.3923/pjbs.2006.2450.2455

[12] Nahvi I, Almodares A, Motocalemi M. 1994. Comparing ethanol production from sweet sorghum juice and sugar beet molasses. In: Proceeding of the 1st Iranian Congress on Chemical Engineering (Eds. I Nahvi, A Almodares and M. Motocalemi). Tehran, Iran.

[13] Nahvi I, Almodares A, Rasuli A. 1994. The effect of environmental factors on ethanol production from sweet sorghum juice. In: Proceeding of the 1st Iranian Congress on Chemical Engineering (Eds. I Nahvi, A Almodares and M. Motocalemi). Tehran, Iran.

[14] Anglani C. Sorghum carbohydrates – a review. Plant Foods for Human Nutrition 1998; 52(1): 77–83. [PMID: 9839837]

[15] Almodares A, Taheri R, Adeli S. Stalk yield and carbohydrate composition of sweet sorghum [(*Sorghum bicolor* (L.) Moench) cultivars and lines and different growth stages. Journal of Malaysian Applied Biology Journal. 2008, 37(1):31–36.

[16] Jacques K, Lyons TP, Kelsal DR. The Alcohol Textbook. Nottingham University Press: United Kingdom 1999. 3rd ed. p. 388.

[17] Liu SY, Lin CY. Development and perspective of promising energy plants for bioethanol production in Taiwan. Renewable Energy. 2009; 34: 1902–1907. DOI: 10.1016/j.renene.2008.12.018

[18] De Vries SC, Vandeven GWJ, Ittersum MKV, Giller KE. The production-ecological sustainability of cassava, sugarcane and sweet sorghum cultivation for bioethanol in Mozambique. GCB Bioenergy 2012; 4 (1): 20–35. DOI: 10.1111/j.1757-1707.2011.01103.x

[19] Xuan TD, Phuong NT, Khang DT, Khanh TD. Influence of sowing times, densities and soils to biomass and ethanol yield of sweet sorghum. Sustainability 2015; 7(9): 11657–11678. DOI: 10.3390/su70911657

[20] Uchino H, Watanabe T, Ramu K, Sahrawat KL, Marimuthu S, Wani SP, Ito O. Effects of nitrogen application on sweet sorghum (*Sorghum bicolor* (L.) Moench) in the semi-arid tropical zone of India. The Japan Agricultural Research Quarterly 2013; 47(1): 65–73. DOI: 10.6090/jarq.47.65

[21] Turgut I, Bilgili U, Duman A, Acikgoz E. Production of sweet sorghum (*Sorghum bicolo*r L. Moench) increases with increased plant densities and nitrogen fertilizer levels. Acta Agriculturae Scandinavica, Section B: Soil & Plant Science 2005; 55 (3): 236–240. DOI: http://dx.doi.org/10.1080/09064710510029051

[22] Reddy PS, Reddy BVS, Kumar AA, Rao S. Standardization of nitrogen fertilizer rate for sugar yield optimization in sweet sorghum. Journal SAT Agricultural Research 2008; 6, 1–4.

[23] Rao S, Kumar CG, Malapaka J, Kamal A, Reddy BVS. Effect of micronutrient treatments in main and ratoon crops of sweet sorghum cultivar ICSV 93046 under tropical conditions. Sugar Tech 2012; 14: 370. DOI: 10.1007/s12355-012-0172-y

[24] Ismail S, Arbad BK, Syed JJ. Impact of integrated nutrient management practices on yield, juice quality and nutrient uptake in sweet sorghum [*Sorghum biocolor* (L.) Moench] grown on vertisol. Asian Journal of BioScience 2009; 4(1): 100–102.

[25] Stewart CE, Follett RF, Pruessner EG, Varvel GE, Vogel KP, Mitchell RB. N fertilizer and harvest impacts on bioenergy crop contribution to SOC. GCB Bioenergy 2016; 8: 1201–1211. DOI: 10.1111/gcbb.12326

[26] Wilson H, Ovid A. Influence of fertilizers on cassava production under rainfed conditions. Journal of Plant Nutrition 1994; 17 (7): 1127–1135. DOI: 10.1080/01904169409364793.

[27] Olugbemi O, Ababyomi YA. Effects of nitrogen application on growth and ethanol yield of sweet sorghum (*Sorghum bicolor* L. Moench) varieties. Advances in Agriculture 2016; Article ID 8329754, 7 pages. DOI: 10.1155/2016/8329754

[28] Martinelli LA, Filoso S. Expansion of sugarcane ethanol production in Brazil: environmental and social challenges. Ecological Applications 2008; 18 (4) 885–898. [PMID: 18536250]

6

Catalytic Biomass Valorization

Aiguo G. Wang, Danielle Austin and Hua Song

Abstract

Biomass is a significant non-conventional energy reserve, which has been considered as a promising alternative over other renewable sources such as solar, wind or hydroelectric storage due to its comparatively ample availability. A variety of biomass types can be converted into useful products via bioenergy technologies. The deep understanding and knowledge of these processes are necessary for optimization and advancement in a cost-effective way. A comprehensive comparison and discussion is conducted with respect to biochemical and thermochemical conversion technology such as microbic digestion and fermentation, pyrolysis, liquefaction and gasification. Pyrolysis is the process of converting biomass into bio oil, charcoal and gaseous factions by heating anaerobically to above 500°C. Liquefaction is a low temperature (LT) and high-pressure thermochemical process to produce marketable liquid over suitable catalysts under hydrogen or reductive environment. Gasification is the conversion of biomass into preferred combustible gas mixture (syngas) via the partial oxidation at high temperature, typically in the range of 800–900°C. The product gas is more versatile and can be burned in gas turbine for electricity production or synthesis of high-value chemicals. The parametric impact, mechanism, development status and future direction have been summarized for each of these technologies with the aim to pave the way for optimization of future investigation.

Keywords: biomass, pyrolysis, gasification, liquefaction

1. Introduction

The declining reserves and fluctuating prices of fossil fuels necessitate a switch from conventional to renewable power sources, for instance, solar, wind, biomass and hydroelectric generation [1]. Biomass is considered the fourth largest source of energy in the world, currently supplying more than 10% of primary energy. It also provides a potential source of valuable chemicals, such as reducing sugars, furfural, ethanol and other products by using biochemical or chemical and thermochemical procedures [2, 3]. Due to its widespread availability, renewable nature and neutral relation to global warming, the potential of biomass to meet the need for the world energy has been widely recognized. Furthermore, biomass utilization has an advantage over other renewable sources on account of its topographical independence and easy storage and transportation.

Biomass is the term for all biologically produced matter, including land and water-based vegetation, as well as organic wastes, which can be used directly or indirectly by converting it into liquid or gaseous fuel. As the necessity for a renewable and sustainable energy becomes more significant, research interest in biomass conversion to fuel or high value-added chemicals has increased dramatically. The distinction between the energy carriers originated from biomass can be made based on their ability to provide heat, electricity and engine fuels. The famous *Van Krevlen* diagram is usually used for comparing biomass and fossil fuels in terms of their O/C and H/C ratios. The lower the respective ratios, the greater the energy content of the material. More details about this diagram can be obtained from the literature [4].

Currently, there are a number of different approaches used for biomass conversion, such as direct combustion for heat or electricity, thermochemical conversion to bio-oil, chemicals and syngas, and biochemical transformation of biogas or bioethanol. Combustion can be widely used on various scales to obtain heat or electricity from biomass; however, the energy efficiency of this conversion is low. Co-combustion of biomass in coal fire plants is an attractive option to improve the conversion efficiency. Biochemical processing involves anaerobic digestion, the decomposition of biomass via bacterial activity without oxygen to yield biogas and alcoholic fermentation, which is used commercially on a large scale to produce ethanol from sugar and starch crops. In this chapter, more attention has been concentrated on the thermochemical conversion of biomass, including gasification, pyrolysis and liquefaction, due to the diversity of products and high conversion efficiency in the process. The differences of the three process conditions are shown in **Table 1**.

An in-depth survey is carried out regarding the mechanism, parametric impacts, products, techno-economics, development status and future direction for each of the three thermochemical technologies. There are a wealth of articles regarding biomass conversion, and this chapter provides a comprehensive overview with a direct comparison of the current technologies highlighting the benefits and drawbacks of each with emphasis on catalytic biomass conversion.

Process	Temperature, K	Pressure, MPa	Drying	Catalyst
Pyrolysis	650–800	0.1–0.5	Unnecessary	Yes
Liquefaction	525–600	5–20	Necessary	No
Gasification	873–1273	1–2	Necessary	Yes

Table 1. Comparison of thermochemical process conditions [5].

2. Biomass valorization

2.1. Pyrolysis

2.1.1. Mechanism

Pyrolysis is the thermal decomposition of biomass in the absence of oxygen, as illustrated in **Figure 1**. Slow pyrolysis occurs when there is a slow heating rate over a longer residence time (5–30 min). The extended residence time and slow heating rate facilitate the increased production of char products and have been used throughout history for the production of charcoal. Fast pyrolysis occurs when there is a much faster heating rate with much shorter residence times. Increased heating rates and faster residence time result in greater liquid and gas products, making fast pyrolysis more desirable for the purposes of producing bio-oil and valuable chemicals [7]. Typically, decomposition occurs in just a few seconds at atmospheric pressure and relatively low temperatures between 300 and 600°C [8, 9]. The quick reaction time and mild operating conditions make pyrolysis an attractive option for producing bio-oil and valuable chemicals from the cheap feedstock of biomass.

A mixture of a complex feedstock and multiphase products make it difficult to study the exact mechanism of pyrolysis, but there is a general understanding. Initially upon heating, water is removed, resulting in a dry fuel that undergoes primary pyrolysis. Decomposition occurs upon the transfer of heat from the surrounding area to the biomass particles resulting in solid char, bio-oil, gas and water, constituting the primary stage of pyrolysis [8]. An increased

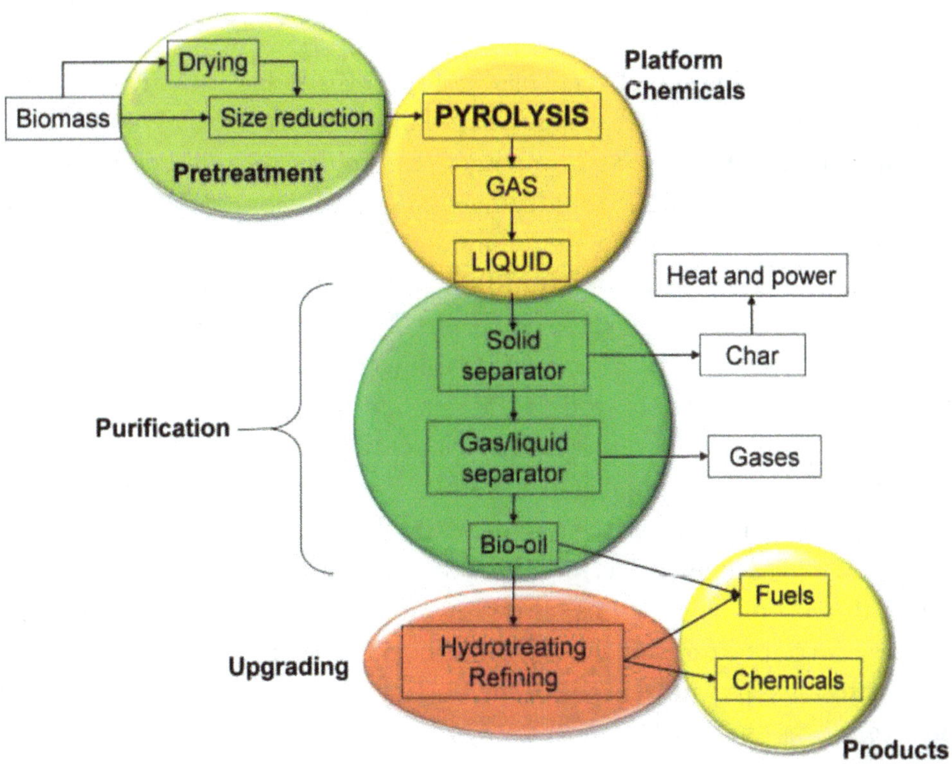

Figure 1. Process schematic of biomass pyrolysis and upgrading [6].

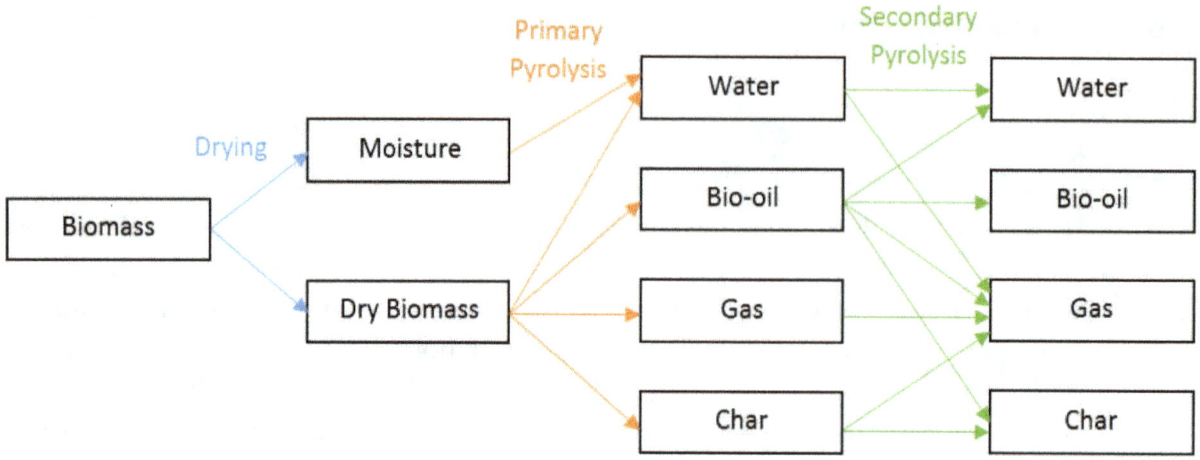

Figure 2. Primary and secondary pyrolysis including arrows indicating reaction pathway [8].

surface area results in better heat transfer, which causes a faster and more complete reaction [9, 10]. For this reason, biomass is typically mechanically ground into a fine powder prior to the reaction. During the secondary stage, further decomposition occurs in addition to various other reactions resulting in an increase in gaseous products. A summary of these processes can be seen in **Figure 2**. Fast pyrolysis can have a biomass to bio-oil conversion of 70–80%. Liquid yield can be maximized by a fast residence time, preventing the secondary decomposition of products. Due to the importance of residence time and heat transfer in pyrolysis, much research has gone into reactor design which is reviewed in reference [9].

2.1.2. Products

Pyrolysis results in a wide range of products that can be divided into the categories of solid char, liquids (water and bio-oil) and gases. Char is the term used for all of the solid products including organic matter (high carbon content) and ash [8]. Water is produced as a product of pyrolysis as well as during the initial drying stage via evaporation [8, 9]. The bio-oil component is a brown, polar liquid comprised of a mix of oxygenated compounds, which differs with feedstock and reaction conditions [9, 10]. At moderate temperatures, gas products are mainly comprised of CO, CO_2 and CH_4, with H_2 and C_XH_Y gases also being produced at higher temperatures [8, 10]. The amount of solid, liquid and gas produced varies greatly depending on the biomass composition as well as the reaction temperature. The temperature has a greater effect on the amount of product in each phase, while the feedstock composition has a greater effect on the bio-oil elemental composition. A higher C/O ratio in the feedstock results in a higher liquid yield and lower gas yield, while the char yield remains relatively unaffected. An increase in hydrogen in the feedstock results in more water formation, more bio-oil and less carbon dioxide and C_xH_y gases [8]. At low temperatures (<350°C) and slow residence times, char is the favoured product [8–10]. Longer residence times allow for the reaction of gas and liquid products to form char [7]. The best yields of liquid product are achieved at moderate temperatures (400−550°C) and fast residence times. These conditions allow for primary pyrolysis to occur, but the fast residence time allows for products to be collected before secondary reactions can take place [8–10]. At very high temperatures (HT) (>800°C), a higher yield of gas products is observed due to the presence of secondary reactions, which further break down the liquid and char components into gases.

The liquid product composition generally mimics the biomass carbon, hydrogen and oxygen composition. Due to the high oxygen content in biomass, the liquid product of pyrolysis is made up of a large portion of oxygen-containing compounds. The oxygen-containing compounds including alcohols and acids make the bio-oil reactive and acidic causing issues for storage and transportation through established pipelines [11–14]. As a result, bio-oil must be upgraded before it can be transported or used as a fuel. An increase in temperature leads to a slight decrease in oxygen content and a slight increase of carbon and hydrogen content in the bio-oil. This results in a small increase of aromatic compounds, which marginally improves the oil quality [8].

2.1.3. Developmental status

The majority of research on pyrolysis involves optimizing the upgrading process to acquire a higher quality product. Hydrodeoxygenation and catalytic cracking are the most popular processes currently being utilized for upgradation [12]. Hydrodeoxygenation involves removing oxygen from the liquid product in the form of water through a hydrogen supply [9, 15]. This process is expensive due to the high operating pressures (up to 20 MPa) and the high cost of hydrogen [9, 15]. Catalytic cracking is a fairly new technology, which uses a catalyst to promote a range of reactions including dehydration, decarboxylation, aromatization, alkylation and polymerization [12]. Zeolite catalysts have found to be particularly effective at aromatization and have been found to remove oxygen in the form of carbon dioxide [9, 14]. The addition of metals such as silver or nickel can increase selectivity and allow for optimized products [12]. Catalytic cracking is significantly cheaper than hydrodeoxygenation due to the relatively mild operating conditions and absence of hydrogen; however, the coking of catalysts is a significant concern and due to the lack of hydrogen supply, the product suffers from a low H/C ratio [12, 16].

At the current time, the combination of pyrolysis and catalytic cracking in one process has been attracting more attention [10]. One research group is combining pyrolysis with catalytic cracking under a methane environment, in a novel process called methanolysis [16, 17]. This process is operated at atmospheric pressure and moderate temperatures on a zeolite catalyst with a supply of methane. Methanolysis is economically attractive due to the relatively low operating conditions and the wide availability and low cost of natural gas of which methane is the major component. The aim of this research is to develop a catalyst that will allow pyrolysis to occur in a methane environment, while at the same time upgrading the liquid product. Promising results were found during feasibility studies, where it was shown that not only was the bio-oil produced of greater quality, but the methane was incorporated into the product [16, 17]. **Table 2** compares pyrolysis in inert gas, hydrogen and methane environments with and without catalysts for a sawdust feedstock. An increased oil yield accompanied by a decrease in oxygen content and an increase in H/C ratio can be observed in a methane environment using a catalyst. It is also clear that the metal used in the catalyst can affect both oil yield and quality. These results are very promising for creating a high quality bio-oil in an economical and environmentally friendly way.

Trials	Oil* yield (wt %)	Water formed (mg/g)	Oil quality		
			H/C ratio	O content (wt %)	O/C atomic ratio
Biomass, inert	5.47	97.0	1.62	5.25	0.226
Biomass, inert, Ag/ZSM-5	4.07	135.6	1.29	0.18	0.009
Biomass, 30% H_2	4.17	73.4	1.46	3.41	0.145
Biomass, 30% H_2, Ag/ZSM-5	3.42	100.2	1.45	0.45	0.024
Biomass, 30% CH_4	4.68	119.0	1.38	0.22	0.009
Biomass, 30% CH_4, Ag/ZSM-5	4.85	128.3	1.76	0.07	0.003
Biomass, 30% CH_4, Ag/P-Ce-ZSM-5	6.89	110.9	2.26	7.35	0.356
30% CH_4, Ag/ZSM-5	0	0	–	–	–

*Moisture-free liquid collections with boiling point less than 150°C.

Table 2. Results of the pyrolysis of sawdust under different environments [16].

2.1.4. *Future direction*

The next steps of pyrolysis research are optimizing product yield and quality, which can be achieved through the use of catalytic cracking. This involves gaining a better understanding of the complex mechanism involving the catalyst by using model compounds. Once there is a clear understanding of the catalyst's role, an industrially viable catalyst and process can be developed to produce high-quality fuels that are cost competitive with oil-derived alternatives. Methanolysis is considered a promising new technology, which can be a sustainable alternative to traditional fuel sources. In addition to developing an optimal catalyst, work must be done to improve methane conversion and incorporate catalyst regeneration into the reaction.

2.2. Liquefaction

2.2.1. *Mechanism*

Liquefaction is the thermochemical conversion of biomass to primarily yield liquid fuel at moderate temperature and high pressures. This is a chemical reforming process of organic matters, usually with a high hydrogen partial pressure and catalysts to enhance the rate of reaction and/or to improve the selectivity of the product. This approach is summarized in **Figure 3**. In the process of liquefaction, the carbonaceous materials of biomass are ultimately converted into liquefied products through a series of complex physical and chemical reactions [18, 19]:

- Micellar-like substructures are formed by solvolysis;
- Smaller and soluble molecules are generated via depolymerization;
- New molecular rearrangements occur through thermal decomposition, including dehydration, decarboxylation, C-O and C-C bonds rupture.
- Hydrogenolysis and hydrogenation of functional groups.

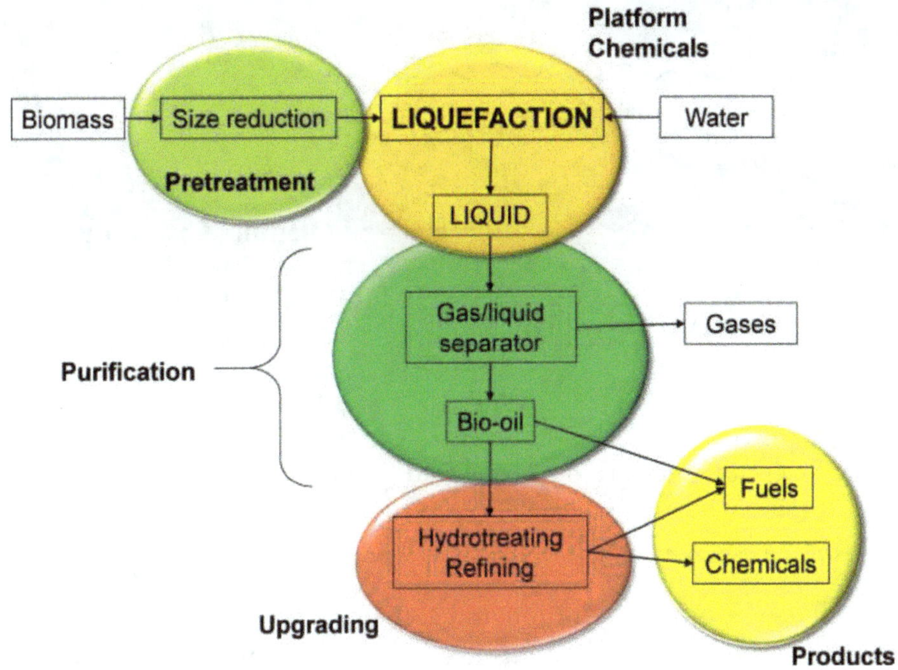

Figure 3. Process schematic of biomass liquefaction and upgrading [6].

Another remarkable feature of liquefaction is that it maximizes the liquid-yields with high quality. One important quality index of product oil is the H/C ratio, which can be increased by removing the oxygen content. The lower oxygen content makes the oil more chemically stable and requires less upgrading to the hydrocarbon product [20]. Generally, biomass contains typically 40–50 wt.% (DAF basis) of oxygen, which can be removed via dehydration and decarboxylation as the form of H_2O and CO_2, respectively. Thermodynamically, removal of H_2O and CO_2 from biomass can be considered as the best option to reduce the oxygen content. Removal of H_2O ultimately forms carbon as the remaining product such as charcoal while removal as CO_2 is prone to generate the product with high H/C ratio. Under hydrothermal upgrading reaction conditions, the decarboxylation selectivity is almost constant, 0.52 [21]. Dehydration and decarboxylation reactions are conceptually described as follows:

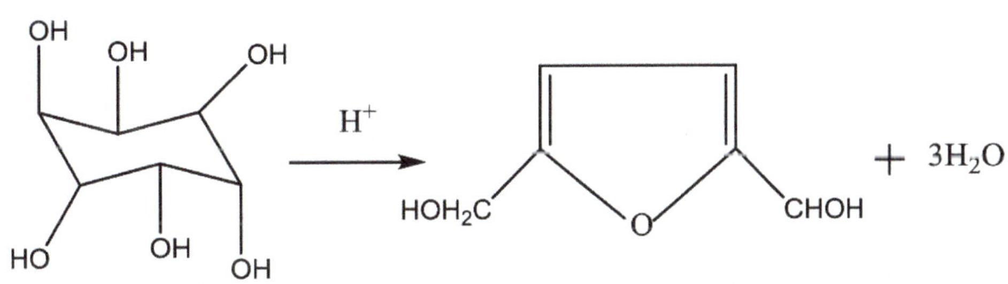

Dehydration of saccharides to 5-hydroxymethylfurfural [22]

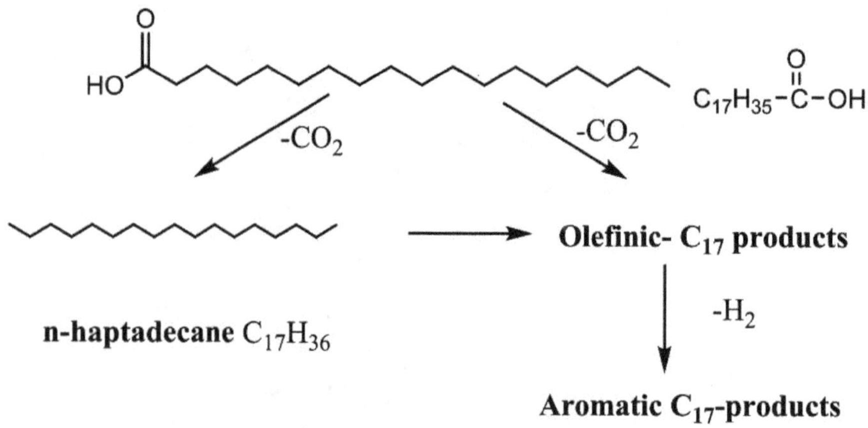

Simplified decarboxylation network of stearic acid to diesel products [23]

2.2.2. Conversion of biomass components

As we know, biomass is a broad definition and contains a wide range of substances with different compositions. The main components are categorized four groups: carbohydrates, lignin, protein and lipids. The protein and lipids are prevalent in the dried distiller grains with soluble (DDGS) and slaughterhouse waste [24, 25], which are not further discussed in this chapter. More concerns are focused on the conversion of carbohydrates and lignin.

2.2.2.1. Carbohydrates

The most abundant carbohydrates contained in the biomass are polysaccharides, cellulose, hemicellulose and starch. Carbohydrates first are hydrolyzed to glucose and other saccharides and further degraded into small molecule compounds at higher temperature. The hydrolysis rate varies with the components. Generally, the hydrolysis rate of hemicellulose and starch is much faster than cellulose, since the cellulose has relatively stable crystalline structure. The conversion of cellulose can achieve 100% at 280°C within 2 min. The addition of carbon dioxide increases cellulose hydrolysis rate, and this is most likely attributed to the formation of carbonic acid, acting as an acid catalyst. However, the beneficial effect is insignificant above 260°C [26]. Hemicelluloses are readily soluble and hydrolyzed above 180°C [27] and close to 100% conversion of hemicellulose is obtained for various wood and herbaceous biomass materials at 230°C within 2 min [28]. Several attempts have been made to reveal the general degradation mechanism of carbohydrates. Glucose is commonly chosen as the model compound because it is one of the main hydrolysates. Even though the employed models are different in details, the overall pattern is quite similar [29, 30]. An overall degradation route for glucose is presented in **Figure 4**.

2.2.2.2. Lignin

Lignin is an aromatic heteropolymer, which is relatively difficult to be degraded chemically or enzymatically [27]. The hydrothermal degradation of lignin can occur at high temperature of 350—400°C, and the main products are catechol, phenols and cresols [37, 38]. However, it can be catalytically hydrolyzed into various phenols and methoxy phenols by cleavage

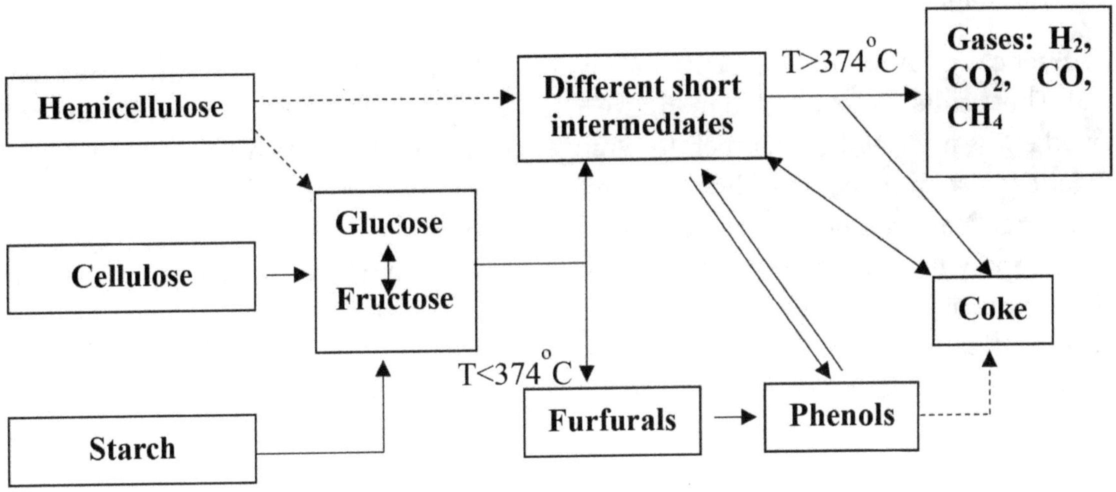

Figure 4. A simplified reaction mechanism for carbohydrate hydrolysis at sub- and supercritical conditions [69].

of ether-bonds under alkaline environment. The reaction temperature for hydrothermal processing of lignin is 200–300°C. The hydrothermal degradation pattern of lignin is shown in **Figure 5**.

2.2.3. *Impact of parameters*

Biomass liquefaction can be influenced by numerous parameters, including temperature, pressure, catalyst, residence time, heating rate, particle size, solvent, reducing agent and the type of biomass. In this chapter, the effect of temperature, pressure, catalyst and residence time on biomass liquefaction will be discussed separately.

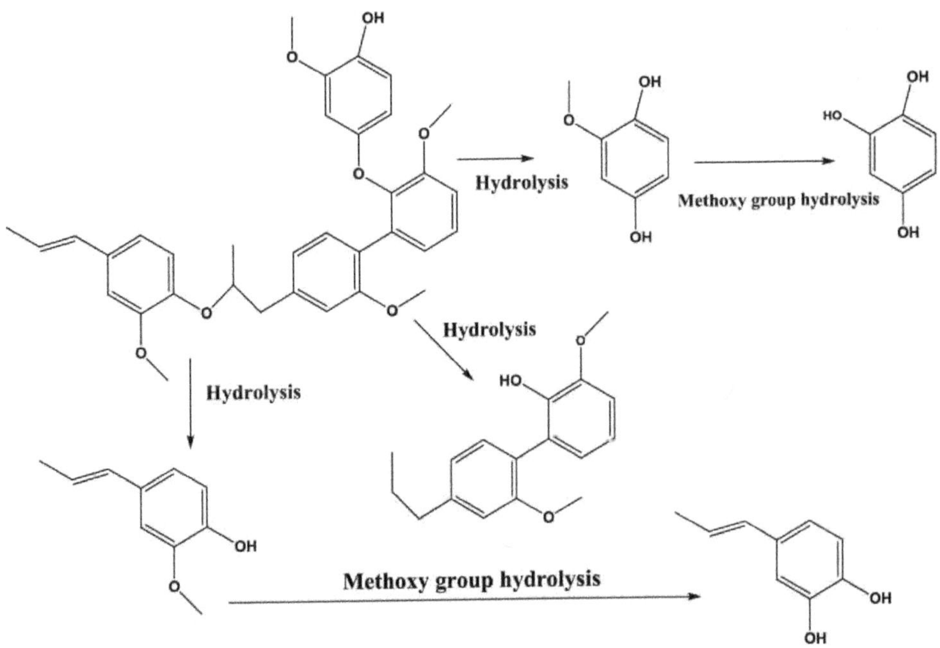

Figure 5. Simplified scheme of lignin degradation [39].

2.2.3.1. *Temperature*

The influence of temperature on the oil yield has been investigated by many researchers [31–36]. The results are illustrated in **Figure 6**. It is obvious that the maximum oil yield can be achieved in the intermediate temperature range. The reason is that temperature enhances the synergetic effect on the yield of liquids due to more biomass fragmentations generated with temperature. When temperature is sufficiently high to provide the activation energy for bond rupture, depolymerization occurs extensively, increasing more free radicals and fragmented species for further repolymerization at higher temperature. Therefore, the dominant reaction is depolymerization of biomass in initial stage, and repolymerization is prevailing in the later stages, and intermediate temperature is expected to yield higher amounts of bio-oil [40].

2.2.3.2. *Pressure*

Pressure is another factor affecting hydrothermal liquefaction (HTL). The hydrolysis rate and biomass dissolution are controlled by maintaining the pressure above the critical pressure of medium, which may enhance the favourable reaction pathways thermodynamically for production of liquid fuels or gas yield [40]. In addition, the density of medium solvent can be increased with pressure, which makes the medium more efficiently penetrates into molecules of biomass components, promoting the decomposition and extraction. However, the influence of pressure on the yield of oil or gas is negligible under supercritical conditions [41–44]. It is reported by Sangon et al. [41] that the oil yield was slightly increased with pressure increase from 7 to 12 MPa during the coal liquefaction in supercritical conditions. Kabyemela et al. [43, 44] observed that pressure rise had a certain degree of adverse effect on the degradation of glucose. The same phenomenon was witnessed in the process of hydrolysis and pyrolysis of cellobiose liquefaction at 400°C. This is because under the supercritical conditions, the increase in pressure results in the cage effect for C–C bonds, which inhibits the breakage of

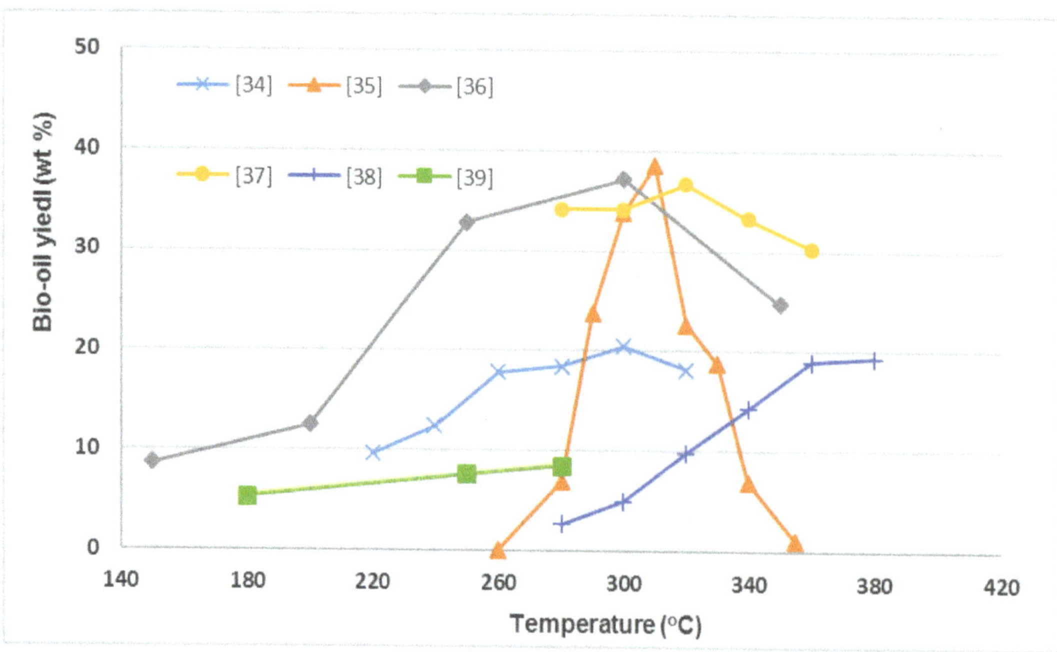

Figure 6. Oil yield dependence of temperature as reported by different research groups [31–36].

bonds and ends up with low degree of biomass fragmentation. Therefore, variation of pressure in supercritical region plays an insignificant role for overall oil yield.

2.2.3.3. *Catalyst*

Generally, homogeneous catalysts, such as alkali salts, organic and inorganic acids, are significantly more active than heterogeneous catalysts in the process of biomass liquefaction. Heterogeneous catalysts, such as Ni-containing catalysts, are frequently used in low-temperature water gasification of biomass. Homogeneous catalysts in the form of alkali salts have been commonly used in biomass liquefaction. For example, sodium carbonate has the catalytic effect of increasing oil yield and reducing char, and potassium carbonate can promote the water-gas shift reaction and lessen solid residue, while the catalysts sodium hydroxide and potassium hydroxide can enhance decarboxylization and increase isomerization of glucose. The heterogeneous catalysts Ni,Ru have the enhancement of H_2 and CH_4 yield in the hydrothermal conditions. According to the studies [45–48], the organic acid catalysts exhibit better catalytic activity than inorganic acid catalysts with respect to the solid residue content. Salts (phosphates, sulphates, chlorides, acetates, carbonates) don't have as good of catalytic performance as alkali (NaOH, KOH) at typical liquefaction temperatures.

2.2.3.4. *Residence time*

The effect of residence time on hydrothermal liquefaction is mainly demonstrated in two aspects: the overall conversion of biomass and the composition of products. According to the study of Boocock and Sherman [49], the bio-oil yield was reduced with increase in residence time. Other researchers [33, 36] witnessed the same results. It is observed by Yan et al. [50] that there is no significant increase in liquid yield for longer residence time. Therefore, the short residence time favours more liquid products in the liquefaction. There may be many reasons to explain the effect of residence time. One convincing explanation is that decrease in oil yield may be as a result of secondary and tertiary reactions for longer residence time when biomass conversion reached the saturation point.

In addition, the duration of liquefaction influences the composition of oil product. The significant difference between the decomposition products obtained for longer and shorter residence time was observed by Karagoz et al. [50]. For instance, some furan carboxaldehydes, benzoic acid and phthalate were gained for 60 min, which were not observed in the case of shorter residence time (15 min). Obviously, the preasphaltenes and asphaltene can be decomposed into lighter products for longer residence time, enhancing the yield of light oil and gases. Therefore, it is necessary to inhibit the further disintegration of lighter products in order to obtain high oil yield. One way to prevent oil decomposition is addition of reducing agents like tetraline, hydrogen and syngas. This strategy not only stabilizes the radicals and lighter products but also favours the high oil yield.

2.2.3.5. *Feedstock*

Theoretically, any biomass can be converted into bio-oil via liquefaction regardless of water content, including forestry, agriculture residues, sewage sludge and food waste. Due to the different behaviours of major biomass components (such as lignin, hemicellulose and cellulose) on the temperature variations, the liquid products and overall yield can be influenced significantly by the heterogeneity of biomass stock. More bio-oil can be obtained from feedstock with a higher content of cellulose and hemicellulose. In addition, the physical properties of bio-oil depend on the feedstock type. The bio-oil produced from liquefaction of the loosely structured biomass usually has poor quality and HHV and a low viscosity. This is because the oxygen and moisture contents are high in the liquid products. Therefore, modifications are necessary before the product can be used in engines, turbines and boilers. It is also unsuitable for long-term storage due to the complexity of the oil [51].

2.3. Gasification

Gasification is a high temperature partial oxidation process to optimize gas production, and it involves the extremely complex catalytic conversion of solid/liquid organic compounds into the gas or solid phase in the presence of a gasifying agent (air, steam, oxygen, CO_2). The gas phase, usually called "syngas", is a mixture of hydrogen, carbon monoxide, carbon dioxide, methane and small hydrocarbons. The solid phase, called "char", is actually composed of unconverted organic fraction, fixed carbon as well as ash. Gasification has been considered as a cost-effective way to increase the use of biomass for energy production, allowing widespread biomass utilization [52]. Especially in the beginning of this century, rising oil price and concerns over climate change resulted in the expansion of more advanced biomass gasification projects around the world. One typical biomass gasification process is shown in **Figure 7**.

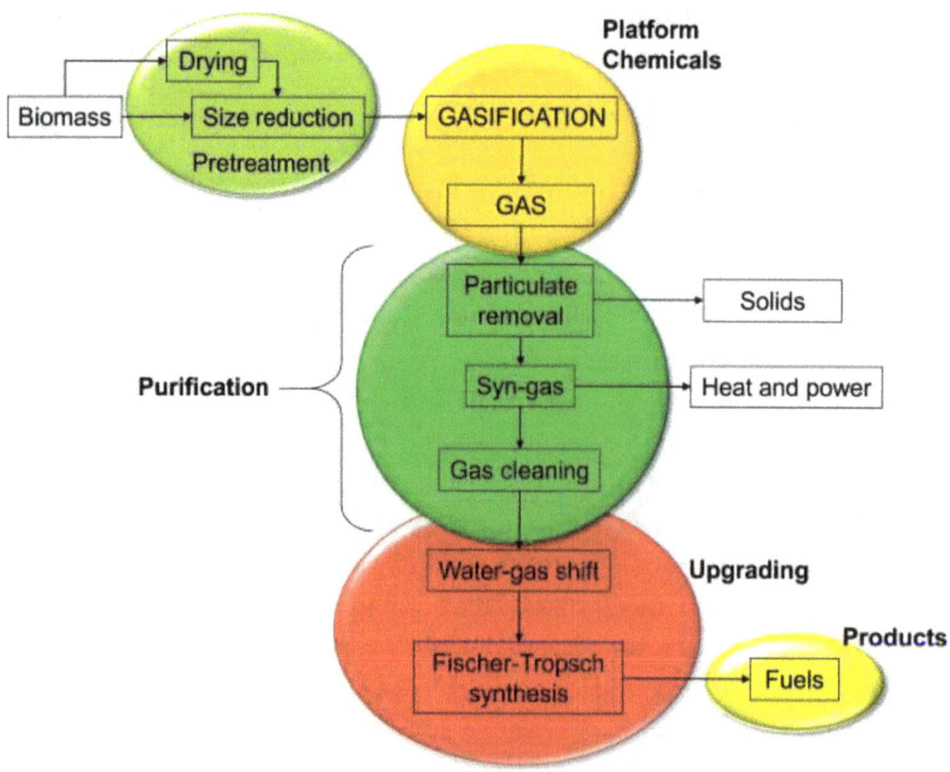

Figure 7. Process schematic of biomass gasification and upgrading [6].

2.3.1. Mechanism

It is believed that gasification involves several sequential and parallel reactions in an oxygen deficient environment. And the primary steps include drying, pyrolysis, reduction and oxidation, shown in **Figure 8**. Initially, the moisture of biomass is evaporated quickly during the drying, and this process is considered complete at 150°C. With the temperature increased, matrix carbonaceous materials, such cellulose and lignin, are decomposed into small species by cracking of chemical bonds within a temperature range of 250–700°C, which can be simply schematized with the following overall reaction:

$$\text{Biomass} \rightarrow H_2 + CO + CO_2 + \text{Hydrocarbons} + H_2O(g) + \text{Tar} + \text{Char}$$

Several complex phenomena, such as heat and mass transfer and reaction kinetics, are involved in this stage. The reaction kinetics is the determining step at low temperatures, while the heat or mass transfer becomes the limiting step at higher temperatures. Then the reduction stage comes into play. It involves all the products of the preceding stages. The final syngas is formed through the complicated reactions between gas mixture and char. The principal reactions include carbonation of char, water gas shift reaction, methanation, tar decomposition and steam reforming [1, 52].

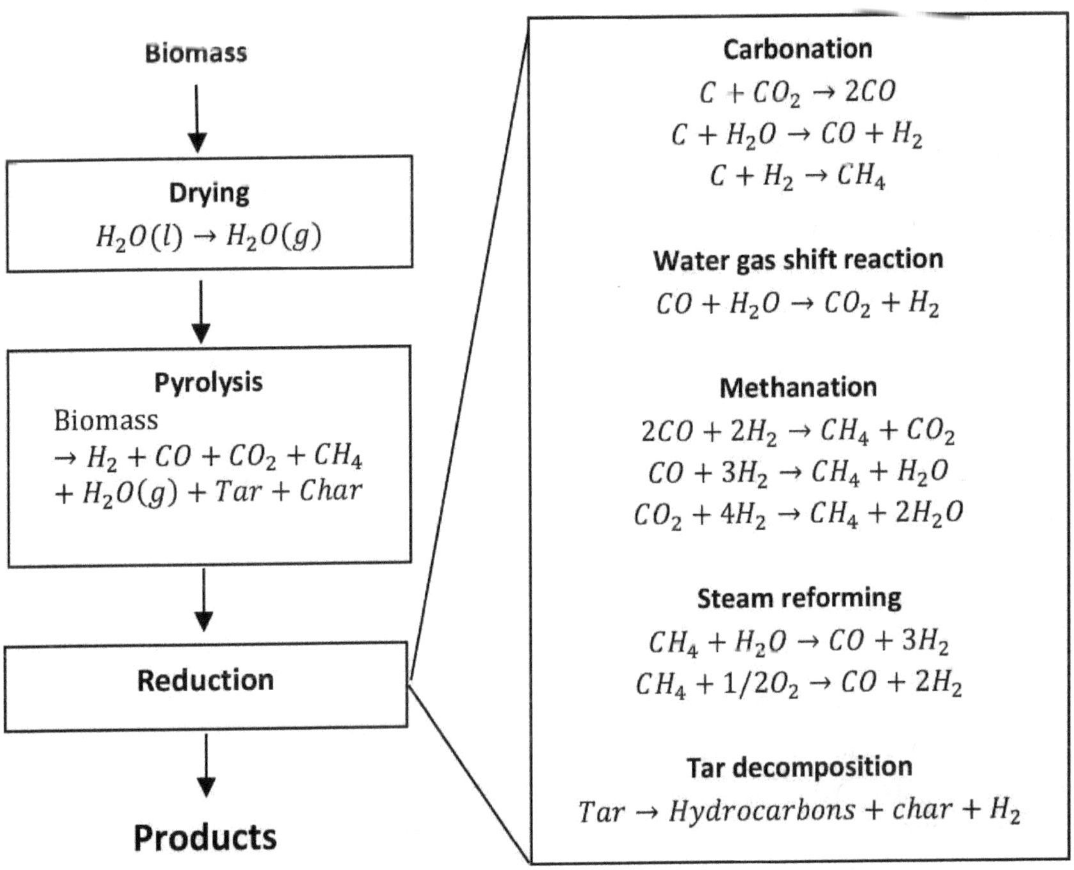

Figure 8. Main stages and the reactions involved in the gasification process [1, 52].

2.3.2. *Parametric impact*

There are many factors influencing the product distribution of biomass gasification, such as feedstock type, quality and moisture, particle size and density, operative conditions, the ratio of steam (or other gas) to biomass and catalysts.

2.3.2.1. *Particle size and density*

The smaller the particle size, the better would be the heat diffusion, which can make the temperature uniform throughout the particle, benefiting the reactions. The effect of particle size on reaction rate is more significant when there is a limitation on size above which heat transfer would be controlling [3]. In addition, the reduction in particle dimension can enhance the carbon conversion and hydrogen yield, improve the syngas efficiency and suppress the formation of tar and char [53, 54]. However, the particle size reduction requires intense energy [55].

Biomass is normally of low density due to the porous structure. The presence of the pores facilitates the heat transfer throughout the pellet, which in turn results in the homogeneous gasification and uniform product composition. As for the dense materials, there is a temperature difference between the exterior and interior of particle, leading to the simultaneous drying, pyrolysis and gasification. Accordingly, a non-homogeneous gas composition is produced.

2.3.2.2. *Operating conditions*

The operating conditions, including pressure, temperature, heating rate and other vital factors, have the potential to affect the gas yield and biomass conversion [54, 55]. The pressure is closely associated with bio-char reactivity and an increase in temperature would involve a faster heating rate, which generates a greater temperature difference over feedstock particles. While the fast and slower heating rate operations largely contribute to the design of gasifier and the desired products. The faster heating rate favours more gas production and less tar yield owing to the volatilization and degradation of the active tars by transforming them to the product gases. The slower heating rate contributes to lower gas yield and higher tar yield because of the recombination of lower-volatility hydrocarbons on the surface of the char particles.

2.3.2.3. *Steam-to-biomass ratio*

The ratio of steam to biomass (S/B) can determine the input energy requirements, outlet gas quality and yield of products. Increasing S/B ratio can positively enhance the reforming, cracking and water gas shift reactions, leading to higher hydrogen yield and consequently generating the syngas with high calorific content [56, 57]. However, there is a threshold limit beyond which increasing S/B will generate excess steam, resulting in the enthalpy loss and reduction in process efficiency [58]. This issue necessitates optimizing the ratio in steam biomass gasification.

2.3.2.4. *Catalysts*

Catalysts can provide an alternative lower-energy pathway for the reactions to follow. The catalysts used in the biomass gasification are categorized alkaline (Na, K) metal, alumina and zeolites, dolomites and limestones, Ni-based, Zn-based, as well as some rarer metals such as platinum- and ruthenium-based, specifically in relation with gasifier design or the type of biomass feed. Among them, alkaline metal oxides, dolomite and Ni-based catalysts have been testified to greatly promote the reformation reaction [59], and alumina silicates are found to enhance the char gasification effectively, whereas Ni-based catalysts are identified to facilitate the conversion of lighter hydrocarbons [60]. It is essential to develop more efficient and economical catalysts with the ability of enhancing the quality and yield of the desired product while minimizing the residual char and tar [1].

2.3.2.5. *Feedstock.*

In the case of gasification, due to the indispensable role these constituents play during the process of gasification, the proportion of cellulose and hemicellulose is closely related to the yield of the gaseous products, while the lignin has significant influence on the production of bio-oil. Generally, the higher the ratio of cellulose (including hemicellulose) to lignin in the feedstock, the higher the gaseous products yields. The moisture content of biomass affects the characteristics of the final gas products significantly. A feedstock with more than 30 wt% moisture would result in less gas production, higher tar content and more energy input. Gasification technology can be a good solution for energy production from renewable sources; however, the water and impurities in the gas products can be problematic in downstream FT processes, where a clean gas feed is required [6].

2.3.3. *Application*

The versatility of gasification manifests in that it can meet the demand for the electricity or thermal energy and industrial purposes. The products obtained from the biomass gasification are shown in **Figure 9**. The gas products, typically including H_2, CO, CO_2, CH_4, H_2O and other gaseous hydrocarbons, can be used to generate heat or electricity by direct firing in diesel engines, gas turbines and boilers. Alternatively, the gas mixture can be further reformed to produce usable fuels such as methanol or hydrogen, which can be transported with high energy densities, enabling the generation of electricity or heat to be centralized based on disperse gasification systems. Renewable biomass is considered as a potential material for gasification to produce syngas, hydrogen or other liquid fuels; however, the economics of current processes favour the use of light hydrocarbons (in natural gas) and coal owing to the characteristic of high oxygen content in the biomass.

On the other hand, bio-syngas originated from biomass gasification can act as the raw material for Fischer-Tropsch Synthesis (FTS) to produce green liquid fuel (diesel) or high value-added chemicals. It is needed that the ratio of H_2 to CO in syngas is adjusted to an appropriate range through water gas shift reaction prior to FTS, and then the syngas is catalytically converted to liquid fuels

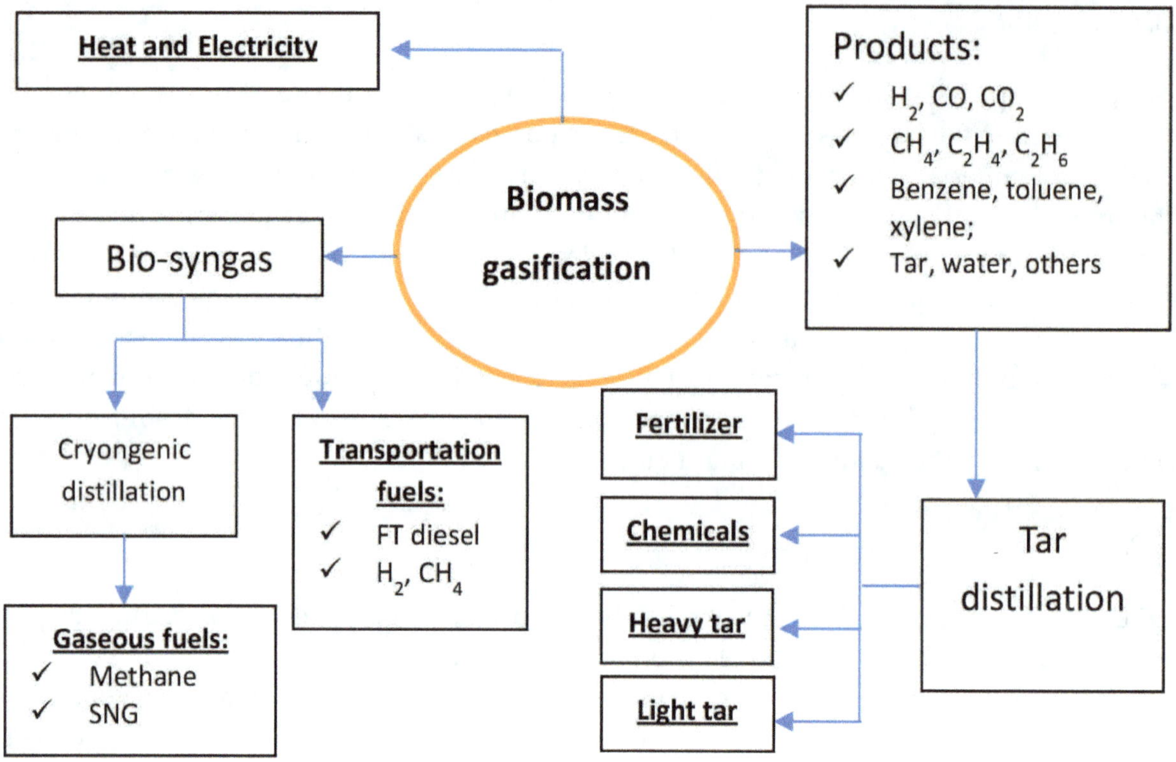

Figure 9. Products from gasification process [61].

and/or chemicals via FT process over Co-based or Fe-based catalysts. This type of synthetic FT diesel fuels has excellent quality. For instance, the emissions can be reduced significantly when fuels are employed in ignition combustion engines because of low content of nitrogenous and aromatic species. In addition, the FT fuels are sulphur-free, which is vital for lowering emissions of exhaust pollutants. Therefore, biomass gasification coupled with FT process provides a promising alternate for mitigating the burden on conventional transportation fuel; however, it requires significant development and scale-up efforts for commercial-scale installations [62].

The poly-generation concept has been developed with advancement in biomass gasification. For instance, co-generation of thermal power with electricity; poly-generation of heat, fertilizer and bio-char; poly-generation of heat, SNG/bio-fuels; and poly-generation of H_2 with heat and electricity. All these approaches not only optimize the thermal efficiency of the process but also provide flexibility and sustainability, thereby enhancing the economic advantage in the long run [3].

2.4. Biochemical transformation

Biochemical conversion of biomass involves the use of bacteria, microorganisms and enzymes to break down biomass into gaseous or liquid fuels, such as biogas or bioethanol. These processes are essentially microbic digestion and fermentation, which are most suitable for the high moisture herbaceous plants, marine crop and manure. The main products in the process are methanol, ethanol and gas mixture (CH_4, CO and CO_2) [63]. This biocatalytic process is considered to be a suitable alternative to chemical catalysis. Some important considerations when selecting the biocatalyst used for the upgrading of biomass are listed in **Table 3**.

(A) Source of biocatalyst	
Whole cell or free enzyme	Whole cells can also be used in permeabilized form
Wild type or mutant enzyme produced by rDNA technology	Some organisms are frequently used as the source for industrial enzymes
Refolded from IBs or use of IBs itself	Choice of the host is important. For enzymes where activity depends upon post-translational modification, higher organisms are used as host expression systems
(B) Free or immobilized form	
Soluble conjugates	For insoluble substrates (which is often the case with biomass), soluble conjugates or enzymes in membrane reactors are preferred
Carrier free or insoluble support	Enzyme aggregates like CLEA or CLEC have high volumetric activity
(C) Operational stability	This may be different from storage stability
(D) Reaction medium	The use of co-solvents is under exploited
(E) Normal or promiscuous activity	As more enzymes, engineered for better promiscuous activity, become available, this application may increase

Table 3. Summary of important parameters of biocatalysts [64].

2.5. Techno-economic analysis

Techno-economic analyses are essential for determining the economic viability of biomass conversion technologies in the research or development stage. On the basis of current performance and targets established for process improvements, evaluation of the production costs can allow us to assess the potential economic feasibility of a process configuration. These results are significant for investing in the technology with high potential for deployment success.

The economic feasibility of fast pyrolysis and upgrading facilities employing two bio-oil upgrading pathways have been evaluated based on the scale of 2000 metric tons/day and 65% biomass conversion in the literature [65]. The results showed that the hydrotreating followed by fluid catalytic cracking (FCC) pathway with hydrogen production via natural gas reforming is the best option with highest IRR (internal rate of return) of 13.3% when compared with single stage hydrotreating. According to the sensitivity analysis, product yield, fixed capital investment (FCI) and biomass cost have significant impacts on the facility IRR. Variation of 5% bio-oil yield can lead to product value (PV) of \$2.60–3.89/gal [66]. The hydrotreating/FCC pathway is proved to be of relatively low risk for project investment based on the Monte Carlo analysis.

Techno-economic analysis was implemented to assess the feasibility of developing a commercial large-scale woody biomass hydrothermal liquefaction (HTL) and upgrading plant with the scale of 2000 dry metric ton/day in the literature [67]. Two cases were involved: state-of-technology (SOT) and goal case considering future improvements. The annual production rate was estimated to be \$42.9 and \$69.9 million gallon gasoline-equivalent (GGE) for SOT and goal case, respectively. The minimum fuel-selling price was estimated to be $4.44/GGE and $2.52/GGE. Obviously, the liquid fuel cost of current HTL and upgrading technologies (SOT) is not competitive with the conventional petroleum-based gasoline. For advancing SOT to goal case, future improvements of HTL competitiveness should be focused on reducing organic loss to obtain higher final yield and lowering wastewater treatment cost. Based on the sensitivity analysis, the key factors that influence the production cost significantly are plant scale, yield, upgrading equipment cost and feedstock cost. Although the goal case technology is promising for future liquid fuel production via woody biomass HTL, there are still financial risks resulting from lack of knowledge and experience.

The techno-economic analysis of green fuel production via BG-FT (biomass gasification and Fischère Tropsch) with a scale of 2000 metric tons per day of corn stover was conducted in the literature [68]. Two scenarios are considered: high temperature (HT) gasification and low temperature (LT) gasification. The results showed that the total capital investment required for nth plant scenarios are $610 million and $500 million for HT and LT, respectively. Product value (PV) for two scenarios were estimated to be $4.30/GGE and $4.80/GGE, but the HT scenario was more desirable due to its higher carbon efficiency and subsequent higher fuel yield. Based on the sensitivity analysis, the capital investment and feedstock cost are chiefly responsible for affecting the PV, whereas the conversion of FT process, feedstock moisture and catalyst cost are least influential parameters.

3. Conclusion

Pyrolysis and liquefaction are inclined to obtain more liquid products (bio-oil), but the stability of bio-oil produced in the pyrolysis process is worse than that generated in the liquefaction process. However, the interest in liquefaction is considerably lower due to the reactors and feed systems, which are more complex and expensive than that for the pyrolysis process. Therefore, upgrading bio-oil by lowering the oxygen content, removing alkalis by means of hydrogenation and catalytic cracking are required for further application of pyrolysis conversion. Catalytic cracking pyrolysis gasification has shown promising application, since the syngas can be used for the production of methanol and hydrogen, which may have a future as fuel for transportation. The economic studies show biomass gasification plants can be as economical as conventional coal fired plants. Compared with chemical catalysts in the process of biomass conversion, biocatalysts have higher selectivity to products and cheaper capital investment. However, the practical industrial application of biocatalysts is limited due to slow conversion process, high dependence of environment and feedstock and difficult recycling of the biocatalysts.

It is noteworthy that one novel technology, achieved by a green catalysis research group at the University of Calgary, is catalytic coal or biomass upgrading using natural gas for liquid chemical or fuel productions. This process directly employs methane as reducing agent instead of hydrogen for removing oxygen from biomass under the facilitation of specially developed low-cost supported catalyst, leading to significantly reduced operation cost.

Acknowledgements

We gratefully acknowledge the financial supports from Natural Sciences and Engineering Research Council of Canada (NSERC, RGPIN/04385-2014).

Author details

Aiguo G. Wang, Danielle Austin and Hua Song*

*Address all correspondence to: sonh@ucalgary.ca

University of Calgary, Calgary, Canada

References

[1] Vineet S. S., Ming Z., Peter C., Joseph Y., Xia Z., Mohammad Z. M., Nilay S., Edward J. An overview of advances in biomass gasification. Energy & Environmental Science. 2016; 9; 2939-2977. DOI: 10.1039/c6ee00935b

[2] Kucuk M. M., Demirbas A. Biomass conversion process. Energy Conversion and Management. 1997;**38**(2):151–165.

[3] Kirubakaran V., Sivaramakrishnan V., Nalini R., Sekar T., Premalatha M., Subramaniam P. A review on gasification of biomass. Renewable and Sustainable Energy Reviews. 2009;**13**:179–186. DOI: 10.1016/j.rser.2007.07.001

[4] Peter M. Energy production from biomass (part 2): Conversion technologies. Bioresource Technology. 2002;**83**(1):47–54. DOI: 10.1016/S0960-8524(01)00118-3

[5] Ayhan D. Biomass resource facilities and biomass conversion processing for fuels and chemicals. Energy Conversion and Management. 2000;**42**(11):1357–1378. DOI: 10.1016/S0196-8904(00)00137-0

[6] Alonso D. M., Bond J. Q., Dumesic J. A. Catalytic conversion of biomass to biofuels. Green Chemistry. 2010;**12**(9):1493–1513. DOI: 10.1039/C004654J

[7] Bahng M., Mukarakate C., Robichaud D. J., Nimlos M. R. Current technologies for analysis of biomass thermochemical processing: A review. Analytica Chimica Acta. 2009;**651**:117–138. DOI: 10.1016/j.aca.2009.08.016

[8] Neves D., Thunman H., Matos A., Tarelho L., Gomez-Barea A. Characterization and prediction of biomass pyrolysis products. Progress in Energy and Combustion Science. 2011;**37**:611–630. DOI: 10.1016/j.pecs.2011.01.001

[9] Bridgwater A. V. Review of fast pyrolysis of biomass and product upgrading. Biomass and Bioenergy. 2012;**38**:68–94. DOI: 10.1016/j.biombioe.2011.01.048

[10] Bridgwater A. V., Meier D., Radlein D. An overview of fast pyrolysis of biomass. Organic Geochemistry. 1999;**30**:1479–1493. DOI: 10.1016/S0146-6380(99)00120-5

[11] Mettler M. S., Vlachos D. G., Dauenhauer P. J. Top ten fundamental challenges of biomass pyrolysis for biofuels. Energy and Environmental Science. 2012;**5**:7797–7809. DOI: 10.1039/c2ee21679e

[12] French R., Czernik S. Catalytic pyrolysis of biomass for biofuels production. Fuel Processing Technology. 2010;**91**:25–32. DOI: 10.1016/j.fuproc.2009.08.011

[13] Su-Ping Z., Yong-Jie Y., Zhengwei R., Tingchen L. Study of hydrodeoxygenation of bio-oil from the fast pyrolysis of biomass. Energy Sources. 2003;**25**:57–65. DOI: 10.1080/00908310290142118

[14] Corma A., Huber G. W., Sauvanaud L., O' Conner P. Processing biomass-derived oxygenates in the oil refinery: Catalytic cracking (FCC) reaction pathways and role of catalyst. Journal of Catalysis. 2007;**247**:307–327. DOI: 10.1016/j.jcat.2007.01.023

[15] Elliott D. C. Historical developments in hydroprocessing bio-oils. Energy & Fuels. 2007;**21**:1792–1815. DOI: 10.1021/ef070044u

[16] He P., Song H. Catalytic conversion of biomass by natural gas for oil quality upgrading. Industrial & Engineering Chemistry Research. 2014;**53**:15862–15870. DOI: 10.1021/ie502272j

[17] Xiao Y., He P., Cheng W., Liu J., Shan W., Song H. Converting solid wastes into liquid fuel using a novel methanolysis process. Waste Management. 2016;**49**:304–310. DOI: 10.1016/j.wasman.2015.12.017

[18] Esteban C., Ralph P. O. Fundamentals of thermochemical biomass conversion. In: R. P. Overend, T. A. Milne, L. K. Mudge, editors. Biomass Liquefaction: An Overview. 1st ed. New York: Springer Netherlands; 1985. pp. 967–1002. DOI: 10.1007/978-94-009-4932-4_54

[19] Zhang Y., Riskowski G., Funk T. Thermochemical conversion of swine manure [dissertation]. Urbana: University of Illinois at Urbana-Champaign; 1999. 46 p.

[20] Morf P. O. Secondary reactions of tar during thermochemical biomass conversion [dissertation]. Swiss: Federal Institute of Technology Zurich; 2001.

[21] Goudriaan F., van de Beld B. Thermal efficiency of the HTU process for biomass liquefaction. In: A. V. Bridgwater, editor. Progress in Thermochemical Biomass Conversion Conference; September 18–21; Tyrol, Austria; 2000. pp. 1312–1325.

[22] Feridoun S. A., Hiroyuki Y. Dehydration of fructose to 5-hydroxymethylfurfural in sub-critical water over heterogeneous zirconium phosphate catalysts. Carbohydrate Research. 2006;**341**(14):2379–2387. DOI: 10.1016/j.carres.2006.06.025.

[23] Snare M., Kubickova I., Maki-Arvela P. Production of diesel fuel from renewable feeds: Kinetics of ethyl stearate decarboxylation. Chemical Engineering Journal. 2007;**134**(1–3): 29–34. DOI: 10.1016/j.cej.2007.03.064

[24] Heddle J. F. Activated sludge treatment of slaughterhouse wastes with protein recovery. Water Research. 1979;**13**(7):581–584. DOI: 10.1016/0043-1354(79)90004-6

[25] Youngmi K., Nathan S., Mosiera R. H. Composition of corn dry-grind ethanol by-products: DDGS, wet cake, and thin stillage. Bioresource Technology. 2008;**99**(12):5165–5176. DOI: 10.1016/j.biortech.2007.09.028

[26] Tim R., Kaiyue L., Tobias A. Hydrolysis kinetics of biopolymers in subcritical water. The Journal of Supercritical Fluids. 2008;**46**(3):335–341. DOI: 10.1016/j.supflu.2007.09.037

[27] Ortwin B. Hydrothermal degradation of polymers derived from plants. Progress in Polymer Science. 1994;**19**(5):797–841. DOI: 10.1016/0079-6700(94)90033-7

[28] William S., Lai M., Antal M. J., Jr. Uncatalyzed solvolysis of whole biomass hemicellulose by hot compressed liquid water. Industrial & Engineering Chemistry Research. 1992;**31**(4):1157–1161. DOI: 10.1021/ie00004a026

[29] Andrea K., Palanikumar M., Franziska S. Influence of proteins on the hydrothermal gasification and liquefaction of biomass. 2. Model Compounds. Industrial & Engineering Chemistry Research. 2007;**46**(1):87–96. DOI: 10.1021/ie061047h

[30] Kruse A., Gawlik A. Biomass conversion in water at 330–410°C and 30–50 MPa. Identification of key compounds for indicating different chemical reaction pathways. Industrial & Engineering Chemistry Research. 2003;**42**(2):267–279. DOI: 10.1021/ie0202773

[31] Dong Z., Liang Z., Shicheng Z., Hongbo F., Jianmin C. Hydrothermal liquefaction of Macroalgae Enteromorpha prolifera to bio-oil. Energy Fuels. 2010;**24**:4054–4061. DOI: 10.1021/ef100151h

[32] Sudong Y., Ryan D., Matt H., Zhongchao T. Subcritical hydrothermal liquefaction of cattle manure to bio-oil: Effects of conversion parameters on bio-oil yield and characterization of bio-oil. Bioresource Technology. 2010;**101**(10):3657–3664. DOI: 10.1016/j.biortech.2009.12.058

[33] Motoyuki S., Hirokazu T., Katsumi H., Kiyoshi M. Hydrothermal liquefaction of plantation biomass with two kinds of wastewater from paper industry. Journal of Materials Science. 2008;**43**(7):2476–2486. DOI: 10.1007/s10853-007-2106-8

[34] Yixin Q., Xiaomin W., Chongli Z. Experimental study on the direct liquefaction of Cunninghamia lanceolata in water. Energy. 2003;**28**(7):597–606. DOI: 10.1016/S0360-5442(02)00178-0

[35] Shupinga Z., Yulong W., Mingde Y., Imdad K., Chun L., Junmao T. Production and characterization of bio-oil from hydrothermal liquefaction of microalgae Dunaliella tertiolecta cake. Energy. 2010;**35**(12):5406–5411. DOI: 10.1016/j.energy.2010.07.013

[36] Selhan K., Thallada B., Akinori M., Yusaku S., Md. Azhar U. Low-temperature hydrothermal treatment of biomass: Effect of reaction parameters on products and boiling point distributions. Energy Fuels. 2004;**18**(1):234–241. DOI: 10.1021/ef030133g

[37] Wahyudiono, Takayuki K., Mitsuru S. Decomposition of a lignin model compound under hydrothermal conditions. Chemistry Engineering and Technology. 2007;**30**(8):1113–1122. DOI: 10.1002/ceat.200700066.

[38] Bo Z., Hua-Jiang H., Ramaswamy S. Reaction kinetics of the hydrothermal treatment of lignin. Applied Biochemistry and Biotechnology. 2008;**147**(1):119–131. DOI: 10.1007/s12010-007-8070-6

[39] Aiguo L., YoonKook P., Zhiliang H. Product identification and distribution from hydrothermal conversion of Walnut Shells. Energy Fuels. 2006;**20**(2):446–454. DOI: 10.1021/ef050192p

[40] Javaid A., Nor A., Saidina A. A review on process conditions for optimum bio-oil yield in hydrothermal liquefaction of biomass. Renewable and Sustainable Energy Reviews. 2011;**15**(3):1615–1624. DOI: 10.1016/j.rser.2010.11.054

[41] Sangon S., Ratanavaraha S., Ngamprasertsith S., Prasassarakich P. Coal liquefaction using supercritical toluene–tetralin mixture in a semi-continuous reactor. Fuel Processing Technology. 2006;**87**(3):201–207. DOI: 10.1016/j.fuproc.2005.07.007

[42] Sascha R. A., Kersten Biljana P., Wolter P., Wim P. M., Van S. Gasification of model compounds and wood in hot compressed water. Industrial & Engineering Chemistry Research. 2006;**45**(12):4169–4177. DOI: 10.1021/ie0509490

[43] Kabyemela B. M., Takigawa M., Adschiri T., Malaluan R. M., Arai K. Mechanism and kinetics of cellobiose decomposition in sub- and supercritical water. Industrial & Engineering Chemistry Research. 1998;**37**(2):357–361. DOI: 10.1021/ie9704408

[44] Bernard M. K., Tadafumi A., Roberto M. M., Kunio A. Kinetics of glucose epimerization and decomposition in subcritical and supercritical water. Industrial & Engineering Chemistry Research. 1997;**36**(5):1552–1558. DOI: 10.1021/ie960250h

[45] ElGayar M., McAuliffe, Shellsol C. As a processing liquid in biomass liquefaction. Energy Sources. 1997;**19**(7):665-676. DOI: 10.1080/00908319708908881

[46] Mun S., Hassan E., Hassan M. Liquefaction of lignocellulosic biomass with dioxane/polar solvent mixtures in the presence of an acid catalyst. Journal of Industrial and Engineering Chemistry. 2004;**10**(3):473–477.

[47] Funazukuri T., Cho J., Wakao N. Effect of adding Na_2CO_3, HCl and or CO during liquefaction of lignin sulfonate with water. Fuel. 1990;**69**(10):1328–1329. DOI: 10.1016/0016-2361(90)90298-5

[48] Alma M., Maldas D., Shiraishi N. Liquefaction of several biomass wastes into phenol in the presence of various alkalis and metallic salts as catalysts. Journal of Polymer Engineering. 1998;**18**(3): 161-177.

[49] Boocock D. G. B. Further aspects of powdered poplar wood liquefaction by aqueous pyrolysis. The Canadian Journal of Chemical Engineering. 1985;**63**(4):627–631.

[50] Yongjie Y., Jie X., Tingchen L., Zhengwei R. Liquefaction of sawdust for liquid fuel. Fuel Processing Technology. 1999;**60**(2):135–143. DOI: 10.1016/S0378-3820(99)00026-0

[51] Akhtar J., Saidina Amin N. A review on process conditions for optimum bio-oil yield in hydrothermal liquefaction of biomass. Renewable and Sustainable Energy Reviews. 2011;**15**(3):1615–1624. DOI: 10.1016/j.rser.2010.11.054

[52] Antonio M., Simeone C., Dino M. Biomass gasification technology: The state of the art overview. Journal of Energy Chemistry. 2016;**25**(1):10–25. DOI: 10.1016/j.jechem.2015.11.005

[53] Lv P. M., Xiong Z., Chang J., Wu C. Z., Chen Y., Zhu J. X. An experimental study on biomass air-steam gasification in a fluidized bed. Bioresource Technology. 2004;**95**(1):95–101. DOI: 10.1016/j.biortech.2004.02.003

[54] Siyi L., Bo X., Xianjun G., Zhiquan H., Shiming L., Maoyun H. Hydrogen-rich gas from catalytic steam gasification of biomass in a fixed bed reactor: Influence of particle size on gasification performance. International Journal of Hydrogen Energy. 2009;**34**(3):1260–1264. DOI: 10.1016/j.ijhydene.2008.10.088

[55] De Lasa H., Salaices E., Mazumder J., Lucky R. Catalytic steam gasification of biomass: Catalysts, thermodynamics and kinetics. Chemical Review. 2011;**111**(9):5404–5433. DOI: 10.1021/cr200024w

[56] Prakash P., Sheeba K. N. Hydrogen production from steam gasification of biomass: Influence of process parameters on hydrogen yield: A review. Renewable Energy. 2014;**66**:570–579. DOI: 10.1016/j.renene.2013.12.025

[57] Ian N., Alberto O., Maria P. A., José C. Biomass gasification with air in an atmospheric bubbling fluidized bed: Effect of six operational variables on the quality of the produced raw gas. Industrial & Engineering Chemistry Research. 1996;**35**(7):2110–2120. DOI: 10.1021/ie9507540

[58] Shweta S., Pratik N. S. Air–steam biomass gasification: Experiments, modelling and simulation. Energy Conversion and Management. 2016;**110**:307–318. DOI: 10.1016/j.enconman.2015.12.030

[59] Meng N., Leung D. Y. C., Michael K. H., Leung K. S. An overview of hydrogen production from biomass. An overview of hydrogen production from biomass. Fuel Processing Technology. 2006;**86**(5):461–472. DOI: 10.1016/j.fuproc.2005.11.003

[60] Corte P., Lacoste C., Traverse J. P. Gasification and catalytic conversion of biomass by flash pyrolysis. Journal of Analytical and Applied Pyrolysis. 1985;**7**(4):323–335. DOI: 10.1016/0165-2370(85)80104-2

[61] Mustafa B., Mehmet B., Elif K., Havva B. Main routes for the thermo-conversion of biomass into fuels and chemicals. Energy Conversion and Management. 2009;**50**(12):3158–3168. DOI: 10.1016/j.enconman.2009.08.013

[62] Mustafa B., Mehmet B., Elif K., Havva B. Main routes for the thermo-conversion of biomass into fuels and chemicals. Part 1: Pyrolysis systems. Energy Conversion and Management. 2009;**50**(12):3147–3157. DOI: 10.1016/j.enconman.2009.08.014

[63] Mustafa B. Biomass energy and biochemical conversion processing for fuels and chemicals. Energy Sources, Part A: Recovery, Utilization, and Environmental Effects. 2006; 28(6): 517–525. DOI: 10.1080/009083190927994

[64] Mukherjee J., Gupta M. N. Biocatalysis for biomass valorization. Sustainable Chemical Process. 2015;3(1):3–7. DOI: 10.1186/s40508-015-0037-2

[65] Zhang Y., Brown T. R., Hu G., Brown R. C. Techno-economic analysis of two bio-oil upgrading pathways. Chemical Engineering Journal. 2013;**225**:895–904. DOI: 10.1016/j.cej.2013.01.030

[66] Wright M. M., Daugaard D. E., Satrio J. A., Brown R. C. Techno-economic analysis of biomass fast pyrolysis to transportation fuels. Fuel. 2010;**89**:S2–S10. DOI: 10.1016/j.fuel.2010.07.029

[67] Zhu Y., Biddy M. J., Jones S. B., Elliott D. C., Schmidt A. J. Techno-economic analysis of liquid fuel production from woody biomass via hydrothermal liquefaction (HTL) and upgrading. Applied Energy. 2014;**129**(15):384–394. DOI: 10.1016/j.apenergy.2014.03.053

[68] Swanson R. M., Platon A., Satrio J. A., Brown R. C. Techno-economic analysis of biomass-to-liquids production based on gasification. Fuel. 2010;**89**(1):S11–S19. DOI: 10.1016/j.fuel.2010.07.027

[69] Saqib Sohail T., Lasse R., Andreas R. Hydrothermal liquefaction of biomass: A review of subcritical water technologies. Energy. 2011;**36**(5):2328–2342. DOI: 10.1016/j.energy.2011.03.013

7

Determination of the Biomass Content of End-of-Life Tyres

Leticia Saiz Rodríguez, José M. Bermejo Muñoz,

Adrien Zambon and Jean P. Faure

Abstract

Studies have been conducted in France and Spain for (1) the validation of sampling methods to achieve representative samples of end-of-life tyre (ELT) materials and (2) the comparison and validation of test methods to quantify their biomass content. Both studies conclude that the ^{14}C techniques are the most reliable techniques for determining the biomass content of end-of-life tyres. Indeed, thermogravimetry and pyrolysis-GC/MS do not lead to results consistent with the theoretical content of biogenic materials present in tyres, and results in both cases differ considerably from the known natural rubber content of the reference samples studied using thermogravimetric analysis. Furthermore, in the two last techniques, natural isoprene cannot be distinguished from synthetic isoprene. Results obtained with radiocarbon analysis based on ^{14}C contents could be used as reference values of the biomass content of the ELTs: in the ranges of 18–22% for passenger car tyres and 29–34% for truck tyres, in line with actual natural rubber and other components content. Additionally, the presence of textile fibres and stearic acid, which are known sources of biomass in the tyre, cannot be evaluated by thermogravimetry and pyrolysis-GC/MS techniques.

Keywords: end-of-life tyres, biomass, natural rubber, reference values, alternative fuel, pyrolysis-GC-MS, thermogravimetry, radiocarbon analysis, isoprene

1. Introduction

End-of-life tyre (ELT) is any pneumatic tyre removed from any vehicle and not selected to be mounted on a vehicle again [1]. Because the end-of-life tyre is a non-reusable tyre in its original form, it enters a waste management system based on product/material recycling and energy recovery. For that, it is important to know the composition of the tyres and their properties in order to recycle them or recover their energy.

Tyres could be classified according to the type of vehicle in which they are suitable to be mounted, so the main groups established by the industry are passenger car, truck and bus, and agro tyres, all of them included in this chapter.

Tyres are composite materials, essentially made of rubber, metal wire and textile fibres, in Ref. [1]. The rubber component is based on a vulcanized mixture of elastomers with some different chemicals (see **Table 1**), such as carbon black, silica and extender oils which also constitute a high percentage of the tyres.

Their compositions and combustion properties are similar to, or even better than those of coal (see **Table 2**). Due to their high carbon content (60–70%), they have become an interesting alternative fuel with a net calorific value in the range of 26.4–30.2 MJ/kg [2].

The use of secondary fuels is progressively increasing, not only because of its economic benefits, but also because of the environmental advantages of using solid recovered fuels [3]. These include natural resource savings, the preservation of fossil fuels such as petroleum coke, and above all the reduction in net emissions of CO_2 due to the biogenic origin of some components of the tyre, mainly natural rubber. Indeed, according to Directive 2003/87/EC [4], emissions associated with biomass fraction are considered to be neutral with regard to the greenhouse effect.

It also leads to the reduction in other pollutants [5] such as SO_x, mainly because the sulphur content in tyres (1–2%), used for the vulcanization process, is in any case lower than the quantity in most fossil fuels (see **Table 2**).

ELTs contain a fraction of biogenic carbon that mainly comes from their natural rubber content. This is not, however, the only source of biogenic carbon. Most tyre formulations also include stearic acid in small quantities, used as activator of the vulcanization reaction, and

Material	Passenger car tyre (%)	Trucks tyre (%)
Rubber/elastomers	43	42
Carbon black and silica	28	24
Metal	13	25
Textile	5	–
Zinc oxide	2	2
Sulphur	1	1
Accelerators/antidegradants	8	–
Stearic acid	1	n.a
Oils	7	n.a

Table 1. Main components of the passenger car tyre and truck tyre.

Content in % mass				
Major elements (>1%)				
	Passenger car	**Truck**	**Coal**	**Petroleum coke**
Carbon (C)	68 to 70%	60 to 63%	63.9%	84 to 97%
Iron (Fe)	11 to 12%	25 to 27%		0 to 0.2%
Hydrogen (H)	6 to 6.3%	5.3 to 5.6%	3.6%	0 to 5%
Oxygen (O)	3.3 to 3.8%	1.5 to 2.2%		
Silica (Si)	1.5 to 1.9%	0.3 to 0.5%	2%	Nd
Zinc (Zn)	1.3 to 1.5%	1.3 to 1.8%		
NCV (MJ/kg)	30.2	26.4	26	32
Sulphur (S)	1 to 1.5%	1.2 to 1.6%	1.3%	0.2 to 6%
Biomass fraction (% mass)	18.3%	29.1%	0%	0%

Table 2. Elemental analysis of tyres.

also smaller quantities of rayon, a natural fibre used as a reinforcement material in the manufacturing of some tyre carcasses. Nowadays, cotton could only be found in carcasses of older tyres [6, 7].

Despite the range of variation in the formulation of tyres, in practice their composition hardly varies, and thus they are one of the more dependable fuels, coming from wastes.

The greatest variation in biomass content is found among tyres of different types: passenger car tyres, truck tyres or agro vehicle tyres [1].

Each tyre particles obtained from treatment by shredding is quite heterogeneous in terms of biomass content. This intrinsic heterogeneity of tyres particles at the microscopic level is related to their composition. For example, elastomer mixtures are not the same in each part of the tyre (see **Figure 1**).

Although the heterogeneity does not appear at the industrial scale (consumption of around one ton per hour), the microscopic heterogeneity is of importance when it is necessary to take representative laboratory samples and to prepare them for analysis, in order to prevent different results.

In order to quantify the biomass content of an ELT, three analytical techniques have been identified. The first one is a method that determines the biogenic carbon content by measuring the activity of the ^{14}C isotope, a technique employed in archaeology to date organic materials [8–10]. Another widely used technique for determining the composition of vulcanized elastomers is thermogravimetric analysis (TGA) [11]. This method is based on the measurement of the weight variation of a sample when it is submitted to a progressive increase in temperature in a controlled atmosphere. The third method, pyrolysis-gas chromatography/

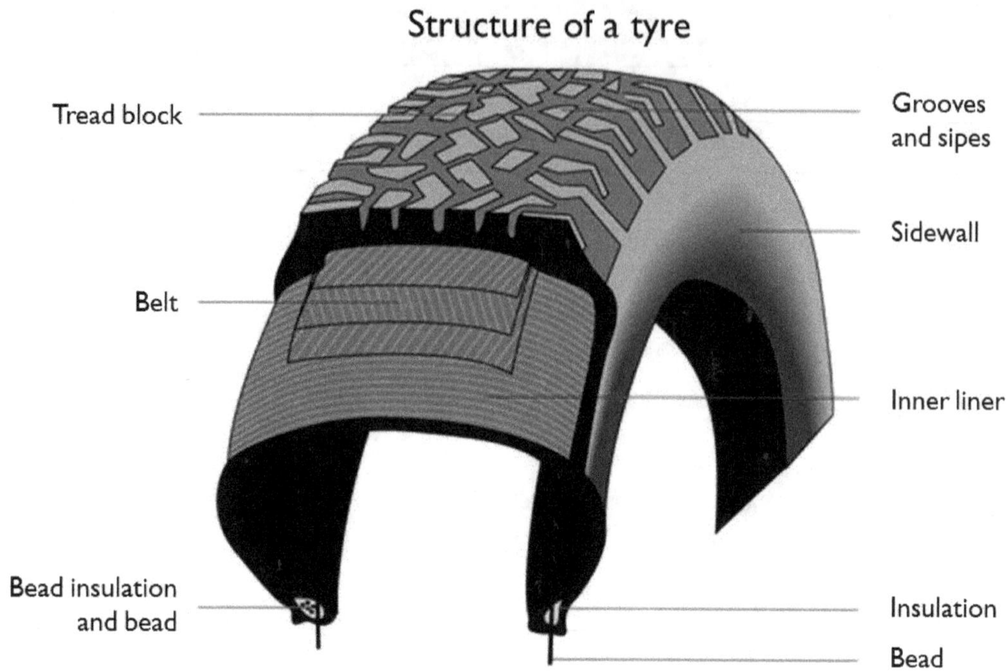

Figure 1. Different layers of a tyre.

mass spectrometry (Py-GC/MS), has been used extensively for qualitative and quantitative identification of polymer blends [12].

The results obtained under sampling, testing and analysis conditions respecting good practices with regards to heterogeneous materials show a remarkable stability in the measured parameters.

This chapter provides first the appropriate methodology to make a selection of test samples from tyres, and secondly, novel information on the use of different technologies for the determination of their biomass content.

Two cases studies conducted independently by ALIAPUR and SIGNUS, in France and Spain, respectively, have probed which techniques are appropriate to measure biomass content in tyres and which are not. Also in both studies reference values of biogenic content have been established, and results are quite close to theoretical values.

This chapter then offers an analysis of the differences between the results of the three techniques as well as the advantages, disadvantages and problems.

2. Sampling procedure for the estimation of biogenic fraction of end-of-life tyres

The management of ELTs to be used as secondary fuel mainly consists of the shredding of relevant quantities of different types of tyres coming from a diverse range of origins.

Shredded material can be stocked in piles, in which each particle contains parts of different layers of the tyre, each with a particular composition. So, the first problem to solve is how to estimate the biomass content taking random portions of materials. If this process is not performed carefully, there is an important risk not to be representative enough of the total stock. This is especially critical if we take into account that any of the used analytical techniques hardly needs a few milligrams of material.

One of the key points for this calculation of biomass content is then the design of a sampling plan which is representative of the big sized lots. In the case of samples taken on the shredding site, the samples will be representative of one to several days of production. Depending on the size of the facility, the sample could represent tens or even hundreds of tons of material.

In the case of samples taken during loading or unloading processes at the storage site, the samples will be representative of several weeks of production; the stocks in this case could even be of thousands of tons of material.

2.1. General criteria for the definition of sampling procedure

The minimum mass to produce a representative laboratory sample from the lot should be determined by the following formula [13]:

$$m_m = \frac{\pi}{6} \cdot d_{95}^3 \cdot s \cdot \lambda \cdot g \cdot \frac{(1-p)}{(Cv)^2 \cdot p} \tag{1}$$

where,

- m_m is the mass of the minimum sample size, in grams as received.
- d_{95} is the nominal top size of a particle (a mass fraction of 95% of the particles are smaller than d_{95}), in mm. This value is measured by means sieves following the method described within CEN/TS 15415:2006 [14].
- s is the shape factor, in mm³/mm³; reference value of 1.0 in the case of granular materials with nominal size smaller than 50 mm [13].
- λ is the average particle density of the particles in the solid recovered fuel, in g/mm³ as received [15].
- g is the correction factor for distribution in the particles size. Its value is related to the superior nominal size d_{95} and the minimum size of the particle d_{05}.
- p is the fraction of the particles with a specific characteristic (such as a specific contaminant), in g/g, and is equal to 0.1 [13].
- Cv is the coefficient of variation. Its value is 0.1 [13].

2.2. General criteria for the preparation of laboratory samples

Samples were prepared using a riffle splitter with 14 slots of 27 mm in width, until the mass of the sample is greater than the minimum size of the laboratory sample necessary to guarantee total representativeness for few milligrams. This must be calculated according to the third-power law [16], which for granular materials is expressed as:

$$m > \alpha \cdot d_{95}{}^{3} \tag{2}$$

where m is the mass retained after each sample division step in grams, d_{95} is the nominal top size in millimetres, and α is a constant over the whole sample preparation procedure for a particular material in g/mm^3.

Depending on the size of the laboratory sample for each method, a particular number of tests are needed to guarantee the representativeness of the sample.

2.3. Methodology for taking representative samples from ELTs

The preparation of the sample consists in a reduction made in several stages of fragmentation/quartering until the different subpopulations (rubbers, metal wires, textile fibres) are obtained (see **Figure 2**). The standard means and procedures for reducing test samples are carried out in a laboratory with the appropriate equipment (sample division, cryogenic mill…). **Figure 3** shows the flow diagram of the whole process for collection and preparation of the representative sample to obtain a test portion.

Taking into account the size of the average production lots, it was estimated that 1.5 t of whole tyres is the minimum necessary quantity of ELTs. For the Spanish case, four different samples were taken in order to estimate the biomass content by type of tyre.

Starting with 1.5 t of tyres, the first step is the shredding of tyres to reduce the size of the sample and to obtain a mix of particles from different tyres. This first-size reduction can be carried out in a primary shredder to obtain pieces within the interval of 35–200 mm.

The sample must be taken at the production platform, using a tool of the open rectangular shovel type by completely cutting the flow of falling material. Samples are considered valid if a total quantity of at least 25 kg is taken per increment to represent a production of 1.5 t.

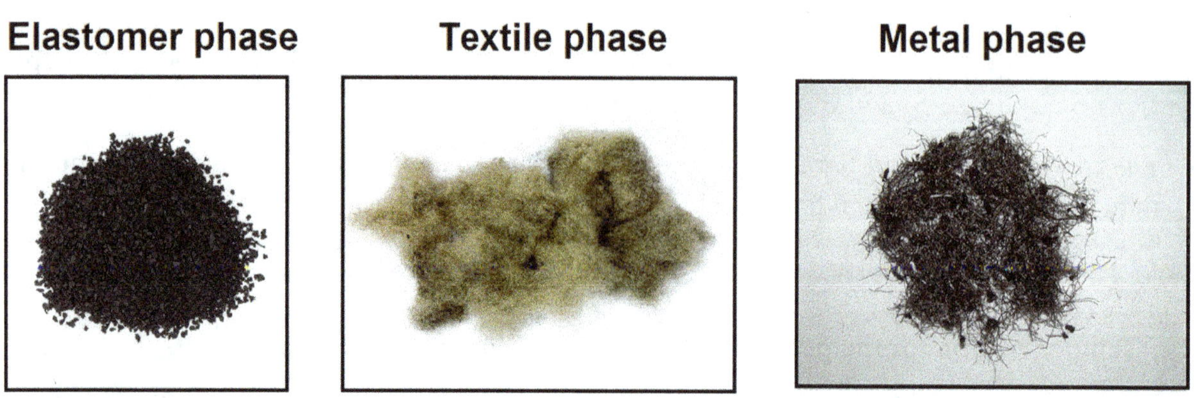

Figure 2. Different fractions of the tyre.

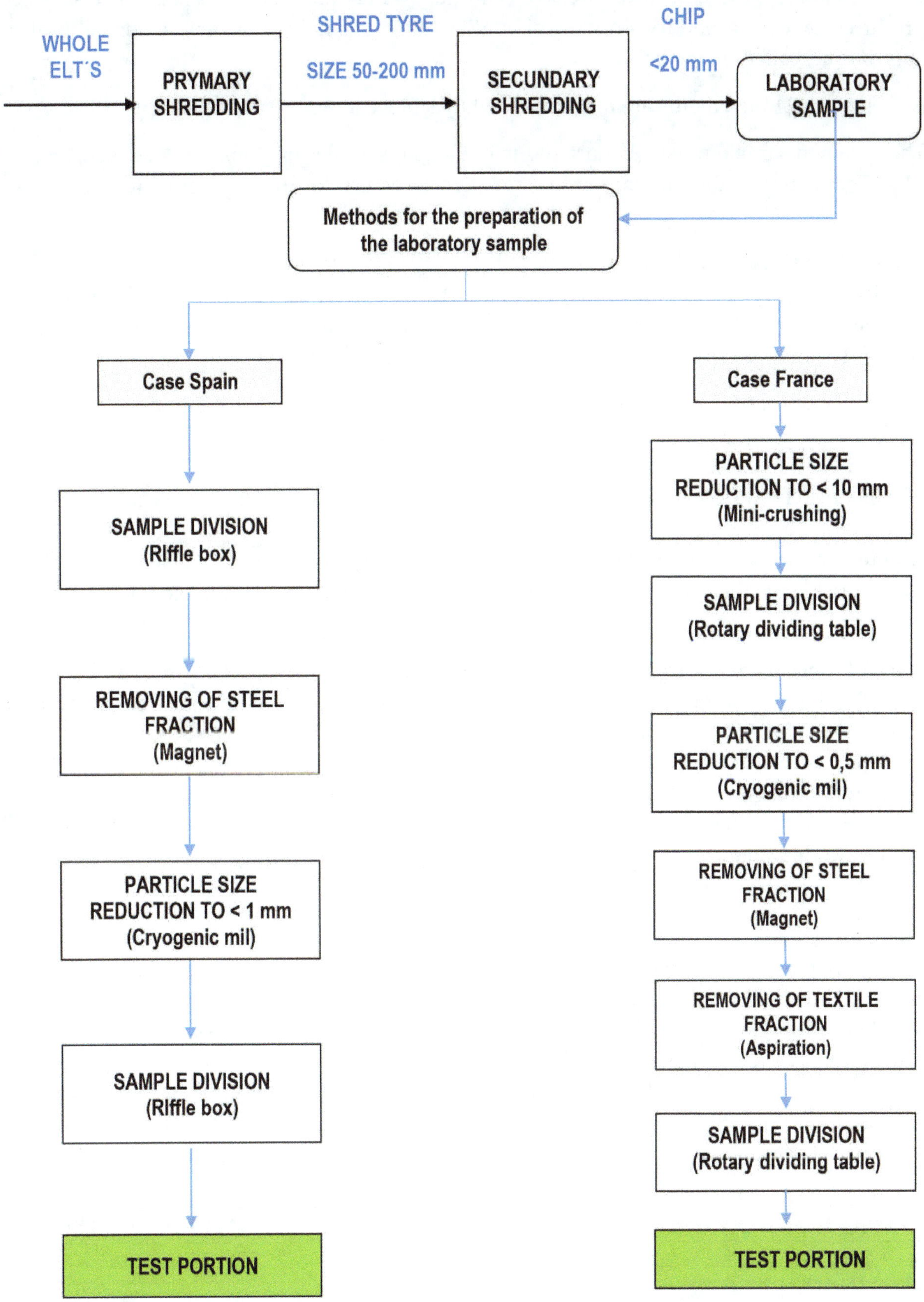

Figure 3. Complete scheme for test sampling preparation.

In a second step, those particles are reduced under a maximum size of 20 mm. The samples produced by this secondary shredder should be taken in smaller portions, at constant intervals of time, called increments.

No less than 24 increments are recommended to achieve a representative sample.

The obtained sample must be quartered to obtain subsamples of a quantity between 1.5 and 3 kg. That is bigger than the minimum quantity necessary for this type of material (around 1.0 kg).

Then the steel fraction should be removed from the sample using a magnet, taking special care in leaving the rest of material (rubber and textile fibres) that should be reduced in particles sized under 1 mm. Different methods could be used for this purpose, especially cryogenics. The obtained product should then be quartered to obtain a test portion.

For the French case, the procedure is similar to the above described (see **Figure 3**).

3. Test methods for the determination of biomass content in tyres

Different test methods were identified in the literature for the determination of natural rubber content in blends that can thus be indirectly used for the quantification of biomass content in tyres [17].

Some of those methods are based on the determination of content in elastomers and are particularly able to distinguish the presence of isoprene, the main component of the natural rubber, among other different elastomers of the rubber blend. This natural rubber has been identified as the main source of biogenic carbon in a tyre; nonetheless, this component is not the only bio-based one. Thus, some of the test methods identified for this chapter are not really conclusive about the total content of biomass in tyres. On the other hand, there are significant differences in the time to analyse the samples (see **Table 3**).

3.1. Pyrolysis-GC/MS

This method is based on the degradation of a sample in an electric furnace at 500–600°C and keeping the sample within this temperature range [18]. This temperature range is recommended to obtain rapid pyrolysis without excessive degradation or carbonization of the rubber sample. However, a temperature of 550°C is advised to obtain the maximum quantity of pyrolysate for NR, IR, BR, SBR, IIR, BIIR and CIIR that are the major elastomeric components of a tyre.

	TGA	Py-GC/MS	^{14}C	
			^{14}C (LSC)	^{14}C (BI)
Test duration	A few hour	A few days	2 months	A few days

Table 3. Test duration of the different methods.

This pyrolysis must be performed passing a stream of nitrogen through the pyrolysis reactor. Nitrogen serves to displace air, prevents oxidation and facilitates the transfer of the pyrolysis products to the gas chromatographer.

The gas chromatographer is equipped with 30-m-long capillary chromatographic column in a fused non-polar-type silica.

The gas chromatographer is coupled to a mass spectrometer operating in scan mode. It detects and registers certain decomposition substances between 35 and 550 atomic mass units (amu).

The pyrolysis-GC/MS carried out in one of the studies is based on ISO standard 7270-2 and requires a calibration curve by pyrolysing the samples with known styrene/butadiene/isoprene ratios.

The approach of this method is to evaluate the natural rubber content in a sample of tyres: It is possible to calculate the total concentration of elastomers in samples and also the concentration in natural elastomers by reporting the result on the previously produced calibration curve.

The authors of this study observed several problems during the application of this method; the main one being related to the non-possibility to distinguish natural isoprene from synthetic isoprene.

One major drawback is that determining the content comparing results with a curve made with different ratio of known samples of styrene/butadiene/isoprene rubbers only gives relative values inside the elastomeric fraction and not in the whole sample. An unrealistic composition on biomass could then be reported with this method.

Another problem derived from the use of PY-GC/MS is the non-detection of other biogenic components of the rubber. A variability of results with pyrolysis temperature and the extraction time in solvents before pyrolysis is also reported. Finally, the presence of brominated butyl could also disturb the results.

Taking into account all these issues, Pyrolysis-GC/MS is then not considered as a valid technique for the evaluation of biogenic content.

3.2. Thermogravimetric analysis (TGA)

This method is based on the continuous measurement of the weight loss of a sample submitted to a ramp of temperature in a controlled atmosphere [11, 19].

In the case of TGA, each type of elastomer has a particular temperature at which the loss of mass occurs. When a sample of vulcanized rubber is tested, some particular peaks appear at specific temperatures.

At lower temperatures, below 300°C, moisture, volatile components derived from plasticizers and other simple chemicals of the rubber blend, volatize.

In the range from 300 to 525°C most of the elastomers in a tyre rubber blend are degraded by the heat. The first thermal decomposition corresponds to natural rubber NR, and the maximum weight-loss rate occurs in the 300–400°C interval. Styrene-butadiene SBR maximum weight-loss rate occurs between 420 and 550°C.

One of the studies tried to produce a reference calibration curve based on different binary NR/SBR rubber samples of known composition and the intensity—or height—of the peaks of the DTG curves. Each measurement produces a typical graph of weight loss as a function of the increase in temperature. **Figure 4** shows an example of a weight loss curve for a (NR 75%, SBR 25%) blend and its derivative weight loss curves. This curve shows one minimum per elastomer, H_{NR} and H_{SBR}.

The calibration curve in **Figure 5** represents the relationship between the peak height ratio H_{NR}/H_{SBR} and the NR content of each of the reference samples. The resulting values are adjusted to a second degree polynomial equation [19]:

$$r = \left(\frac{NR}{NR + SBR}\right)\% = a\left(\frac{H_{NR}}{H_{SBR}}\right)^2 + b\left(\frac{H_{NR}}{H_{SBR}}\right) + c \quad (3)$$

$$a = -22.331;\ b = 84.503;\ c = 3.0782$$

where r is the percentage of NR in the elastomeric fraction (NR+SBR), H_{NR} is the maximum rate of weight loss in the area where NR decomposes, and H_{SBR} is the maximum rate where SBR decomposes.

Using this calibration curve, the value of r of an unknown sample can be then determined, based on the height of the peaks for NR and SBR.

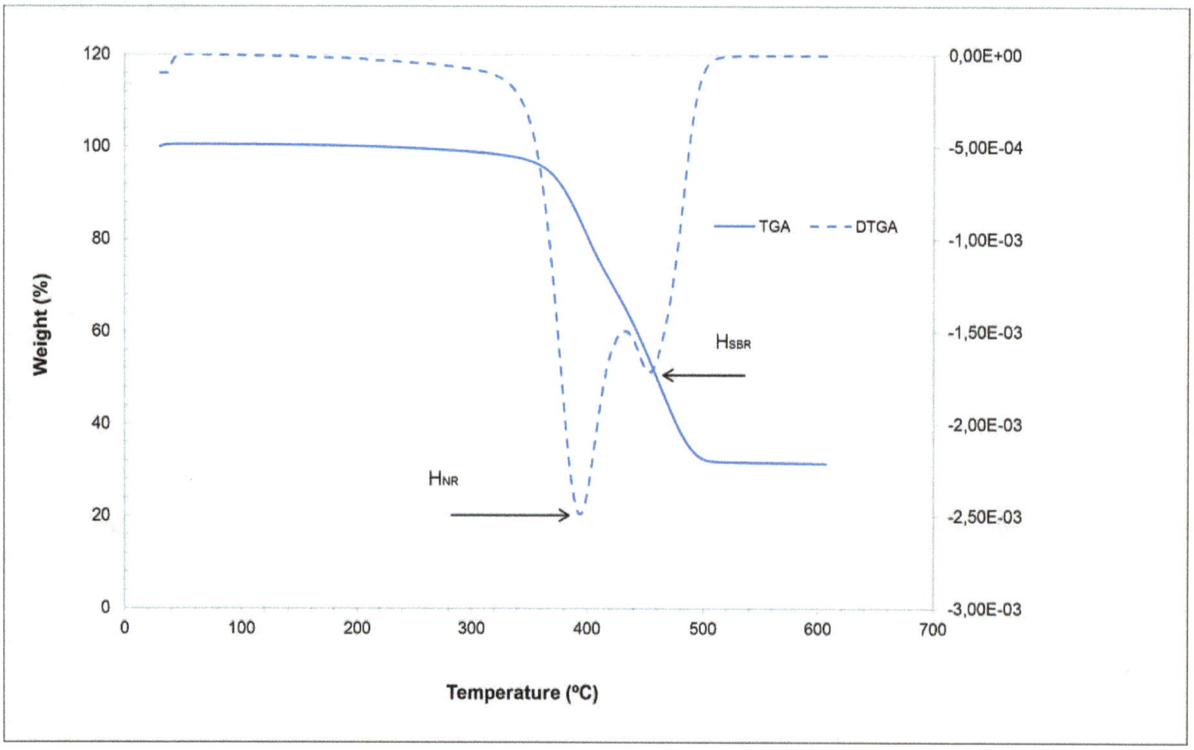

Figure 4. Weight loss and derivative weight loss curve for a polymer blend (NR75%/SBR25%).

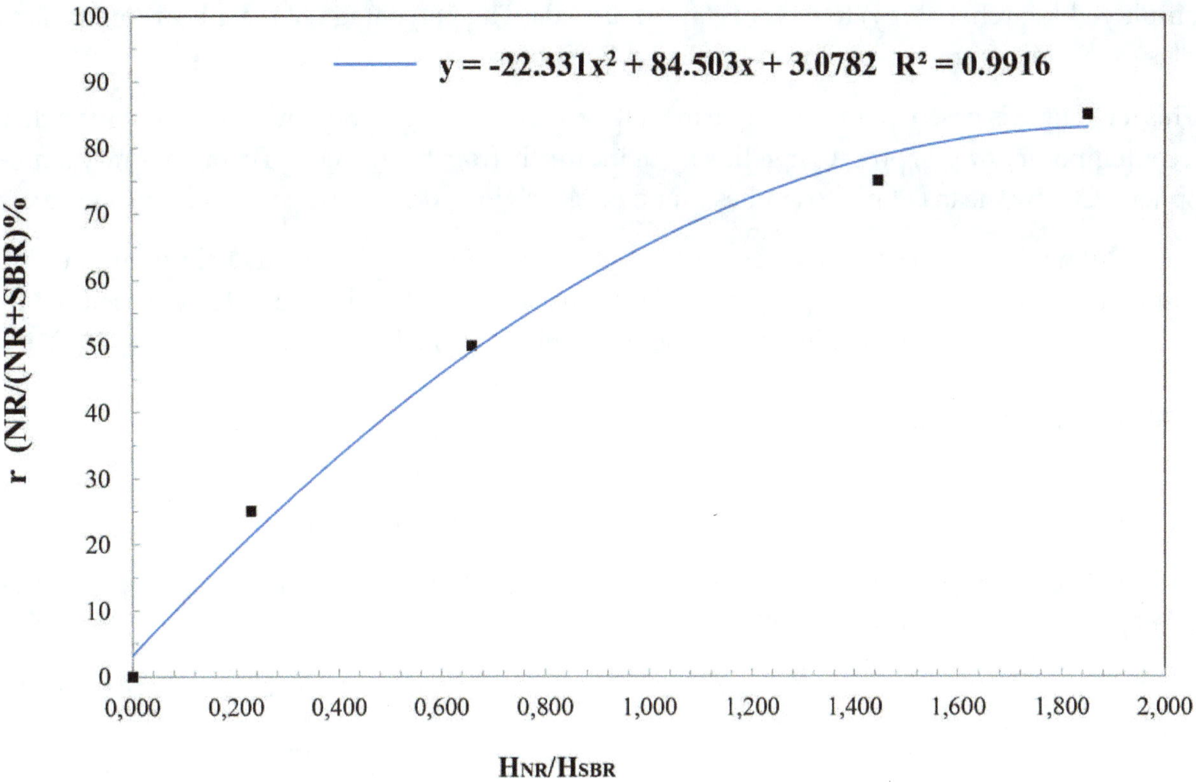

Figure 5. Plot of r (NR/(NR+SBR)%) vs. the peak heights ratio (H_{NR}/H_{SBR}) to determine the biomass content in the tyres studied.

However, it has been reported by authors of this chapter [17] that the correlation of the result for NR obtained by TG analysis and the actual content of NR in samples with known quantities of this elastomer is very bad. Like in the case of PY-GC/MS, this analysis technique is only valid for an estimation of the natural rubber content, independently of the accuracy of the results. Indeed, there is not any possibility to distinguish natural isoprene from synthetic isoprene.

Furthermore, problems derived from the use of this technique start when there is a combination of more than two elastomers in the sample; in such case, the identification and quantification of the elastomers are very difficult because of the overlapping of peaks. Finally, other biogenic components of the rubber, such as cotton or stearic acid, are not detectable by the use of this technique.

Taking into account all these issues, thermogravimetric analysis is then not considered as a valid technique for the evaluation of biogenic content.

3.3. Radiocarbon analysis: ^{14}C methods

The determination of biomass content in different materials using ^{14}C methods is based on analytical procedures used for the determination of the age of carbon contained in materials [8, 9].

Three well-known methods for the determination of ^{14}C content are described in the literature, two of which have been used in the studies made by the authors of this chapter. Those methods are commonly accepted for the determination of the age of objects, especially employed in archaeology, to date organic materials. The percentage of recent carbon in samples gives an accurate idea of non-fossil carbon content.

This technique is based on the principle that all the carbon atoms in organic materials have either a contemporary origin, proceeding directly or indirectly from the fixation of contemporary atmospheric CO_2 by means of photosynthesis, or a fossil origin and were fixed millions of years ago.

Every living organism contains a quantity of ^{14}C proportional to the relative abundance of ^{14}C in the atmosphere. Thus, the percentage of biomass in a material is directly proportional to its ^{14}C content. Fossil fuels, however, do not contain ^{14}C, as its half-life is 5,700 years [20, 21].

3.3.1. $^{14}C/^{12}C$ *determination by beta-ionization (BI)*

One of the studies used the analysis of the carbon concentration of bio-based origin (Beta ionization method) focusing on the biomass ^{12}C assay, considered to be more accurate.

The test has been developed according to standard ASTM D6866-08 and has determined the biogenic carbon content specifically for this purpose.

Particularly, in this case, this test method has been adapted and used for measuring ^{14}C, in elastomeric fraction and textile fraction.

3.3.2. *Liquid scintillation spectrometer (LSC)* ^{14}C *determination*

Another alternative is the ^{14}C determination by liquid scintillation spectrometer LSC using butyl-PBD, as scintillation agent, added to benzene (C_6H_6) samples, previously prepared by the following chemical reactions:

$$2CO_2 + 10Li \rightarrow Li_2C_2 + 4Li_2O \tag{4}$$

$$Li_2C_2 + 2H_2O \rightarrow C_2H_2 + 2LiOH \tag{5}$$

$$3C_2H_2 \rightarrow C_6H_6 \tag{6}$$

CO_2 is then produced in a combustion chamber by burning an appropriate sample of rubber coming from tyre. It was ensured that the reacting CO_2 was only coming from the sample.

The ^{14}C activity was corrected by the isotopic fractionation according to the directives of the ASTM D6866-05 standard test methods for the determination of biomass content [10]. To do this, the $^{12}C/^{13}C$ ratio was established in the stable isotopes laboratory of reference.

4. Discussion of results of the biomass content in tyres

Two different studies have been conducted recently, one in France and the other Spain for the quantification of the biomass content in tyres.

	Spanish case (number of tests)		French case (number of tests)	
Description of the sample	TGA	^{14}C (method C)	Py-GC/MS	^{14}C (method B)
Passenger car tyres	8	4	8	8
Truck and bus	4	2	4	4
Agro tyres	4	2	–	–
Total number of tests	12	8	12	12

Table 4. Tests are conducted for each method.

In both cases, the purpose of the studies was not only the quantification of the biomass content in tyres but also the validation or not of different techniques for this purpose.

In paragraph 3 of this chapter, three techniques have been compared: PY-GC/MS and TGA were discarded, and techniques related with ^{14}C determination have been accepted and are highly recommended. **Table 4** shows the number of tests conducted for each method.

Table 5 indicates the content of each sample of tyres in the recycling plant selected to conduct this study in Spain, taking into account the Spanish shared market, in terms of sizes. Lot coded PT140611-1 represents a sample of motorcycle, passenger car, SUV and vans tyres representative in percentage of the market in Spain. The other sample PA140611-2 was randomly taken from a stocked pile of those types of tyres. In addition, two other samples, one corresponding to truck tyres, and other sample with agro tyres were prepared.

Tyres were shredded and reduced to granules following the procedure described in paragraph 2 of this chapter. Samples were appropriately divided using a riffle box with the appropriate number of slots.

After reducing the size of the particles below 20 mm, the steel wires were removed from samples. This process was carried out with a magnet and the percentage of steel in samples is given in **Table 6**. Some rubber particles remain attached to the steel samples, so a calcination of these samples was performed to calculate the actual content of metal wires in the representative samples.

In the same way, the rubber granulates samples were reduced to fine powder, under 1 mm.

Sample code	Description of the sample	Number of tyres	Total weight in tons
PT140611-1	Passenger car tyres of a representative sample of the market	200	1.62
PA140611-2	Random sample of passenger car tyres	190–210	1.58
MC190711-3	Truck and bus	30	1.67
MA190711-4	Agro tyres	30	1.76

Table 5. Samples of different types of tyres (Spanish case).

	% in weight of the steel fraction	% percentage of steel after calcination	% real content per sample
PT140611-1	23	70.25	15.9
PA140611-2	24	62.22	15.1
MC190711-3	36	77.35	27.9
MA190711-4	6	90.15	5.2

Table 6. Percentage of steel fraction (Spanish case).

The French study used a specific laboratory equipment that permits a complete separation of the three different fractions of materials with no contamination between them, starting from shredding stage. So, the phase of calcination of steel fraction was not used. Summary of content of different materials are listed in **Tables 7** and **8**. Then the rubber and textile fractions were also reduced to particles under 1 mm.

In both studies, a cryogenic laboratory mill was used to achieve this particle size total reduction in the samples.

4.1. Results of the ^{14}C analysis

According to the sampling plan, the sample for laboratory test is representative of the lot if quantity is over 66 mg. In the case of liquid scintillation spectrometry, the mass of the sample should be between 10 and 12 g and then, only one single test per sample is necessary. Nevertheless, all the samples have been tested twice to ensure the repeatability of the results.

Results of ^{14}C per sample of rubber and textile using this LSC are shown in **Table 9**.

The percentage of modern carbon is calculated with the following expression.

Sample	Rubber	Metal	Textile
VL1	80.90	12.16	6.93
VL2	82.54	10.05	7.41
VL3	82.30	13.15	4.55
VL4	80.30	13.57	6.13
VL5	83.59	10.77	5.64
VL6	83.08	11.25	5.68
VL7	84.54	12.58	2.89
VL8	84.03	12.28	3.68
Average reference values	82.7	12.0	5.4

Table 7. Summary of content of different fractions per sample in French case (passenger car tyres).

Sample	Rubber	Metal	Textile
PL1	78.74	21.26	0.00
PL2	79.67	20.33	0.00
PL3	73.66	25.50	0.84
PL4	72.55	27.45	0.00
Average reference values	73.1	26.5	0.42

Table 8. Summary of content of different fractions per sample in French case (truck and bus tyres).

$$pMC = \frac{A_{SN}}{A_{ON}} \tag{7}$$

where A_{SN} is the isotopic activity of the sample, standardized for isotopic division, A_{ON} is the activity of the oxalic acid reference, also standardized for isotopic division and pMC is the percentage of modern carbon. Using this Eq. (7), the results of the biomass content in the Spanish study are summarized in **Table 10**.

In general terms, duplicated samples show similar values, all of them are under the limits of tolerance established for this analytical technique.

Table 10 shows the average value of the percentage of biomass in the rubber-textile fraction for each sample. The final result of biomass in the tyres sample, taking into account the content of steel in each one, is written in the **Table 11**.

As a conclusion, the biomass content in the sample corresponding to the passenger car tyres both random and representative to the Spanish share market are exactly the same with an average value of 22.2%.

Sample code	Weight of the sample burned (g)	Yield		Weight of C_6H_6 obtained (g)	Weight of C_6H_6 analyzed (g)	Isotopic division ($^0/_{00}$)
		C_2H_2%	C_6H_6%			
PT140611-1.1	12,1	99,2%	94,5%	7,4	6,188	-26,6 ± 0,2
PT140611-1.2	9,7	99,9%	97,5%	7,8	6,077	-26,6 ± 0,2
PT140611-2.1	11,2	99,6%	97,8%	8,1	6,123	-27,2 ± 0,2
PT140611-2.2	10,1	99,5%	97,9%	7,6	6,121	-27,7 ± 0,2
MC190711-3.1	10,8	99,3%	97,4%	7,7	6,112	-26,7 ± 0,2
MC190711-3.2	10	99,6%	97,6%	7,7	6,100	-26,7 ± 0,2
MA190711-4.1	9,7	97,9%	98,4%	7,8	6,103	-26,0 ± 0,2
MA190711-4.2	9,9	96,0%	97,5%	7,6	6,127	-26,0 ± 0,2

Table 9. Summary of synthetized samples for ^{14}C analysis (Spanish case).

Sample	Rubber and textile fraction	
	pMC (%)	Biomass (%)
PT140611-1.1	29,2 ± 0,7	27,2 ± 0,7
PT140611-1.2	27,5 ± 0,7	25,6 ± 0,7
PA140611-2.1	27,8 ± 0,7	25,8 ± 0,7
PA140611-2.2	28,5 ± 0,7	26,5 ± 0,7
MC190711-3.1	50,2 ± 0,7	46,7 ± 0,7
MC190711-3.2	50,9 ± 0,7	47,3 ± 0,7
MA190711-4.1	29,7 ± 0,7	27,6 ± 0,7
MA190711-4.2	30,1 ± 0,7	28,0 ± 0,7

Table 10. Results of ^{14}C analysis for the biomass content in tyres (Spanish case).

Sample	% biomass of the rubber-textile fraction	% biomass of the total sample	Range of biomass content in tyres (%)
Passenger car tyres of a representative sample of the market	26.4	22.2	21.5–22.9
Random sample of passenger car tyres	26.15	22.2	21.9–22.5
Truck and bus	47.0	33.9	33.7–34.1
Agro tyres	27.8	26.4	26.2–26.5

Table 11. Results of biomass content in tyres samples (Spanish case).

Biomass values obtained for tyres samples corresponding to truck and bus tyres resulted in an average percentage of 33.9% by weight, and the average value for biomass content of agro tyres resulted in a 26.4% by weight.

On the other hand, the study carried out in France on several tyre samples, both passenger car, truck and bus tyres, for the determination of biomass content have been tested by the beta-ionization technique.

Results obtained for different types of samples are shown in **Tables 12** and **13**. The value of biomass for passenger car tyre samples is in average 18.3 and 29.1% for the truck and bus samples.

In any case, the magnitude order of the biomass value obtained in the French case are in accordance with approximate theoretical values of natural rubber and other chemical substances with bio-based content in tyres. More specifically, the different values of natural rubber content in car and truck tyre have been obtained and are in line with theoretical values.

Sample	Biomass content (%)	Standard deviation
VL1	17	1.69
VL2	18.9	0.36
VL3	17.0	1.69
VL4	18.4	0.01
VL5	20.3	4.00
VL6	18.1	0.04
VL7	17.3	1.00
VL8	19.4	1.21
Average reference values	18.3	

Table 12. Biomass content of passenger car tyres measured by BI (French case).

Sample	Biomass content (%)	Standard deviation
PL1	30.7	0.36
PL2	31.4	1.69
PL3	29.7	0.16
PL4	28.6	2.25
Average reference values	29.1	

Table 13. Biomass content of truck and bus tyres measured by BI (French case).

5. Conclusions

A tyre is a complex product with a biomass content heterogeneously distributed inside it. Furthermore, there is also a huge variability in composition inside the market, by brands and types, each one having a different biomass content. Therefore, to carry out a proper study of biomass content of a batch or a sample representing a significant number of tyres, it is absolutely necessary to make a good selection of the laboratory sample by means of statistical methods to ensure representativeness of the lot. In this chapter, the procedure for managing tyres samples to obtain representative samples by continuous size reduction and quartering has been described.

Furthermore, it should be taken into account that necessary quantities for the biomass content determination are in the order of a few grams or even milligrams, so the sampling step is crucial to obtain proper results.

Several methods have been explored for determining the biomass fraction of tyres, but some of them are not completely reliable and conclusive, mainly for the following reasons:

The results of the thermogravimetric method differed considerably from the known natural rubber content of the reference samples as well as results obtained from the ^{14}C technique that is because the synthetic isoprene cannot be distinguished from natural isoprene. The pyrolysis-GC-FID method is neither considered as a reliable method, mainly for the same reason and because results are affected by temperature and extraction time.

In addition, the presence of textile fibres and stearic acid in the tyre, a well-known source of biomass, cannot be evaluated for both techniques (pyrolysis-GC-MS and TGA),

French and Spanish studies conclude that the ^{14}C techniques are the most reliable for determining the biomass content of end-of-life tyres. Both methods, beta-ionization (BI) and liquid scintillation spectrometry (LSC), lead to results close to the actual biomass content in tyres.

Finally, the reference values of the biomass content of the end-of-life tyres have been established. Average content for passenger car tyres is 18.3% for the French market and 22% for the Spanish market.

In the case of truck and bus tyres, content in both cases are in the range from 29.1% in the French case to 34% in the Spanish case. Apart from errors of different analytical techniques and laboratories, differences found in both studies could be related to different market share in term of sizes and brands in both countries.

In the Spanish case, a reference value has also been established for agro tyres with an average biomass content of 26.4%.

Author details

Leticia Saiz Rodríguez[1*], José M. Bermejo Muñoz[1], Adrien Zambon[2] and Jean P. Faure[2]

*Address all correspondence to: lsaiz@signus.es

1 Signus Ecovalor S.L, Madrid, Spain

2 Aliapur, Lyon, France

References

[1] R&D ALIAPUR. Life cycle assessment of 9 recovery methods for end-of-life tyres [Internet]. 2010. Available from: http://www.etrma.org/uploads/Modules/Documentsmanager/aliapur_lca-reference-document-june-2010.pdf [Accessed: 2016-03-10]

[2] R&D ALIAPUR. Using used tyres as an alternative source of fuel [Internet]. 2009. Available from: https://www.aliapur.fr/sites/default/files/files/RetD/Extract_-_Used_tyres_as_an_alternative_source_of_fuel_July_09.pdf [Accessed: 2016-03-10]

[3] W. Staber, S. Flamme, J. Fellner: Methods for determining the biomass content of waste. Waste Management & Research. 2008; 26: 78–87. doi:10.1177/0734242X07087313

[4] European Parliament and of the Council. Directive 2003/87 of the European Parliament and Council, Establishing a scheme for greenhouse gas emission allowance trading within the community and amending Council Directive 96/61/EC [Internet]. 2003. Available from: http://eur-lex.europa.eu/legal-content/EN/TXT/PDF/?uri=CELEX:02003L0087-20140430&from=EN [Accessed: 2016-03-10]

[5] S. Singh, W. Nimmo, B.M. Gibbs, P.T. Williams: Waste tyre rubber as a secondary fuel for power plants. Fuel. 2009; 88: 2473–2480. doi:10.1016/j.fuel.2009.02.026

[6] M.N. Chukwu, I. C. Madufor et al: Effect of stearic acid level on the physical properties of natural rubber vulcanisate. The Pacific Journal of Science and Technology. 2011; 12: 344–350. url:www.akamaiuniversity.us/PJST12_1_344.pdf

[7] B. Lah, D.Klinar, B. Likozar: Pyrolysis of natural, butadiene, styrene–butadiene rubber and tyre components: modelling kinetics and transport phenomena at different heating rates and formulations. Chemical Engineering Science. 2013; 87: 1–13. doi:10.1016/j.ces.2012.10.003

[8] CEN (European Committee for Standardization). CEN/TS 15747–SRF, "^{14}C-based methods for the determination of the biomass content" [Internet]. 2008. Available from: http://standards.globalspec.com/standards/detail?docid=1166222&familyid=EUPDNCAAAAAAAAAA [Accessed: 2016-03-10]

[9] CEN (European Committee for Standardization). EN 14440–SRF, "Methods for the determination of biomass content" [Internet]. 2011. Available from: http://standards.globalspec.com/std/1465300/cen-en-15440 [Accessed: 2016-03-10]

[10] ASTM International Standard. D 6866-05. Determining the biobased content of natural range materials using radiocarbon and isotope ratio mass spectrometry analysis. Book of Standards. 2005; 0803. doi:10.1520/D6866-16

[11] S. R. Shield, G. N. Ghebremeskel and C. Hendrix: Pyrolysis-GC/MS and TGA as tools for characterizing blends of SBR and NBR. Rubber Chemistry and Technology. 2001; 74: 803–813. doi:10.5254/1.3547654

[12] Y.S. Lee, W. Lee, S. Cho, I. Kim, C. Ha, J.: Quantitative analysis of unknown compositions in ternary polymer blends: a model study on NR/SBR/BR system. Journal of Analaytical and Applied Pyrolysis. 2007; 78: 1, (85–94). doi:10.1016/j.jaap.2006.05.001

[13] CEN (European Committee for Standardization). EN 15413:2011–"Solid recovered fuels. Methods for the preparation of the test sample from the laboratory sample". [Internet]. 2011. Available from: http://www.aenor.es/aenor/normas/normas/fichanorma.asp?tipo=N&codigo=N0049240#.V9ktiPmLSJA [Accessed: 2016-03-10]

[14] CEN (European Committee for Standardization). EN 15415-1:2011, "Solid recovered fuels. Determination of particle size distribution. Screen method for small dimension particles". [Internet]. 2011. Available from: http://shop.bsigroup.com/ProductDetail/?pid=000000000030219013 [Accessed: 2016-03-10]

[15] CEN (European Committee for Standardization). CEN/TS 15401:2010, "Solid recovered fuels. Determination of bulk density". [Internet]. 2010. Available from: http://shop.bsigroup.com/ProductDetail/?pid=000000000030209997 [Accessed: 2016-03-10]

[16] CEN (European Committee for Standardization). EN 15443:2011–"Solid recovered fuels. Methods for the preparation of the Laboratory sample". [Internet]. 2011. Available from: http://www.aenor.es/aenor/normas/normas/fichanorma.asp?tipo=N&codigo=N0048493#.V9kuy_mLSJA [Accessed: 2016-03-10]

[17] L.Saiz, J.M. Bermejo et al: Comparative analysis of ^{14}C and TGA techniques for the quantification of the biomass content of end-of-life tires. Rubber Chemistry and Technology. 2014; 87: 4 (664–678). doi:10.5254/rct.14.85993

[18] ISO 7270-2_2012 "Rubber—Analysis by pyrolytic gas-chromatographic methods—part 2: determination of styrene/butadiene/isoprene ratio". [Internet]. 2012. Available from: http://www.iso.org/iso/catalogue_detail.htm?csnumber=53025 [Accessed: 2016-03-10]

[19] M.J. Fernández-Berridi, N. González, A. Mugica, C. Bernicot: Pyrolysis-FTIR and TGA techniques as tools in the characterization of blends of natural rubber and SBR. Thermochimica Acta. 2006; 444: 65–70. doi:10.1016/j.tca.2006.02.027

[20] J. Mohn, S. Szidat, J. Fellner, H. Rechberger, R. Quartier, B. Buchmann, L. Emmenegger: Determination of biogenic and fossil CO_2 emitted by waste incineration based on $^{14}CO_2$ and mass balances. Bioresource Technology. 2008; 99: 6471–6479. doi:10.1016/j.biortech.2007.11.042

[21] S.W.L. Palstra, H.A.J. Meijer: Carbon-14 based determination of the biogenic fraction of industrial CO_2 emissions—application and validation. Bioresource Technology. 2010; 101: 3702–3710. doi:10.1016/j.biortech.2009.12.004

8

Modeling the Calorific Value of Biomass from Fruit Trees using Elemental Analysis Data

Borja Velázquez-Martí, Isabel López-Cortés,
Domingo Salazar-Hernández and
Ángel Jesús Callejón-Ferre

Abstract

Pruning of fruit trees produces a great quantity of biomass each year that can be used for energy production. For this purpose, it is necessary to carry out an energy characterization of these pruned wastes, where the determination of heating value is significant. This value is usually measured by an adiabatic or isoperibolic calorimeter, which causes high economic costs and wastes time. The present study is focused on the development of indirect models for heating value prediction of biomass from orange trees *Citrus × sinensis* Osbeck, almond trees *Prunus dulcis* (Mill) D.A. Webb, and olive trees *Olea europaea* L. from an elemental analysis in order to reduce the time of determination as well as the economic costs. Residual biomass was classified and characterized according to CEN regulations such as received, without drying. Also, moisture content wet basis, bark ratio, density, heating value, and elemental composition (carbon, hydrogen, nitrogen, and sulfur) were measured. The influence of these variables on the heating value was analyzed. Finally, mathematical models were developed to predict this value for this studied species. These models showed coefficients of determination between 0.83 and 0.97, being suitable for industrial use.

Keywords: bioenergy, economical studies, wood residues, higher heating value

1. Introduction

A number of researchers have published mathematical models to predict the higher heating value of different biomass materials from the concentration of the main elements present, such as percentage of carbon, percentage of hydrogen, and percentage of nitrogen together with others [1–6]. On the other hand, other models have used proximate analysis [7–10] or structural analysis [9–11]. Indirect calculation of the higher heating value by means of these types of models is justified by the expensive cost of the use of calorimeters [3, 12]. The aim of this chapter is to compare the resources used in direct heat value determination with that used in indirect calculation from elemental analysis by means of prediction models in three common lignocellulosic materials coming from pruning Mediterranean fruit trees. In this work, predictive models of the heating value of biomass from pruning of *Citrus* × *sinensis* Osbeck (orange tree), *P. dulcis* (Mill.) D.A. Webb (almond tree) and *O. europaea* L. (olive tree) have been developed; the calorific value determined directly using the isoperibolic calorimeter was compared with that determined from the elemental composition of materials. The influence of the percentage of leaves, wood moisture content, bark percentage, and percentage content of C, H, N, and S was analyzed.

The heat value is an invariable parameter for a material with constant composition, defined by the empirical form $CH_wO_xN_yS_z$, where w is the number of moles of hydrogen per mole of carbon, x is the number of moles of oxygen per mole of carbon, y is the number of moles of nitrogen per mole of carbon, and z is the number of moles of sulfur per mole of carbon. The moles of each element are obtained by multiplication of the sample mass with its ratio and dividing by the atomic weight of each element. The values w, x, y, and z are obtained from the division of the moles of each element in the sample by the moles of carbon.

The capacity of retaining water in the biomass, caused by its porosity, must be considered in the measurement of the calorific value. The moisture content in the material changes its molecular formula, and therefore the gravimetric percentages of C, H, O, and N. For this reason, standards to determine the calorific value for a particular material, such as UNE 164001:2005 EX [13], require the material to be obtained in the anhydrous state, or with a known moisture content.

The problem in industrial applications is that the biomass materials received for the combustion in boilers show variation in their moisture content, they are sometimes mixed with leaves and even with other materials. Under these conditions, the heat properties are not constant. In order to measure the calorific value instantly, regression models are analyzed in this paper from the percentage contents of C, H, and N of the materials as received. Currently, there are devices capable of measuring these elements in 5 min [5, 6].

When there is variability in the composition, and uncertainty of the type of materials or proportion of the mixture thereof, the mathematical models for the indirect determination of the calorific value are only applicable in the scope where they were developed. This uncertainty is quite common in industry. Thus researchers, such as Francis and Lloyd [14], Ebeling and Jenkins [15], and Kathiravale et al. [16], have provided models for different types of mixtures. However, none of them showed an economical study to justify the use of the model. In this chapter, the development of models specific for the studied material is shown; in addition, an analysis was carried out to compare the cost of direct determination with an indirect mathematical model from elemental analysis.

In industrial facilities, it is very difficult to find biomass received for combustion without some moisture. Because of processes increase production costs, they are rarely used in the production of energy wood. On the other hand, open air drying rarely decreases the moisture content below 20% [17, 18]. Moreover, it is normal that industrial boilers do not work with a well-defined type of material but with mixtures of different types of biomass. These reasons make the composition of the biomass used in industry variable that directly affects the expected calorific value. So, calorific determination before the introduction of the materials in the boiler is useful to understand the energy performance of the combustion. If this direct determination by the bomb calorimeter is more expensive than the indirect determination from their composition, developing predictive models is fully justified. This is studied in this work.

2. Materials and methods

2.1. Vegetal material

The species studied in this work were *Citrus* × *sinensis* Osbeck (orange tree), *P. dulcis* (Mill) D.A. Webb (almond tree), and *O. europaea* L. (olive tree). These three species are widely grown in the Mediterranean region; they represent 10% of the total cultivated area in Spain [19].

The studied orange variety was "Valencia Late". This is one of the latest varieties most cultivated in Spain. The harvesting begins from March, and it is usually pruned in summer after harvesting. Its main use is fresh consumption due to its high juice content. It also has high chances of industrialization [20].

The almond variety studied was "Blaquerna". It is a self-fertile cultivar that is currently replacing other varieties of lower production. Almond pruning can be annual or biennial according to the development of the tree [21]. During the first 4 years, winter pruning is carried out focused on defining the architecture of the vegetation. From the fourth production year, pruning is performed to remove unproductive branches and to improve the fruit quality [22].

The studied olive tree variety was "Villalonga", which is the variety most widely used in Valencia, with a total cultivation area of 23,550 ha. Its main use is in the manufacture of oil and it is also used for fresh consumption [23].

To define these raw materials as biomass for biofuel, the standard UNE EN 14961-1:2011[24] was used. According to this standard, the analyzed samples in this work were classified by their origin and sources of solid biofuels: 1. Wood biomass → 1.1. Wood biomass from forest or plantation → 1.1.4. Residues from cuttings → 1.1.4.1. Fresh/green, hardwood (including leaves). Therefore, it would be "cutting hardwood waste" (1.1.4.1). Following the mentioned standard, the specifications of the properties of the studied biofuel are defined in **Table 1**.

Origin: Cutting hardwood waste (1.1.4.1)		Wood biomass (1.1) - *Citrus × sinensis* Osbeck - *Prunus dulcis* (Mill.) D.A. Webb - *Olea europaea* L.
Marketed form		Wood logs, firewood
Sizes (cm)		
Length (*L*) cm		
Citrus × sinensis Osbeck *L* 100 (max 100 cm ± 5 cm)	*Prunus dulcis* (Mill.) D.A. Webb *L* 100+ (max 153 cm)	*Olea europaea* L. *L* 100+ (max 182 cm)
Diameter (*D*) cm		
Citrus × sinensis Osbeck *D*10 (2 cm ≤ *D* ≤ 10 cm)	*Prunus dulcis* (Mill.) D.A. Webb *D*10 (2 cm ≤ *D* ≤ 10 cm)	*Olea europaea* L. *D*10 (2 cm ≤ *D* ≤ 10 cm)
Moisture content, M (% as it is received) CEN/TS 15149-1, CEN/TS 15149-2		
Citrus × sinensis Osbeck M 45 (≤45%)	*Prunus dulcis* (Mill.) D.A. Webb M 35 (≤35%)	*Olea europaea* L. M 40 (≤40%)
Volume or weight, m^3, kg, loose or piled as received		15–20 kg of each specied.
Volume ratio of split logs		Trunk without cuttings, whole branches
Cut surface[a]		Surface smooth and regular cut
Rust and rot		None of the samples has mold

[a] The use of chainsaw is considered to produce a smooth and regular surface.

Table 1. Specification of the properties of the energy wood pieces (EN 14961-1:2011) [22].

2.2. Sample preparation

The branches of each species, obtained from pruning, were divided into six size classes (0–1, 1–2, 2–3, 3–4, 4–5, and >5 cm). Five samples of each class were taken for analysis; therefore, it resulted in 30 samples per species. For sample preparation, the methods defined by the UNE-EN 14780:2012 [25] were followed. The main purpose of sample preparation was to reduce the size of the branches in test portions with the same initial composition, being representative of the original sample. Initial wet basis moisture content average in wood was about 42.24%. In **Figure 1**, the process of the preparation of the sample is shown. First step was to separate the leaves and wood of the 30 branches of each species arrived at the laboratory (**Figure 2a–c**). These leaves were crushed with hammer mill and stored in airtight jars with identification labels. On the other hand, the wood was milled until their sizes were lower than 3 mm. Special care was taken to avoid loss of fine particles and moisture during milling and other operations. Average wet basis moisture content in tested particles was 29.85%. The devices used for the sample preparation were as follows:

- Manual saw. In order to prevent the moisture loss at the border, cuttings of the central part was used for the analytical determinations. The pieces obtained using this device are shown in **Figure 2(d)**.

- Hammer mill crusher of stainless steel, equipped with a 3 mm screen.

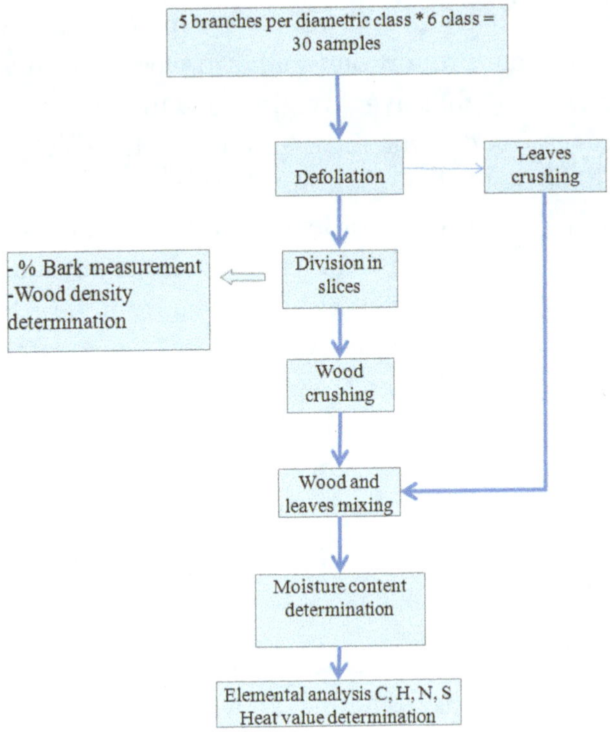

Figure 1. Sample preparation process.

Figure 2. Images of sample preparation process.

Once the samples were crushed, leaves and wood were mixed in defined proportions in each group of size class. The different proportions analyzed were from 10% weight leaves/weight wood in the first sample, 20% weight leaves/weight wood in the first sample in the second, etc. up to 50% of mixture leaves/wood to the fifth sample. So, the effect of leaves on the calorific value was measured. This is very important in this study because pruned material used in boilers has always got a high percentage of leaves, and this influences the moisture content and composition. In many publications, the calorific value calculated referred to that of completely dry and bare material, but this condition is far from the actual applications. In this work, the analysis was focused on samples with variable elemental composition which is usually obtained due to variation in the percentage of leaf and bark content (caused by different size class) and the moisture content (obtained after outdoor drying). For this reason, neither the wood nor the leaf fractions were dried rigorously, but they only experience the natural loss of moisture content during the transport and storage prior to analysis.

2.3. Measurement process

Higher heating value, wet basis moisture content, and elemental composition (C, H, and N) of each sample were measured. For this, standards shown in **Table 2** were used.

Reference of standard	Standard
CEN/TS 14779	Solid biofuels—sampling—methods for preparing sampling plans and sampling certificates
CEN/TS 14780	Solid biofuels—methods for sample preparation
EN 14774-3	Solid biofuels—determination of moisture content—Stufe drying method part 3 moisture content analysis for overall sample analysis
EN 14918	Solid biofuels—determination of calorific value
CEN/TS 15104	Solid biofuels—determination of total carbon, hydrogen, and nitrogen—instrumental methods
CEN/TS 15290	Solid biofuels—determination of major elements
EN 14961-1	Solid biofuels—fuel specifications and classes—part 1: general requirements

Table 2. Methods used to sample and measure parameters.

The calorific value was measured by means of a LECO AC-500 isoperibolic calorimeter. Before analyzing the samples, the calorimeter was calibrated by the combustion of a reactive standard of a known calorific power (benzoic acid, 1 g). Subsequently, each sample was prepared with a mass between 0.1 and 1 g. This was introduced into a combustion vessel where a fuse wire caused ignition. Note that 10 ml of distilled water was added. Then, combustion vessel was sealed, and oxygen with a pressure of 3000 kPa was introduced inside the calorimeter . This container was placed in a bucket of water which was surrounded by an insulating layer to maintain a constant temperature. During analysis, the water temperature was measured by an electronic thermometer with an accuracy of 1/10,000 degree. In order to control energy exchange, the temperatures of the cuvette and the insulating layer were continuously monitored. With this, the device applies a correction to the result.The water temperature was monitored by a microprocessor in every 6 seconds. The difference between the water temperature before ignition and after ignition was processed through the calorimeter software, for obtaining the calorific value.

The weight percentage of carbon, hydrogen, and nitrogen was measured by means of a LECO TruSpec CHN analyzer. According to EN 14918, samples between 0.1 and 1 g were weighed. Then, they were wrapped in titanium sheets that is completely inorganic. These were placed in a feeding carrousel. The analysis cycle consists of three phases: purging, combustion, and analysis. In the purge phase, the sample is casted into the load compartment, which is sealed and atmospheric gases are removed. In the second phase, the sample is casted into a compartment at 950°C, and oxygen is injected for rapid and complete combustion. The gases pass through an afterburner at 850°C to oxidize and remove particles. The combustion gases are collected in the ballast (gas collection vessel). During the analysis phase, combustion gases are homogenized in the ballast. Subsequently, an aliquot of 3 cm^3 is purged with helium through infrared detectors of CO_2 and H_2O. Another gas loop aliquot is transported through hot copper to remove O_2 and transform the NO_X to N_2. Then, in order to remove CO_2 and H_2O, they are allowed to flow through the tubes containing Anhydrone Lecosorb. The N content is determined on a thermal conductivity cell. The results are shown as percent or kg/mg.

In order to calculate the cost of analysis all inputs were counted. Market prices of nine enterprises were checked. Consumables, labor maintenance, and amortization were considered. Cost of technical labor was estimated in 20 €/h. Residual value of the device was considered to be 10% of investment. Time of analysis was measured.

2.4. Percentage of bark

The percentage of bark in the branches was calculated after their classification according to six diameter classes: 0–1, 1–2, 2–3, 3–4, 5–6, and >6 cm:. The diameter influences the bark ratio [26]. For each diameter, class five samples were taken, so 30 samples in total were analyzed in each species. In each branch, diameter with bark and diameter without bark were measured using of a digital caliper with 0.01 mm accuracy, as shown in **Figure 3**. From these diameters, the percentage of bark was calculated by Eq. (1) [26], where *Bark* (%) is the percentage of bark; ϕ_{cc} is the diameter over bark; ϕ_{sc} is the diameter without bark:

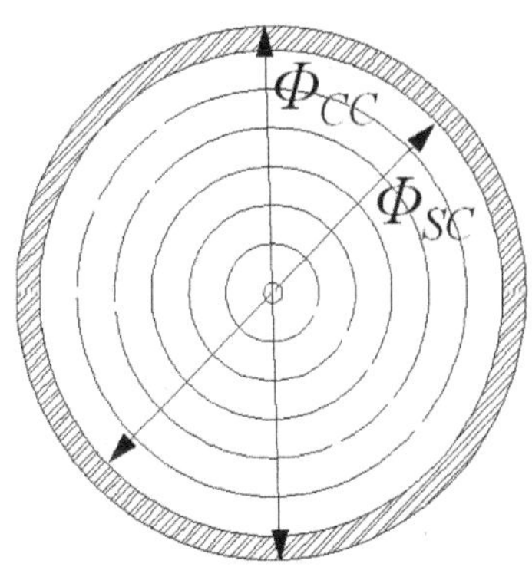

Figure 3. Measurements to calculate the percentage of bark.

$$\text{Bark (\%)} = \frac{\varphi_{cc}^2 - \varphi_{sc}^2}{\varphi_{cc}^2} \cdot 100 \tag{1}$$

2.5. Determination of wood density

Wood density is expressed as the mass of dry wood per unit volume. To calculate the density, first the samples were immersed in a beaker with water, calibrated to 250 mL. The volume was measured by the difference between the water level before and after immersing. Then the dry weight of the samples was determined; for this purpose, samples were placed in a drying oven at a constant temperature of 105 ± 2°C for 24 h. Mean and standard deviation for the densities are obtained by Eq. (2) [26]:

$$\rho_m = \frac{P_s}{V_v} \tag{2}$$

where ρ_m is the wood density (g·cm^{-3}), P_s is the dry weight of the sample (g), and V_v is the volume of the sample (cm^3).

2.6. Prediction model of heat value

To obtain predictive models by regression up to three variables (C, H, and N) have been used. For evaluating the models, the coefficient of determination (R^2), root mean square of the errors (RMS), and mean absolute error (MAE) were obtained. The coefficient of determination, denoted by R^2, is a number that indicates the proportion of the variance in the dependent variable that is predictable from the independent variable. RMS represents the sample standard deviation of the differences between predicted values and observed values. The mean absolute error is an average of the absolute errors $|e_i|=|f_i-y_i|$, where f_i is the prediction and y_i is the true value. The model with the best fit had highest R^2, minimum MAE, and RMS.

For all equations, 30 data were used to develop the models, and another 15 independent data were used for validation. The statistical program used was Statgraphics Centurion XV©14, even for the calculation of the significance of the variables of the mathematical prediction models by the beta coefficients and Student's *t*-test. The data observed in the new experiments and predicted by the model were compared with paired-sample test based on the *t*-Distribution.

3. Results and discussion

Several tests were initially applied to check the normal distribution of the data, such as Shapiro-Wilk test [27, 28], Anderson-Darling test, and the Lilliefors test [29, 30]. In **Table 3**, statistical description of the studied variables is shown for each species. It is observed that the coefficient of skewness and kurtosis are between −2 and +2. This indicates that the observed samples are normally distributed.

Variable	Species	Average	Standard deviation	Coef. skewness	Coef. kurtosis	Max.	Min.
HHV (kJ/kg)	Almond	15840.63	234.752	−0.775	−0.435	17458.54	13744.85
	Orange	12653.35	230.03	−0.23	−1.044	14108.26	10781.85
	Olive	15234.97	243.7	−0.281	−0.717	16861.92	13354.64
C %	Almond	40.026	2.261	−0.382	−0.683	44.050	35.300
	Orange	31.192	2.414	−0.274	−1.158	34.800	26.650
	Olive	37.974	2.359	0.192	−0.706	41.800	33.600
H %	Almond	5.661	0.740	1.740	−0.239	7.410	4.875
	Orange	7.179	0.291	−0.284	−0.485	7.590	6.540
	Olive	6.911	0.275	−0.429	−1.061	7.285	6.390
N %	Almond	1.132	0.535	0.147	−1.415	1.935	0.349
	Orange	0.430	0.143	1.383	0.474	0.800	0.222
	Olive	0.549	0.178	−0.792	−0.871	0.812	0.219
S %	Almond	0.128	0.024	0.705	−0.055	0.182	0.088
	Orange	0.061	0.054	1.802	−0.898	0.166	0.006
	Olive	0.031	0.013	1.100	−0.031	0.063	0.015
Moisture content %	Almond	19.009	4.870	0.967	−0.392	30.314	12.291
	Orange	34.679	4.496	−0.457	−1.084	41.094	26.770
	Olive	31.085	7.659	−0.020	−0.637	46.070	18.490
Bark %	Almond	24.741	6.485	0.548	−1.070	36.282	13.830
	Orange	15.038	3.688	1.936	0.176	23.960	10.135
	Olive	16.469	4.297	1.586	1.235	28.260	10.500
Leaves %	Almond	30.231	14.434	0.235	−1.346	50	10
	Orange	30.417	14.590	−0.156	−1.325	50	10
	Olive	31.905	14.703	−0.483	−1.202	50	10

HHV, high heat value; C, carbon; H, hydrogen; N, nitrogen; S, sulfur.

Table 3. Statistical analysis of the studied variables in each species.

The calorific values are clearly influenced by the moisture content and the leaf content in the sample. The obtained calorific value for the three species was between 12 and 16 MJ/kg, which were relatively lower than values cited in the literature. This is because, in this work, we have studied samples of diverse diameters with different percentage of bark, without any drying process, and mixtures of wood and leaves in different proportions, so smaller high heat values (HHVs) were obtained. For example, González et al. [31] gave values of HHV for biomass from orange tree pruning as 16–18 MJ/kg. Yin [32] also analyzed mixtures of biomass and got HHV values 18 MJ/kg, but all these were measured on a dry condition basis.

To compare the calorific value of the three species studied, analysis of variance was performed. In **Figure 4**, LSD intervals are shown at 95% confidence level. It can be seen that the

HHV of almond and olive trees were similar, but that of orange was significantly lower, which may be due to the characteristics of wood, leaves, and bark with moisture.

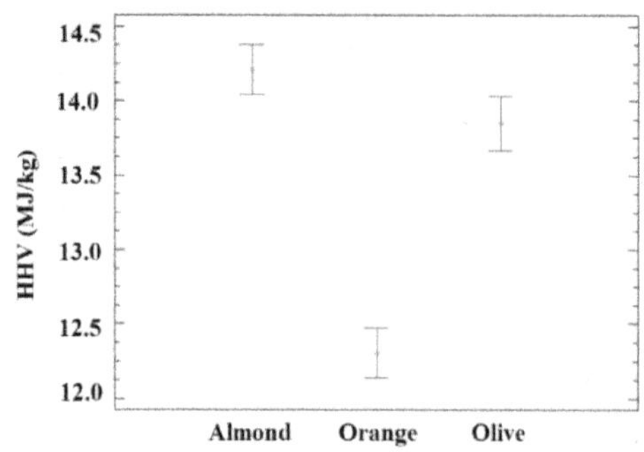

Figure 4. Intervals LSD for the calorific value of the species studied at 95% level of confidence.

Prunus dulcis	C	H	N	S	% leaves	% bark	% moisture content	HHV (kJ/kg)
C		−0.576*	0.1515	−0.3196	0.2535	−0.0076	−0.9844*	0.9739*
H	−0.576*		0.1712	0.397*	−0.0914	−0.1514	0.6393*	−0.5449*
N	0.1515	0.1712		0.4904*	0.8992*	−0.2699	−0.2055	0.2928
S	−0.3196	0.397*	0.4904*		0.3773	−0.3122	0.2873	−0.2427
% leaves	0.2535	−0.0914	0.8992*	0.3773		−0.09	−0.301	−0.3658
% bark	−0.0076	−0.1514	−0.2699	−0.3122	−0.09		0.0612	−0.0737
% moisture	−0.9844*	0.6393*	−0.2055	0.2873	−0.301	0.0612		−0.9786*
HVV (kJ/kg)	0.9739*	−0.5449*	0.2928	−0.2427	0.3658	−0.0737	−0.9786*	
Citrus × sinensis	C	H	N	S	% leaves	% bark	% moisture content	HHV (kJ/kg)
C		−0.8699*	−0.1595	0.5819*	−0.4233*	−0.5698*	−0.9551*	0.916*
H	−0.8699*		−0.1615	−0.5637*	0.165	0.4785*	0.8811*	−0.8249*
N	−0.1595	−0.1615		0.2159	0.7504*	0.2578	−0.0793	−0.1708
S	0.5819*	−0.5637*	0.2159		−0.173	−0.3339	−0.6484*	0.6292*
% leaves	−0.4233*	0.165	0.7504*	−0.173		0.1614	0.2708	−0.3111
% bark	−0.5698*	0.4785*	0.2578	−0.3339	0.1614		0.4213*	−0.7104*
% moisture	−0.9551*	0.8811*	−0.0793	−0.6484*	0.2708	0.4213*		−0.8464*
HVV (kJ/kg)	0.916*	−0.8249*	−0.1708	0.6292*	−0.3111	−0.7104*	−0.8464*	
Olea europaea	C	H	N	S	% leaves	% bark	% moisture content	HHV (kJ/kg)
C		−0.2761	0.8297*	0.0164	0.7945*	0.3936	−0.9715*	0.9752*
H	−0.2761		0.1347	0.2282	0.0321	−0.3109	0.3303	−0.279
N	0.8297*	0.1347		0.0287	0.9104*	0.1444	−0.7907*	0.8261*
S	0.0164	0.2282	0.0287		0.0894	0.0508	0.0596	−0.0737
% leaves	0.7945*	0.0321	0.9104*	0.0894		0.0024	−0.727*	−0.7955*
% bark	0.3936	−0.3109	0.1444	0.0508	0.0024		−0.4414*	−0.421
% moisture	−0.9715*	0.3303	−0.7907*	0.0596	−0.727*	−0.4414*		−0.9411*
HVV (kJ/kg)	0.9752*	−0.279	0.8261*	−0.0737	0.7955*	0.421	−0.9411*	

C, carbon; H, hydrogen; N, nitrogen; S, sulfur; HHV, calorific value.
* Pairs of variables with p-values below 0.05 (95% of significance level).

Table 4. Correlation coefficients of Pearson between the studied variables per species.

Table 4 shows the correlation between analyzed variables. Significant negative influence of the percentage of moisture in the calorific value (HHV) is observed, with −0.97 Pearson coefficient. On the other hand, it is noted that a higher percentage of C increases the HHV (+0.97). It is observed that the percentage of H is associated with the highest moisture content (+0.63) and obviously decreases the percentage of carbon in the sample and the calorific value (−0.54). It is observed that the percentages of sulfur, leaves, and bark do not have a clear influence on the calorific value of almonds but have a negative influence on the olive and orange trees.

Specie	**Model**	**R^2 adj. (%)**	**RMS**	**MAE**
Mixing three species	HHV = 717.79 + 380.85·C + 7.61% leaves−14.00% bark	97.352	278.77	195.74
	HHV = 863.61 + 383.93·C − 15.257% bark	96.976	297.90	218.00
	HHV =769.58 + 371.80·C + 8.02% leaves	97.147	289.35	201.07
	HHV = 928.66 + 374.30·C	96.733	309.62	223.64
Prunus dulcis	HHV = 7624.11 + 237.76·C + 228.75·N − 83.37%w	97.131	166.21	125.38
	HHV = −1024.04 + 412.99·C + 272.54·N	96.743	177.08	133.28
	HHV = −1106.84 + 422.77·C	94.630	227.38	173.87
Citrus × sinensis	HHV = 4323.54 + 301.46·C − 72.74% bark	88.127	331.30	250.40
	HHV = 1254.39 + 365.79·C	83.17	394.46	303.26
Olea europaea	HHV = −790.92 + 421.37·C	94.858	231.17	118.44

HHV, high heat value (kJ/kg); MAE, mean absolute error; C, carbon; N, nitrogen; %w, moisture content in wet basis.

Table 5. Prediction models proposed for the indirect calculation of calorific value.

Calorific models from elemental analyses are proposed in **Table 5**. All models show a high R^2 and relatively low standard error (RMS) and mean absolute error (MAE). All models are considered valid for calculating the calorific value, as an alternative to the calorimeter, reducing the time and cost of analysis. The p-values for all explanatory variables were less than 0.05. It is observed that the difference between the simplest models, whose explanatory variable is only carbon, and the more complicated is very small. Therefore, we recommend using the simplest, common to all species Eq. (3). In the variance analysis of the regression models, the p-value was less than 0.01 for all variables. This means that there is a significant relationship between variables and the volume for a confidence level of 99%:

$$HHV = 28.66 + 374.30 \cdot C \tag{3}$$

Vargas-Moreno et al. [4] conducted a review of models to predict the calorific value for different biomass materials based on elemental analyses. Most of these models reviewed gave determination coefficients between 0.8 and 0.99. For this reason, the correlation coefficients obtained in this work, higher than 0.8, were considered very acceptable (**Table 4**). However, on the other hand, Acda [33] proposed models with higher R^2 for different materials, but their models were obtained for dry materials and without leaves, whereas the models obtained in this study were obtained from mixtures of wood and leaves, and different moisture content, as received. Therefore, they are more applicable in actual situations in industries, where the material cannot be usually dried completely and they have different leaf percentage.

Velázquez-Martí et al. [5] already applied this method to obtain prediction models to predict a high heat value on lignocellulosic waste materials from urban tree pruning. In these works, the determination coefficients obtained are similar to those of the present work, between 0.7 and 0.9. This means that the presence of moisture causes decrease in accuracy of the calorific value calculated by prediction models.

It is observed that a higher carbon content (C) provides bigger calorific values. This coincides with many other studies [2, 34]. It is also shown in **Table 4** with a Pearson correlation coefficient of 0.97 between HHV and C. Because of this, all proposed equations for predicting the gross calorific value (**Table 5**), whether univariate or multivariate, present the variable C.

The sulfur (S) is not high in all species studied (**Table 3**), so it would not be a problem in biomass combustion boilers caused by this element. As it can be seen, in olive and citrus wood the S and N contents are lower than the limits established by the standard EN 14691-part 4 [35], which fix the conditions for chips used as biofuels in %N < 1% and %S < 0.1%. Almond has not got values excessive high. This allows preparing mixtures with materials with low N and S contents.

This discrepancy may be due to the extractive substances of plant biomass (sugars, tannins, sterols, fatty acids, resin acids, oligomeric terpenics, hydrocarbons, etc.) that influence the gross calorific value. Their contents depend on the species, the part of the plant, season, and the growing stage, among other factors [36]. Another explanation could be the botanical family of the studied species [37].

Once calculated using the prediction models, we proceeded to the validation of these data by applying them to new samples, whose observations were compared with the predicted values.

3.1. Comparison cost between direct and indirect measurement of HHV

Measurements with the isoperibolic LECO AC-500 calorimeter showed that the time per analyzed sample was 15 min, which includes sample preparation. As consumables, 6 L of pure oxygen measured at standard conditions (25°C and 1 atm of pressure) and a fuse for ignition of the sample are used. **Table 6** shows the cost of reagents. Device depreciation and calibration cost are respectively shown in **Tables 7** and **8**. The total costs for different number of analyzed samples are shown in **Table 9**.

Consumable	Price per consumable pack		Consumption per sample		Analysis cost (€/analysis)
Standard for calibration ref. 502-208-T Ac. Benzoic	160.0	€/100 tablets	5 tablets/calibration		1.60
Spike for combustion ref. 502-815 Mineral oil	175.7	€/118 mL	0.25	mL	
Fuse for ignition ref. 502-462	38.2	€/375 fuse wire	1	Ud	0.10
Rent gas tank O_2	94.4	€/year			0.019
Gas O_2					2.25
Maintenance	1200	€/year			0.24
Technical labor	20	€/h	20	min/sample	6.67
			100	min/calibration	

Table 6. Cost of the reagents for isoperibolic LECO AC500 calorimeter (12/03/2013).

Acquisition cost	50,000	€
Useful life	10	
Residual value 10%	5000	€
Depreciation cost	0.9	€/sample

Table 7. Depreciation cost per sample of LECO AC-500 calorimeter.

	n	Time (min/analysis)	Total time (min)	Cost (€)
Number of analysis	5	20	100	33.33
Standard for calibration ref. 502-208-T Ac. Benzoic	5			8.00
Fuse wire, ref. 502–462	5			0.51
Gas O_2				11.25
Rent gas tanks + maintenance				1.294
			Total	54.39

n, number of analysis.

Table 8. Calibration cost of LECO AC-500 calorimeter.

Sample number	Number of calibration	Technical labor time (h)	Technical labor cost (€)	Gasses (€)	Wire fuse (€)	Total cost (€)	Cost/sample (€/sample)
1	1	2.00	6.67	1.71	0.10	**64.03**	64.03
2	1	2.33	13.33	3.43	0.20	**73.67**	36.83
3	1	2.67	20.00	5.14	0.30	**83.31**	27.77
4	1	3.00	26.67	6.86	0.40	**92.95**	23.24
5	1	3.33	33.33	8.57	0.50	**102.59**	20.52
6	1	3.67	40.00	10.29	0.60	**112.23**	18.70
7	1	4.00	46.67	12.00	0.70	**121.87**	17.41
8	1	4.33	53.33	13.71	0.80	**131.51**	16.44
9	1	4.67	60.00	15.43	0.90	**141.15**	15.68
10	1	5.00	66.67	17.14	1.00	**150.78**	15.08
11	1	5.33	73.33	18.86	1.10	**160.42**	14.58
12	1	5.67	80.00	20.57	1.20	**170.06**	14.17
13	1	6.00	86.67	22.29	1.30	**179.70**	13.82
14	1	6.33	93.33	24.00	1.40	**189.34**	13.52
15	1	6.67	100.00	25.71	1.50	**198.98**	13.27
16	1	7.00	106.67	27.43	1.60	**208.62**	13.04
17	1	7.33	113.33	29.14	1.70	**218.26**	12.84
18	1	7.67	120.00	30.86	1.80	**227.90**	12.66
19	1	8.00	126.67	32.57	1.90	**237.54**	12.50
20	1	8.33	133.33	34.29	2.00	**247.18**	12.36
21	1	8.67	140.00	36.00	2.10	**256.82**	12.23
22	1	9.00	146.67	37.71	2.20	**266.46**	12.11
23	1	9.33	153.33	39.43	2.30	**276.10**	12.00
24	1	9.67	160.00	41.14	2.40	**285.74**	11.91
25	1	10.00	166.67	42.86	2.50	**295.38**	11.82

Table 9. Total costs for different number of samples analyzed by LECO AC-500 calorimeter.

The time spent for elemental analysis using a TruSpec CHN analyzer is 5 min. The costs of the reagents, calibration, and amortization are respectively shown in **Tables 10–12**.

Consumable	Price per consumable pack		Consumption per sample		Analysis cost (€/ analysis)
Standard for calibration EDTA ref. 502-092	44.8	€/50 g	0.2 × 8	g/calibration	0.18
Tin Foil for solid samples (Large Tin Foil ref. 502-397-400)	65.5	€/400 tin foil	1	Unit/sample	0.16
Rent tank O_2	94.4	€/year			0.019
Rent tank N_2	47.2	€/year			0.009
Rent tank He	47.2	€/year			0.009
Rent tank compressed air	47.2	€/year			0.009
Gas O_2	450.0	€/tank	1	Tank/200 analysis	2.25
Gas N_2	342.9	€/tank	1	Tank/200 analysis	1.71
Gas He	342.9	€/tank	1	Tank/200 analysis	1.71
Compressed air	126.2	€/year			0.03
Maintenance	1200	€/year			0.24
Technical labor	20	€/h	5	min/simple	3.33

Table 10. Cost of the reagents for elemental LECO TRUSPEC CHNS analyzer (12/03/2013) (supposing 5000 analysis per year).

		n	Time (min/analysis)	Total time (min)	Cost
Time of calibration calculation	Blanc test	15	5	75	
	Standard analysis	8	5	40	
				115	38.33 €
Standard for calibration	EDTA ref. 502-092	8			1.43 €
Tin foil for solid samples	Large Tin Foil ref. 502-397-400	8			1.31 €
Gases		8			18.00 €
Rent gas tanks + maintenance		8			2.30 €
				Total	61.37 €

n, number of analysis.

Table 11. Calibration cost of LECO Truspec CHNS analyzer.

Acquisition cost	**125,000**	**€**
Useful life	10	
Residual value 10%	12,500	€
Depreciation cost	2.25	€/sample

Table 12. Depreciation cost per simple of LECO Truspec CHNS analyzer.

In **Table 13**, the total costs of the elemental analysis with the analyzer LECO CHNS TruSpec are shown. When **Tables 9** and **13** are compared, it can be seen that the cost for determining the calorific value indirectly from elemental analysis is 23% cheaper than the direct measurement with AC500 LECO isoperibolic calorimeter for 25 samples. Moreover, time of determination is lower. The possibility to calculate the calorific value of a substance from its elemental composition reduces investment to a single computer, instead of two. This is very important in laboratories with limited resources of small and medium enterprises.

No. analysis	Calibration	Calibration cost	Analysis time (h)	Technician labor	Gases	Tin foil	Total	Cost/sample (€/sample)
1	1	61.37	2.00	1.67	2.25	0.16	**67.99**	67.99
2	1	61.37	2.08	3.33	4.50	0.33	**74.61**	37.30
3	1	61.37	2.17	5.00	6.75	0.49	**81.23**	27.08
4	1	61.37	2.25	6.67	9.00	0.66	**87.85**	21.96
5	1	61.37	2.33	8.33	11.25	0.82	**94.46**	18.89
6	1	61.37	2.42	10.00	13.50	0.98	**101.08**	16.85
7	1	61.37	2.50	11.67	15.75	1.15	**107.70**	15.39
8	1	61.37	2.58	13.33	18.00	1.31	**114.32**	14.29
9	1	61.37	2.67	15.00	20.25	1.47	**120.93**	13.44
10	1	61.37	2.75	16.67	22.50	1.64	**127.55**	12.76
11	1	61.37	2.83	18.33	24.75	1.80	**134.17**	12.20
12	1	61.37	2.92	20.00	27.00	1.97	**140.79**	11.73
13	1	61.37	3.00	21.67	29.25	2.13	**147.40**	11.34
14	1	61.37	3.08	23.33	31.50	2.29	**154.02**	11.00
15	1	61.37	3.17	25.00	33.75	2.46	**160.64**	10.71
16	1	61.37	3.25	26.67	36.00	2.62	**167.26**	10.45
17	1	61.37	3.33	28.33	38.25	2.78	**173.87**	10.23
18	1	61.37	3.42	30.00	40.50	2.95	**180.49**	10.03

No. analysis	Calibration	Calibration cost	Analysis time (h)	Technician labor	Gases	Tin foil	Total	Cost/sample (€/sample)
19	1	61.37	3.50	31.67	42.75	3.11	**187.11**	9.85
20	1	61.37	3.58	33.33	45.00	3.28	**193.73**	9.69
21	1	61.37	3.67	35.00	47.25	3.44	**200.34**	9.54
22	1	61.37	3.75	36.67	49.50	3.60	**206.96**	9.41
23	1	61.37	3.83	38.33	51.75	3.77	**213.58**	9.29
24	1	61.37	3.92	40.00	54.00	3.93	**220.20**	9.17
25	1	61.37	4.00	41.67	56.25	4.09	**226.81**	9.07
26	1	61.37	4.08	43.33	58.50	4.26	**233.43**	8.98
27	1	61.37	4.17	45.00	60.75	4.42	**240.05**	8.89
28	1	61.37	4.25	46.67	63.00	4.59	**246.67**	8.81
29	1	61.37	4.33	48.33	65.25	4.75	**253.29**	8.73
30	1	61.37	4.42	50.00	67.50	4.91	**259.90**	8.66
31	1	61.37	4.50	51.67	69.75	5.08	**266.52**	8.60
32	1	61.37	4.58	53.33	72.00	5.24	**273.14**	8.54
33	1	61.37	4.67	55.00	74.25	5.40	**279.76**	8.48
34	1	61.37	4.75	56.67	76.50	5.57	**286.37**	8.42
35	1	61.37	4.83	58.33	78.75	5.73	**292.99**	8.37
36	1	61.37	4.92	60.00	81.00	5.90	**299.61**	8.32
37	1	61.37	5.00	61.67	83.25	6.06	**306.23**	8.28
38	1	61.37	5.08	63.33	85.50	6.22	**312.84**	8.23
39	1	61.37	5.17	65.00	87.75	6.39	**319.46**	8.19
40	1	61.37	5.25	66.67	90.00	6.55	**326.08**	8.15
41	1	61.37	5.33	68.33	92.25	6.71	**332.70**	8.11
42	1	61.37	5.42	70.00	94.50	6.88	**339.31**	8.08
43	1	61.37	5.50	71.67	96.75	7.04	**345.93**	8.04
44	1	61.37	5.58	73.33	99.00	7.21	**352.55**	8.01
45	1	61.37	5.67	75.00	101.25	7.37	**359.17**	7.98
46	1	61.37	5.75	76.67	103.50	7.53	**365.78**	7.95
47	1	61.37	5.83	78.33	105.75	7.70	**372.40**	7.92
48	1	61.37	5.92	80.00	108.00	7.86	**379.02**	7.90
49	1	61.37	6.00	81.67	110.25	8.02	**385.64**	7.87
50	1	61.37	6.08	83.33	112.50	8.19	**392.26**	7.85

Table 13. Total cost of analysis by LECO Truspec CHNS analyzer.

4. Conclusions

The advantage of using indirect methods for determining the heat value based on regression models from the analysis of C, H, and N elements has been proven this paper. Time to determine the high heat value indirectly is 40% lower than the time taken using the calorimeter directly. The cost of indirect method is 23% cheaper. Along with the cost savings, reduced analysis time is associated with a lower environmental impact linked to the reagents used.

It is proved that a higher carbon content (C) provides bigger calorific values. However, hydrogen (H) has a negative influence, with a negative Pearson's coefficient. This is due to the fact that hydrogen is associated with the water content. As it is known, the moisture content decreases the high heat value of the biomass, therefore, hydrogen presents inverse proportionality with the heat obtained from combustion. The colinearity between the moisture content and the hydrogen ratio justifies that both were rarely considered in the same model to predict the heat value.

In this paper, models for determining the calorific value in samples of *Citrus × sinensis* Osbeck (orange tree), *P. dulcis* (Mill) D.A. Webb (almond tree), and *O. europaea* L. (olive tree) are proposed using an elemental analysis. The accuracy is high, obtaining coefficients of determination higher than 0.95, an average error of 223.64 kJ/kg, and a RMS of 309.62 kJ/kg.

According to the thermochemical characterization of plum wood, the residual biomass from pruning can be used as chips for bioenergy. These species did not have significant differences in C and H composition, between 30 and 40% C, between 5 and 7% H. However, small differences exist with respect to N and S. Olive and citrus wood have S and N contents lower than the limits established by the standard EN 14691-part 4 [32], which fix the conditions for chips used as biofuels in %N < 1% and %S < 0.1%. Almond has not got values excessive high but still it exceeds. This leads to the preparation of mixtures with materials with low N and S contents to decrease their content.

Author details

Borja Velázquez-Martí[1*], Isabel López-Cortés[2], Domingo Salazar-Hernández[2] and Ángel Jesús Callejón-Ferre[3]

*Address all correspondence to: borvemar@dmta.upv.es

1 Department of Rural and Food Engineering, Universitat Politècnica de València, Valencia, Spain

2 Department of Vegetal Production, Universitat Politècnica de València, Valencia, Spain

3 Department of Engineering, University of Almeria, Almería, Spain

References

[1] Friedl A, Padouvas E, Rotter H, Varmuza K. Prediction of heating values of biomass fuel from elemental composition. Anal Chim Acta. 2005;544:191–198. DOI: 10.1016/j.aca.2005.01.0

[2] Sheng C, Azevedo JLT. Estimating the higher heating value of biomass fuels from basic analysis data. Biomass Bioenergy. 2005;28(5):499–507. DOI: 10.1016/j.biombiec.2004.11.008

[3] Callejón-Ferre AJ, Velázquez-Martí B, López-Martinez JA, Manzano-Agugliaro F. Greenhouse crop residues: energy potential and models for prediction of their higher heating value. Renew Sust Energ Rev. 2011;15(2):948–955. DOI: 10.1016/j.rser.2010.11.01

[4] Vargas-Moreno JM, Callejón-Ferre AJ, Pérez-Alonso J, Velázquez-Martí B. A review of the mathematical models for predicting the heating value of biomass materials. Renew Sust Energ Rev. 2012;16:3065–3083. DOI: 10.1016/j.rser.2012.02.054

[5] Velázquez-Martí B, Sajdak M, López-Cortés I, Callejón-Ferre AJ. Wood characterization for energy application proceeding from pruning *Morus alba* L., *Platanus hispanica* Münchh. and *Sophora japonica* L. in urban áreas. Renew Energy. 2014;62:478–483. DOI: 10.1016/j.renene.2013.08.010

[6] Pérez-Arévalo JJ, Callejón-Ferre AJ, Velázquez-Martí B, Suárez-Medina MD. Prediction models based on higher heating value from the elemental analysis of neem, mango, avocado, banana, and carob trees in Guayas (Ecuador). J Renew Sustain Ener. 2015;7:053122.

[7] Saidur R, Abdelaziz EA, Demirbaş A, Hossain MS, Mekhilef S. A review on biomass as a fuel for boilers. Renew Sust Energ Rev. 2011;15(5):2262–2289. DOI: 10.1016/j.rser.2011.02.015

[8] Khan AA, Jonga WD, Jansens PJ, Spliethoff H. Biomass combustion in fluidized bed boilers: potential problems and remedies. Fuel Process Technol. 2009;90:21–50. DOI: 10.1016/j.fuproc.2008.07.012

[9] Telmo C, Lousada J, Moreira N. Proximate analysis, backwards stepwise regression between gross calorific value, ultimate and chemical analysis of wood. Bioresource Technol. 2010;101(11):3808–3815. DOI: 10.1016/j.biortech.2010.01.021

[10] Demirbas A, Demirbas AH. Estimating the calorific values of lignocellulosic fuels. Energ Explor Exploit. 2004;22:135–143. DOI: 10.1260/0144598041475198.

[11] Demirbas A. Relationship between heating value and lignin, moisture, ash and extractive contents of biomass fuels. Energy Explor Exploit. 2002;20:105–111. DOI: 10.1260/014459802760170420

[12] Callejón-Ferre AJ, Carreño-Sánchez J, Suárez-Medina FJ, Pérez-Alonso J, Velázquez-Martí B . Prediction models for higher heating value based on the structural analysis

of the biomass of plant remains from the greenhouses of Almería. Fuel. 2014;116:377–387. DOI: 10.1016/j.fuel.2013.08.023

[13] UNE-EN 164001:2005 EX. Solids Biofuels. Method for the detemination of calorific value. AENOR: Madrid. Spain; 2005.

[14] Francis HE, Lloyd WG. Predicting heating value from elemental composition. J Coal Qual. 1983;2(2):21–25.

[15] Ebeling JM, Jenkins BM. Physical and chemical properties of biomass. Trans ASAE. 1998;28:898–902.

[16] Kathiravale S, Yunus MNM, Sopian K, Samsuddin AH, Rahman RA. Modeling the heating value of municipal solid waste. Fuel. 2003;82(9):1119–1125. DOI: 10.1016/S0016-2361(03)00009-7

[17] Velázquez-Martí B, Fernández-González E, López-Cortes I, Salazar-Hernández DM. Quantification of the residual biomass obtained from pruning of vineyards in Mediterranean area. Biomass Bioenergy. 2011;35(3):3453–3464. DOI: 10.1016/j.biombioe.2011.04.009

[18] Velázquez-Martí B, Fernández-González E, López-Cortes I, Salazar-Hernández DM. Quantification of the residual biomass obtained from pruning of trees in Mediterranean olive groves. Biomass Bioenergy. 2011;35(2):3208–3217. DOI: 10.1016/j.biombioe.2011.04.042

[19] INE. Instituto Nacional de Estadística. Censo Agrario 2009 [Internet]. 2009 [Updated: 2014]. Available from: http://www.ine.es/censoagrario/censoag.htm

[20] Avanza MM, Bramardi SJ, Mazza SM. Statistical models to describe the fruit growth pattern in sweet orange "Valencia late". Spanish J Agric Res. 2008;6(4):577–585.

[21] Velázquez-Martí B, Fernández-González E, López-Cortes I, Salazar-Hernández DM. Quantification of the residual biomass obtained from pruning of trees in Mediterranean almond groves. Renew Energy. 2011;36:621–626. DOI: 10.1016/j.renene.2010.08.008

[22] López-Cortés I, Salazar-García DC, Salazar DM, Velázquez-Martí B. Fatty Acid, Sterol and Polyphenol content in most Monovarietal Oils of the Spanish East Areas. In: De Leonardis A. (eds.). Virgin olive oil. Production, composition, uses and benefit for the man. New York: Nova Publishers; 2014.

[23] Barranco D, Rallo L. Olive cultivars in Spain. Horttechnology. 2000;10(1):107–110.

[24] UNE-EN 14961-1:2011. Biocombustibles sólidos. Especificaciones y clases de combustibles. Parte 1: Requisitos generales. AENOR: Madrid, Spain; 2011.

[25] UNE-EN 14780:2012. Biocombustibles sólidos. Preparación de muestras. AENOR: Madrid, Spain; 2012.

[26] Husch B, Beers TW, Kershaw JA Jr, editors. Forest Mensuration. John Wiley & Sons. Inc.: Hoboken, NJ; 2003.

[27] Shapiro SS, Francia RS. An approximate analysis of variance test for normality. J Amer Statist Assoc. 1972; 67: 215–216.

[28] Shapiro SS, Wilk MB. An analysis of variance test for normality (complete samples). Biometrika. 1965;52:591–611.

[29] Sheskin DJ. Handbook of Parametric and Nonparametric Statistical Procedures: Fifth Edition. London: Chapman & Hall/CRC Taylor & Francis Group 2011.

[30] Razali NM, Wah YB. Power Comparisons of Shapiro-Wilk, Kolmogorov-Smirnov, Lilliefors and Anderson-Darling Tests. J Stat Model Anal. 2011;2(1):21–33.

[31] González Z, Rosal A, Requejo A, Rodríguez A. Production of pulp and energy using orange tree prunings. Bioresource Technol. 2011;102(19):9330–9334. DOI: 10.1016/j.biortech.2011.07.088

[32] Yin CY. Prediction of higher heating values of biomass from proximate and ultimate analyses. Fuel. 2011;90(3):1128–1132. DOI: 10.1016/j.fuel.2010.11.031

[33] Acda MN. Physico-chemical properties of wood pellets from coppice of short rotation tropical hardwoods. Fuel. 2015;160:531–533. doi:10.1016/j.fuel.2015.08.018

[34] Erol M, Haykiri-Acma H, Kücükbayrak S. Calorific value estimation of biomass from their proximate analysis data. Renew Energy. 2010;35:170–173. DOI:10.1016/j.renene.2009.05.008

[35] EN 14691-part 4. Wood chips—product standard for non-industrial use. CEN-UNE, Madrid, Spain, 2011.

[36] Nhuchhen DR. Prediction of carbon, hydrogen, and oxygen compositions of raw and torrefied biomass using proximate analysis. Fuel. 2016;180:348–356. doi: 10.1016/j.fuel.2016.04.058

[37] Veiga JPS, Valle TL, Feltran JC, Bizzo WA. Characterization and productivity of cassava waste and its use as an energy source. Renew Energy. 2016;93:691–699. doi:10.1016/j.renene.2016.02.078

9

Above-Ground Biomass Estimation with High Spatial Resolution Satellite Images

Adélia M. O. Sousa, Ana Cristina Gonçalves and

José R. Marques da Silva

Abstract

Assessment and monitoring of forest biomass are frequently done with allometric functions per species for inventory plots. The estimation per area unit is carried out with an extrapolation method. In this chapter, a review of the recent methods to estimate forest above-ground biomass (AGB) using remote sensing data is presented. A case study is given with an innovative methodology to estimate above-ground biomass based on crown horizontal projection obtained with high spatial resolution satellite images for two evergreen oak species. The linear functions fitted for pure, mixed and both compositions showed a good performance. Also, the functions with dummy variables to distinguish species and compositions adjusted had the best performance. An error threshold of 5% corresponds to stand areas of 8.7 and 5.5 ha for the functions of all species and compositions without and with dummy variables. This method enables the overall area evaluation, and it is easily implemented in a geographic information system environment.

Keywords: QuickBird, multi-resolution segmentation, crown horizontal projection, forest inventory, regressions

1. Introduction

1.1. Role of forest inventories in biomass estimation

Forest covers a larger area on Earth and provides many products and services. Forest evaluation inventories were initiated when wood shortages arose. It can be said that they are the driving force to acquiring information on forest areas, stand composition and products. In the beginning

(Middle Ages), it was focused on timber volume and forest planning. With time demands for products and services changed, shifting their focus [1].

Forest inventories are based on a sampling design for a given threshold error, with ground plot assessment, providing data sets, which enable the forest evaluation. The need for measuring an increasing number of variables to evaluate products and services expected from forests turns them increasingly expensive and labour intensive. Remote sensing enabling the evaluation of some of those variables (*e.g.* areas, stand composition, crown cover) has been included in forests inventories. Though it cannot replace field work, it can rationalise it, allowing sample efficiency and reducing costs, error and labour intensity [1].

Biomass, both above and below ground, estimation, distribution and dynamics have been acquiring an increasing importance, especially in the last couple of decades. It was powered by the use of wood for bioenergy and by the evaluation of carbon stocks, sequestration and losses. The estimation of biomass can be grouped in two main methods: direct and indirect. The *direct method*, thought the most accurate, requires cutting trees, separating their different components (*e.g.* wood, bark, branches and leaves) and determining their dry weight. It is an expensive and labour demanding method, especially in what roots are concerned, which is feasible for small samples and prohibitive for large ones. The *indirect method* is traditionally based on mathematical relations between biomass (dependent variable) and one or a few easy-to-measure tree variables (independent variables). The biomass estimates attained with the former method are used to fit allometric functions at tree level, whose most frequent explanatory variables are the diameter at breast height and/or total height. Up to 2005 for Europe, North America and Australia, many functions were compiled by several authors [2–6]. More recent ones were found [7–12]. This large number of allometric functions stresses their species-specific and site-specific domain, reason why the selection of the best-suited function is of the utmost importance.

The tree-level biomass functions are frequently used to make estimations at plot scale (usually summing the biomass of all stems). Their estimation for a given area is based on the inventory plots (which depend on the sampling design and intensity for a certain error threshold) and an extrapolation method. In the literature, different extrapolation methods are described [1] with accuracy decreasing with the increase in the forest area, the variability of stand composition and structure, topography, soil and climate [13].

The forest inventories have cycles of 5 or 10 years. In between remote sensing can be used to evaluate the forest dynamics, in general, and biomass in particular. On the other hand, advantages can be gained as it can work at different scales and time frames; all area is under evaluation no requiring extrapolation methods; and maps can be produced in a geographical information system (GIS) environment.

A wide range of studies of biomass estimation derived from remote sensing can be found in the literature. In the next subchapter (2), a brief description of the methods and techniques is given. A case study (subchapter 3) will be used to illustrate the development of a methodology which was aimed to be simple and to be used either by researches or by technicians. The challenge was to produce data from remote sensing images for two evergreen oaks (holm and cork oak), native of the Mediterranean basin, and develop accurate above-ground biomass (AGB) allometric functions whose explanatory variable is crown horizontal projection, for pure and mixed stands of both species, in the latter analysing also the influence of the predictive ability of independent variables that differentiate species and composition.

The innovation of this study is the estimation of above-ground biomass considering or not stand composition, which is identified by the processing of high spatial resolution of satellite image data. In earlier studies, AGB estimation functions using satellite images did not consider the composition as independent variable. In the studies in which medium and low spatial resolution images are used, the pixel size frequently does not allow the separation of the species. High spatial resolution satellite images data enable the identification and delimitation of crowns of different tree species. Nonetheless, the studies with these satellite images seem to be focused in pure stands. The case study presented demonstrates that stand composition improves the accuracy of AGB estimation functions.

1.2. Contribution of remote sensing data to biomass estimation

Remote sensing was under a strong development in the last three decades due to the rapid advancement of remote sensing technology, increasing the availability of satellite imagery with different spectral, spatial, radioactive and temporal resolutions. Their sensor technologies enable a wide range of Earth surface monitoring scales [14] of forest areas distribution, species, and physic and biochemical properties [15]. The data can be derived from two sensors: the passive and the active.

In the ***passive sensor***, the optical spectral reflectance of Earth surface is registered. This multi-spectral information is sensitive to vegetation characteristics such as their shadows and textures, tree density, leaf area index (LAI) and crown size [16]. Good correlations are found between the former and the latter, namely with biomass [17] for a wide range of geographical areas [18]. To develop biomass functions, several methods are used. The most frequent is the regression analysis using forest inventory biomass data and satellite generated data [19], such as spectral reflectance, crown diameter and crown horizontal projection [20–24], original bands and/or vegetation indices [25–28]. The green vegetation canopy has different interactions with specific electromagnetic spectrum ranges, especially the red (R) and the near infrared (NIR) regions, due to the leaves photosynthetic activity and thus with LAI and biomass. The middle infrared (MIR) region, because of its ability to normalise the canopy cover variability, plays also an important contribution [15]. High relations are established between spectral reflectance and/or vegetation indices, and canopy structure, LAI or biomass [16] from tropical [29, 30] to Mediterranean evergreen oaks and shrubland [17, 26, 31, 32] landscapes. Vegetation indices, condensing satellite imagery data in a quantitative numeric form, are related to some forest parameter's estimation, such as the number of trees per hectare [33], canopy cover [26, 34] and volume, basal area and biomass [35]. The aforementioned indices show a better sensitivity when compared with the spectral reflectance [36]. Apart from regression, other methods are described in the literature, such as k-nearest neighbour, neural network, regression tree, random forest and support vector machine [29, 37–40].

Passive sensor, satellite imagery data, depends on its spatial resolution to determine directly the working scale and can be divided into three groups: low, medium and high.

The ***low spatial resolution*** satellite imagery data are characterised by scenes covering large geographical areas. The first studies of biomass estimation at local, regional and global scales used the images from the National Oceanic and Atmosphere Administration (NOAA) with the Advanced Very High Resolution Radiometer (AVHRR) sensor, with 1.1 km of spatial resolution; Moderate Resolution Imaging Spectroradiometer (MODIS), with 500 m [23]; and Satellite *Pour l'Observation de la Terre* (SPOT) Vegetation with 1 km of spatial resolution [41]. The advantage that can be pointed out in the biomass estimation with this spatial resolution is that one scene covers large area allowing working at country and global scales. Conversely, the pixel size makes the image classification more difficult due to the different land cover types encountered in it, resulting frequently in low accuracy biomass estimations [42]. To surpass this disadvantage in some studies, spectral mixture analysis was employed, where each pixel is a physical mixture of the multiple components weighted by the dominant area and the mixture spectrum is a linear combination of the endmember reflectance spectra [43].

The ***medium spatial resolution*** (30 m) satellite imagery data more frequently used result from the processing of Landsat Thematic Mapper (TM), Enhanced Thematic Mapper Plus (ETM +), Advanced Spaceborne Thermal Emission and Reflection radiometer (ASTER) and Wide Field Sensors (WiFS) [23, 25, 44, 45]. The data set is usually combined with forest inventory sample plots data and low spatial resolution imagery. Examples of the first are the volume and biomass estimation of conifer species of boreal forest in northern Europe, with Landsat images and NOAA-AVHRR [46] and Landsat and WiFS [25]; example of the second is AGB estimation for local and regional scales [25, 39, 46].

The data of ***high spatial resolution*** (IKONOS, QuickBird, WordView 2) have had, more recently, a high availability increase. Improved accuracy in biomass estimation is reached when compared with the former two spatial resolutions. The main disadvantage derives from their spatial resolution, which makes the data processing more time-consuming, thus better suited for local or regional scales [27]. In the literature, some references were found for these data sets, but few for biomass estimation, most of which with functions developed with regression analysis methods. Examples with IKONOS image data are the estimation of wet and dry biomass of oil palm plantations in Africa, considering vegetation indices and individual reflectance bands [47] and allometric equations of diameter at breast height and AGB as a function of crown width in tropical forests [48]. Examples with QuickBird are a methodology to estimate and map AGB for black spruce stands in Canada with shadow fraction as explanatory variable [49], AGB functions whose independent variable is crown horizontal projection, applying multi-resolution segmentation and object-oriented classification methods, with regression analysis for *Eucalyptus* sp. and *Tectona grandis* L. [50] and *Quercus rotundifolia* [51].

The three referred spatial resolutions of the optical remote sensing imagery for the same region and their image processing techniques allow the monitoring of AGB with different degrees of detail which are able to decrease the processing time and costs [19, 27, 52].

The ***active sensor*** is based on a mechanism that transmits and receives a portion of energy (microwaves region), interacting with Earth surface, and whose relevance to modelling forest variables depends on how deep the signal penetrates in the canopy. The mostly used data are derived from the Space-borne synthetic Aperture Radar (SAR) and L-band ALOS PALSAR. Two groups of data can be defined: RAdio Detection And Ranging (RADAR) and Light Detection and Ranging (LiDAR). The ***RADAR*** data's main advantage is their independence of atmospheric conditions as wavelength of radiation is much larger that the atmospheric particles [53], reason why it has acquired a marked importance in regions with frequent cloud cover [15]. Several examples of AGB estimation with RADAR were found in the literature [54–59]. The ***LiDAR*** is mainly obtained by airborne discrete-return [60], airborne waveform and from ground-based [61] LiDAR. Several authors developed nonlinear functions for AGB with its derived variables [15, 62] for individual trees, plot and stands, depending on the type of LiDAR system used [16]. LiDAR derives forest structural information (canopy cover), and RADAR backscatter seems to be able to estimate AGB with good accuracy [63].

In summary, the estimation of forest AGB with remote sensing data has advantages and disadvantages. To obtain very good accuracy, these approaches need a heavy field work in order to attain training data sets. However, many studies present methodologies with very good results at local, regional and global scale for AGB estimation. There is a trend towards the combination of several types of remote sensing imagery data to generate vegetation parameters and their relation with forest biomass.

2. Case study

2.1. Introduction

The data resulting from passive sensors, with low and medium spatial resolution, are usually used for the country or global scales. Due to their pixel size, it is not possible to identify and delimitate the tree crowns. Conversely, high spatial resolution satellite images enable it with good accuracy. The active sensors processing is more difficult than the passive sensors. Thus, when developing a methodology to be used both by researches and by technicians, the passive sensors data processing is more straightforward than that of the active sensors. The objectives of this study are the development of allometric functions for the estimation of AGB with crown horizontal projection, obtained with high spatial resolution satellite image, as independent variable with linear regression for (i) cork oak pure stands, (ii) mixed cork oak and holm oak stands (from this point forward referred as evergreen oaks) and (iii) both pure and mixed stands (from this point forward referred as all). Though with similar stand parameters there are some differences between pure stands of cork oak or holm oak and mixed stands of both species. Thus it was considered that the model for both species might be improved with the inclusion of dummy variables for species and composition as independent variables.

Cork oak (*Quercus suber* L.) and holm oak (*Quercus rotundifolia* Lam.) are native of the Mediterranean basin, distributed from Portugal to Greece and Portugal to Syria [64]. These species occur usually in low-density stands in a silvopastoral system, called *montado*, with open heterogeneous canopies, where the main products are bark and fruit, associated with extensive grazing, both in pure and in mixed stands. According to the Portuguese National Forest Inventory, cork oak and holm oak stands account for 22.5 and 13.7% of the forest area, respectively. The biomass for pure cork oak stands is 49.5 t/ha with an error of 5.2% and for holm oak 23.2 t/ha with an error of 7.2%. Mixed stands with cork oak or holm oak as dominant species correspond to 17% of their total area. The biomass estimation for mixed stands is 35.6 t/ha with an error of 13.7% [65]. They are similar in terms of habit, especially due to pruning to enhance fruit production and growth; both are slow growing [64].

2.2. Materials and methods

2.2.1. Study area

The study area is in southern Portugal, region of Mora, with approximately 80 km^2 (**Figure 1**). The area is mainly occupied by forest, composed predominantly by cork and holm oak in both pure and mixed stands. This region has a Mediterranean climate with a hot and dry summer and rainy winter with lower temperature. The terrain is characterised by small variations, with a mean elevation of about 200 m. The used QuickBird satellite image (August 2006) has a spatial resolution of 0.70 m, resulting from a fusion of panchromatic band with the four multi-spectral bands, blue (B), green (G), red (R) and near infrared (NIR).

2.2.2. Processing satellite image

The image was geometric and radiometric corrected using ENVI4.8 [66]. The geometric correction was based on ground control points obtained with Global Navigation Satellite System (GNSS) and geodetic vertices, identified on the ground and in the images, with a Root Mean Square Error (RMSE) of 0.49 m. The pixel values that are the digital numbers of the image were converted to top-of-atmosphere reflectance and finally to soil reflectance, using the dark object subtraction method [67]. The Normalized Difference Vegetation Index (NDVI) [68] was calculated with the latter and used to generate a vegetation mask, isolating tree crowns from other land cover types. For this propose, the multi-resolution segmentation method with the contrast split segmentation algorithm on eCognition, version 8.0.1 was applied [69]. Over this vegetation mask, forest species were identified using the nearest neighbour classification method, producing a map with the delimitation of trees crowns identified by specie [70].

The area was divided into a square grid of 45.5 m× 45.5 m (2070.25 m^2). Grids were classified according to the species present. The crown cover obtained with satellite data (CC_s) was defined as the percentage of area occupied by the crown horizontal projection per grid (CHP_s) in relation to the grid area. It was calculated with the vegetation mask per species using a geographic information system, ArcGIS, version 10 [71]. A grid was considered pure when crown cover of one species was equal or larger than 75% and mixed otherwise. The design of the forest inventory was a random stratified sampling by proportional allocation. Three strata were defined as function of crown cover: (i) 10–30%, (ii) 30–50% and (iii) >50%.

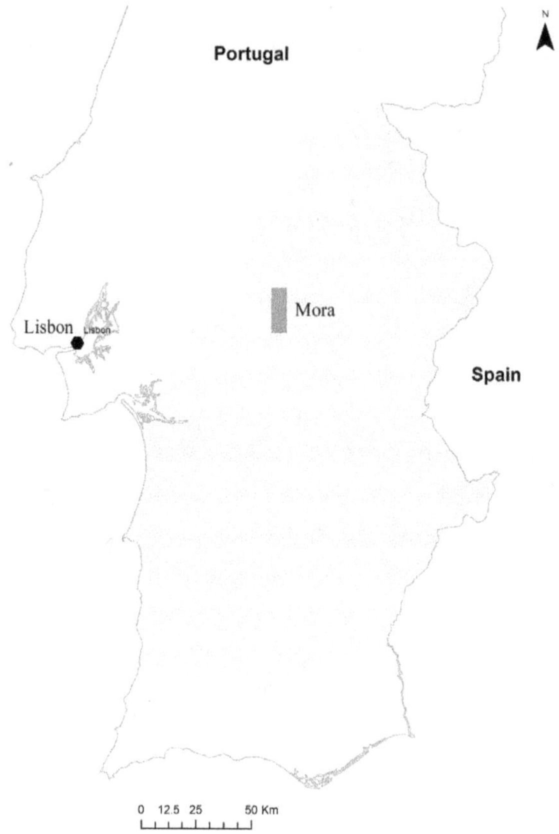

Figure 1. Location of the study area (dark grey) and Alentejo region (light grey).

2.2.3. Forest inventory data and biomass estimation

Forest inventory data set is composed of 17 plots of pure holm oak, 11 plots of pure cork oak and 23 mixed plots of holm oak and cork oak, with a total sampled area of 10.6 ha. In these plots, for all individuals with a breast height diameter ≥6 cm, the diameter at breast height, total height and crown radii in four directions (North, South, East and West) were measured [72]. Each tree geographical location was recorded by GNSS. Tree crown horizontal projection (CHP_p) was calculated as the circle whose radius is the arithmetic mean of the four crown radii measured. The relation (in %) of sum of the tree crown horizontal projections to plot area was used to estimate plot crown cover (CC_p). Ref. [8] developed a biomass allometric function per tree with simultaneous fitting for cork and holm oak. As the stands referred by the aforementioned authors were similar to those of this study, their functions were used to estimate above-ground biomass (*AGB*, Eq. (1)) as the sum of wood (*ww*), bark (*wb*) and crown (*wc*) biomass. Above-ground biomass per plot (AGB_p) was defined as the sum of *AGB* of all the trees in the plot.

$$AGB_t = ww + wb + wc = 0.164185 \times dbh^{2.011002} + 0.600169 \times dbh^{1.355957} + 1.909152 \times dbh^{1.200354} \quad (1)$$

2.2.4. Statistical analysis

Statistical analysis included correlation analysis and linear and multiple regression, implemented in R statistical software [73]. As AGB_p, CHP_p and CHP_s did not have normal distribution (evaluated with Shapiro-Wilk normality test), the Spearman correlation test was used. The linear function was selected as it is especially suited for one layer or without crown closure stands or forests [26]. As it was assumed that null *CHP* corresponded to null *AGB*, the linear regression through the origin (Eq. (2), where β is the slope) was used. The ordinary least square linear method was selected to fit the functions, both for the individual and for their cumulative values (**Figure 2.**) with CHP_s as independent variable. Also, in order to understand whether AGB could be estimated with aforementioned models, a preliminary analysis was carried out for plot inventory data (CHP_{pi}) with the same methodology for cork oak, evergreen oaks and all plots. The preliminary analysis for holm oak was done by [51], and the data were included in the evaluation for all plots in this study.

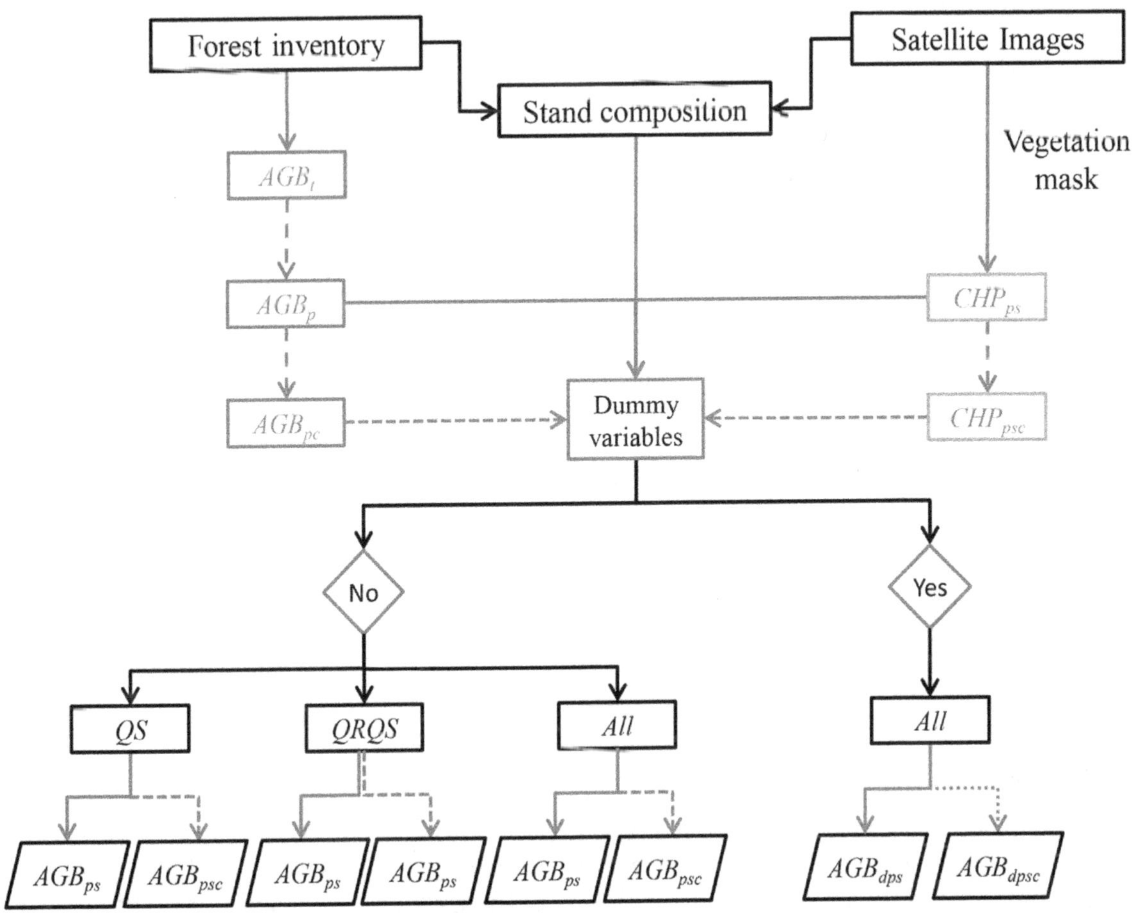

Figure 2. Flow diagram of the models developed (where AGB is the above-ground biomass in kg, CHP is the crown horizontal projection in m^2, t is the tree, p is the plot values, i is the inventory data, c is the cumulative values, s is satellite image data and d is dummy variables).

As the goal was to develop functions able to estimate AGB as function of *CHP* for the two evergreen oaks, in pure and mixed stands, dummy variables were defined to test the species composition contribution in the function predictive ability (**Figure 2**). Three dummy variables were defined. For each dummy variable, 1 indicates that the plot is pure holm oak (*dQR*), pure cork oak (*dQS*) or mixed holm oak and cork oak (*dQRQS*) and 0 otherwise. Multiple linear regression with stepwise method, using Akaike information criterion (AIC) as selection criteria, was used to fit the allometric functions (Eq. (3), where β_i are the regression coefficients, i = 0, 1, 2, 3). Multicollinearity among explanatory variables was analysed with variance inflation factor (VIF), considering that values exceeding 10 are sign of serious multicollinearity [74, 75].

As suggested by several authors [76, 77], the models were studied by the sum of squares of the residuals (SQR), the coefficient of determination (R^2) and the adjusted coefficient of determination (R^2_{aj}). Validation tests entail an independent data set. To overtake the inexistence of an independent data set, Refs. [78–80] suggest using predicted residual error. The sum of its square values, PRESS (Eq. (4)), and the sum of its absolute values, APRESS (Eq. (5)), were used as the validation test. The closer to the null value of residuals, the better is the model. The per cent value of the estimated and calculated AGB defined the error.

$$AGB = \beta \times CHP \tag{2}$$

$$AGB = \beta_0 \times CHP + \beta_1 \times dQS + \beta_2 \times dQR + \beta_3 \times dQRQS \tag{3}$$

$$PRESS = \sum_{i=1}^{n} \left(y_i - \hat{y}_{i,-i} \right)^2 \tag{4}$$

$$APRESS = \sum_{i=1}^{n} \left| y_i - \hat{y}_{i,-i} \right|^2 \tag{5}$$

2.3. Results and discussion

2.3.1. Multi-resolution segmentation and object-oriented classification

Figure 3 presents the vegetation mask resulting from the multi-resolution segmentation and object-oriented classification process (yellow line), over the QuickBird image with false colour composite for an area of Mora. The vegetation index NDVI enables a spectral signature sufficiently different to obtain a good distinction of the two forest species (cork and holm oak), with an agreement between the classification and ground truth using Kappa statistic [81, 82], of 78% and a global precision of 89%. Thus, the simultaneous analysis of these two evergreen oak species in pure and mixed stands can be useful for the estimation of AGB and also because, though they are visually similar, the methodology allows a mapping per species (cork and holm oak) with a good classification accuracy.

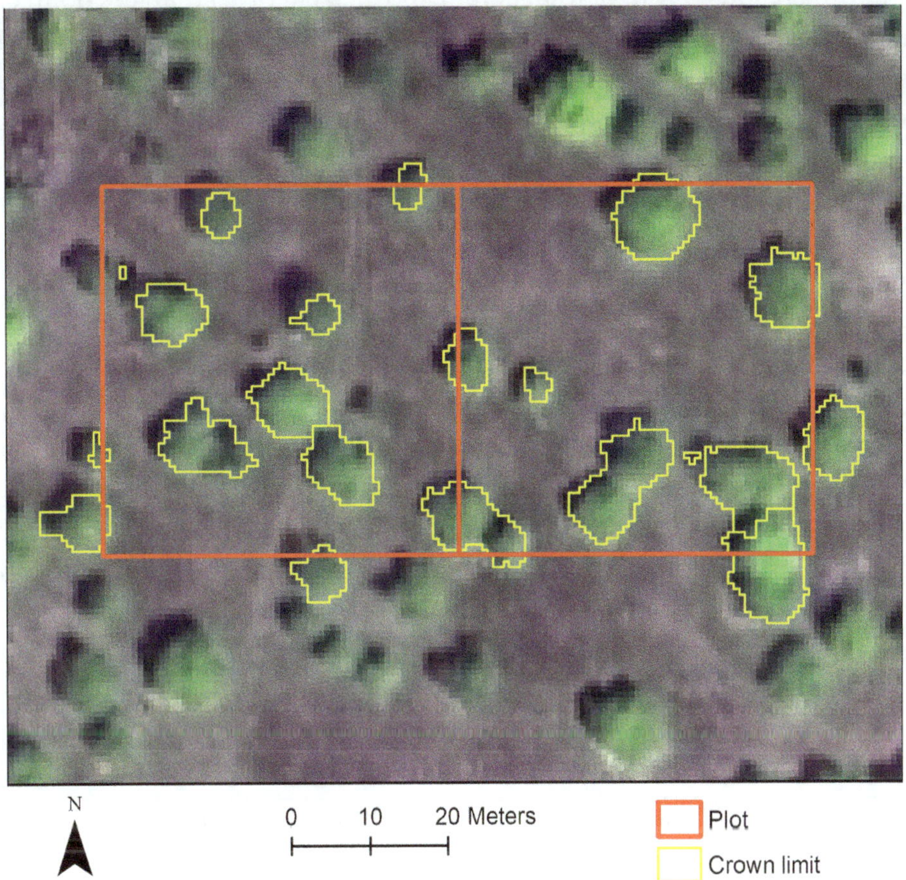

Figure 3. QuickBird image with false colour composite (RGB = red, NIR and blue) and illustration of the results of multi-resolution segmentation process (yellow line).

2.3.2. Individual trees and plot characteristics and above-ground biomass functions with inventory data

The descriptive statistics: minimum (*min*), maximum (*max*), arithmetic mean (*mean*), standard variation (*SD*) and coefficient of variation (*CV*) for all plots (*All*), evergreen oaks mixed plots (*QRQS*) and cork pure plots (*QS*) of the data are presented in **Table 1**. The density measures, number of trees per hectare (*N*), basal area per hectare (*G*), crown cover (CC_s) and above-ground biomass per hectare (*AGB*) were considered for the inventory data, while crown horizontal projection per plot (CHP_{ps}) and crown cover (CC_s) were considered for satellite data. In **Table 1**, the differences between the three groups of plots can be observed. Comparing the descriptive statistics of cork oak pure plots with those of evergreen oak plots (**Table 1**), higher densities and *AGB* are observed for the former. A more detailed analysis at tree level reveals that in the mixed plots about 54% of the trees are holm oak and 46% are cork oak. The differences are, at least partially, explained by the individual tree dimensions. While in the mixed plots mean crown radius is equal for both species (3.1 m) and similar to the cork oak pure plots (3.0 m), the mean total height and the mean quadratic diameter (d_g) are different:

6.8 m, 7.1 m, 35.8 cm and 37.6 cm, respectively. This is also reflected in the AGB_t with a mean larger for the cork oak pure plots (463.9 $kgtree^{-1}$) than for the mixed plots (424.0 $kgtree^{-1}$). When the two species are analysed in the evergreen oak plots, the cork oak individuals have, in average, larger dimensions (7.1 m of total height and 38.4 cm of d_g) than the holm oak ones (6.3 and 33.6 cm), which is also reflected in the mean AGB_t (481.7 and 382.3 $kgtree^{-1}$, respectively). The diameter distribution of the evergreen oak plots covers a wide spam indicative of stands of several cohorts (or age classes). The species spatial distribution visual analysis point outs towards an individual distribution both by species and by cohort. This might be probably due to the mutualistic and competitive interactions [83] and also by the sociability and compatibility [84] of these oaks, with similar growth rates, both slow, and rotation length. The holm oak pure plots [51] have values between the former two compositions. Also, trees with similar crown sizes may have quite different AGB_t. This is expectable considering two factors. First, these species have weak epinastic control, but when in group tend to reduce their lateral development due to branch abrasion or to the effects of high and low shading [83], especially because they are shade intolerant. As in *montado* tree spatial distribution is sparse, though irregular, trees can be found isolated or in clusters and in the latter is the competition phenomenon that explains, at least partially, the variability in crown dimensions. Second, in the *montado* system pruning to enhance fruit production is frequent [64] in mature stands and this reverses temporarily the rate between CHP_t and AGB_t.

	All plots (*All*)					**Evergreen oaks mixed plots (*QRQS*)**					**Cork oak pure plots (*QS*)**				
Variables	**min**	**max**	**mean**	**SD**	**CV**	**min**	**max**	**mean**	**SD**	**CV**	**min**	**max**	**mean**	**SD**	**CV**
N ($treesha^{-1}$)	19	140	70	28	40.5	19	140	61	28	46.5	53	135	88	24	27.7
G (m^2ha^{-1})	2.6	15.4	7.1	2.7	38.1	2.6	13.1	6.2	2.1	34.3	5.2	15.4	9.8	3.0	31.2
CCp (%)	8.7	49.3	25.3	9.2	36.5	8.7	49.3	20.9	8.2	39.3	18.7	42.3	27.5	7.4	26.8
CCs (%)	13.7	70.5	35.0	14.7	41.9	14.3	58.8	28.5	11.2	39.2	23.6	70.5	44.9	13.5	30.0
CHP_{ps} (m^2)	284.2	1460.2	725.3	303.7	41.9	296.9	1216.7	589.3	230.9	39.2	489.0	1460.2	928.7	278.7	30.0
AGB (tha^{-1})	10.4	62.7	30.1	11.1	36.8	10.4	51.5	26.0	8.6	33.1	23.8	62.7	40.7	11.9	29.2

N: number of trees per hectare, G: basal area per hectare, CC_s: crown cover calculated with satellite data, CHP_{ps}: plot crown horizontal projection evaluated with satellite data, AGB: above-ground biomass per hectare, min: minimum, max: maximum, mean: arithmetic mean, SD: standard variation, CV: coefficient of variation.

Table 1. Descriptive statistics.

The strongest positive correlations are found for pure plots of cork oak between AGB_p and CHP_p (0.827), expected due to the smaller variability (**Table 1**). Strong positive correction for pure cork oak plots between AGB_p and CHP_s was attained (0.627). The difference between AGB_p and CHP_s and CHP_p can be, at least partially, explained by the stems spatial distribution, the shade casted by trees and understory. In tree clusters, multi-resolution segmentation process is not able to isolate tree crowns, including mixed pixels of vegetation, soil and/or shadow inside the polygon that delimitates the cluster. Some types of understory vegetation, near tree crowns, have similar spectral signature, and when delimiting the crown those pixels are included [85, 86]. The weakest correlations were observed for the evergreen oak plots between AGB_p and CHP_p (0.482) and CHP_s (0.552), which is expected due to the variability increase in those parameters. When all the plots are considered, strong positive correlations are found between AGB_p and CHP_p (0.671) and CHP_s (0.746).

Model	Plots	Equation	SQR	R^2	R^2_{aj}	PRESS	APRESS
		Inventory data					
1	QS	$AGB_{pi\_QS} = 14.7087 \times CHP_{pi\_QS}$	23576825	0.972	0.969	0.00000019	0.00101252
2	QRQS	$AGB_{pi\_QRQS} = 11.5889 \times CHP_{pi\_QRQS}$	74401387	0.899	0.894	0.00000065	0.00313966
3	All	$AGB_{pi\_All} = 11.4139 \times CHP_{pi\_All}$	187602462	0.916	0.915	0.00000113	0.00631091
4	QS	$AGB_{pic\_QS} = 14.8889 \times CHP_{pic\_QS}$	16220130	0.9995	0.999	0.00000002	0.00034940
5	QRQS	$AGB_{pic\_QRQS} = 11.75374 \times CHP_{pic\_QRQS}$	292087434	0.997	0.997	0.00000004	0.00049897
6	All	$AGB_{pic\_All} = 11.75374 \times CHP_{pic\_All}$	3066829502	0.998	0.998	0.00000001	0.00049770
		Satellite image data					
7	QS	$AGB_{ps\_QS} = 8.847 \times CHP_{ps\_QS}$	40091932	0.953	0.948	0.00000016	0.00106716
8	QRQS	$AGB_{ps\_QRQS} = 8.5093 \times CHP_{ps\_QRQS}$	72931734	0.901	0.896	0.00000061	0.00317184
9	All	$AGB_{ps\_All} = 8.1309 \times CHP_{ps\_All}$	163946315	0.927	0.925	0.00000120	0.00656152
10	QS	$AGB_{psc\_QS} = 9.2671 \times CHP_{psc\_QS}$	43379144	0.999	0.998	0.00000001	0.00025905
11	QRQS	$AGB_{psc\_QRQS} = 9.037 \times CHP_{psc\_QRQS}$	104841889	0.999	0.999	0.00000005	0.00068011
12	All	$AGB_{psc\_All} = 8.85869 \times CHP_{psc\_All}$	1722352935	0.999	0.999	0.00000001	0.00050963
		Satellite image data with dummy variables					
13	All	$AGB_{dps} = 4.836 \times CHP_{ps} + 2191.3997 \times dQR + 3944.2222 \times dQS + 2532.8074 \times dQSQR$	104975100	0.953	0.949	0.00000151	0.00661458
14	All	$AGB_{dpsc} = 8.6029 \times CHP_{psc} + 6425.9261 \times dQR + 6105.1531 \times dQS + 4827.1112 \times dQSQR$	1235713930	0.991	0.999	0.00000007	0.00070861

AGB: above-ground biomass in kg, CHP: crown horizontal projection in m², p: plot, i: inventory data, c: cumulative values, s: satellite data, QS: cork oak, QSQR: evergreen oaks, All: all plots, d: dummy variable.

Table 2. Properties of the fitted models.

The statistical properties and the PRESS and APRESS statistics of the linear models for plot CHP_{pi} (models 1–3) and CHP_{pic} (models 4–6) are indicative of their good performance (**Table 2**).

When comparing the models for the four sets of plots, the statistical properties of the pure plots are better than those of the evergreen oak plots and all plots (**Table 2**), due to the increase in variability in the latter two. The results attained by [51] for holm oak pure plots are also better when compared with the evergreen oak and all plots of this study. Nonetheless, smaller differences in model performance are found for models 4–6 when compared with models 1–3. The differences might be partially due to the small differences in habit of the two species, the spatial distribution pattern of the individuals of both species and silvicultural practices, especially pruning where focus is put in the crown enlargement more for holm oak than for cork oak. This is also reflected in the overall model error. Larger overall errors are found for the models 2 and 3 (24.3% and 23.7%, respectively) when compared with model 1 (13.6%), and models 5 and 6 (9.8% and 9.2%, respectively) when compared with model 4 (4.4%).

2.3.3. Above-ground biomass allometric functions with satellite image data

The statistical properties and the validation statistics of the linear models for CHP_{ps} and CHP_{psc} for pure cork oak plots (models 7 and 10), for evergreen oak plots (models 8 and 11) and for all plots (models 9 and 12, **Figure 4a**, **b**) are indicative of their good performance (**Table 2**). These results are in accordance with [26] that states that linear regression behaves well in stands with one layer, while [47] refers that in multi-layered stands nonlinear functions are better suited. Noteworthy is that models 10–12 have better statistical properties than the models 7–9. When comparing the models for evergreen oaks with those for pure plots, the statistical properties of the former are worse than of latter (**Table 2**). This, as already referred, is due to the larger variability observed in the mixed plots. When comparing the models fitted with inventory and satellite image data, the performance is better for the former, which can be partially explained by the differences in the calculation of *CHP* with the sets of data. Models for cumulative values have slightly better performance for the inventory (models 4–6) than for the satellite image data (models 10–12), corroborated by the overall error of 3.7, 4.4, 9.8 and 9.2%, and 13.4, 4.8, 6.4 and 5.8%, respectively.

Model 10, for cork oak pure stands, has larger errors up to 4000 m^2 and an irregular trend afterwards. For error thresholds of 10 and 5% *CHP*, areas larger than 4000 and 5000 m^2, respectively, are needed. Considering a mean *CHP* area of 4486 m^2ha^{-1}, the corresponding stand area is 0.9 and 1.1 ha, respectively. Model 11, for evergreen oak stands, shows higher errors than the former models for *CHP* areas up to 1000 m^2, with a continuous decrease, in absolute value, up to 3000 m^2 and stabilisation afterwards. Errors smaller or equal than 10 and 5% are attained for *CHP* areas of 3000 and 7000 m^2, respectively. The mean *CHP* area for these stands is 2847 m^2ha^{-1}, which corresponds to the above-referred errors to stand areas of 1.1 and 2.5 ha, respectively. Model 12, for both pure and mixed stands, shows the same error trend as model 11. Errors below the aforementioned are attained for *CHP* areas of 2000 and 8000 m^2. As the mean *CHP* area is 3503 m^2ha^{-1}, the corresponding stand areas are 0.6 and 2.3 ha. As 92% of the holm oak stands and 89% of the cork oak stands have areas bigger than 2 ha [65], this allometric function can be used at local and regional level.

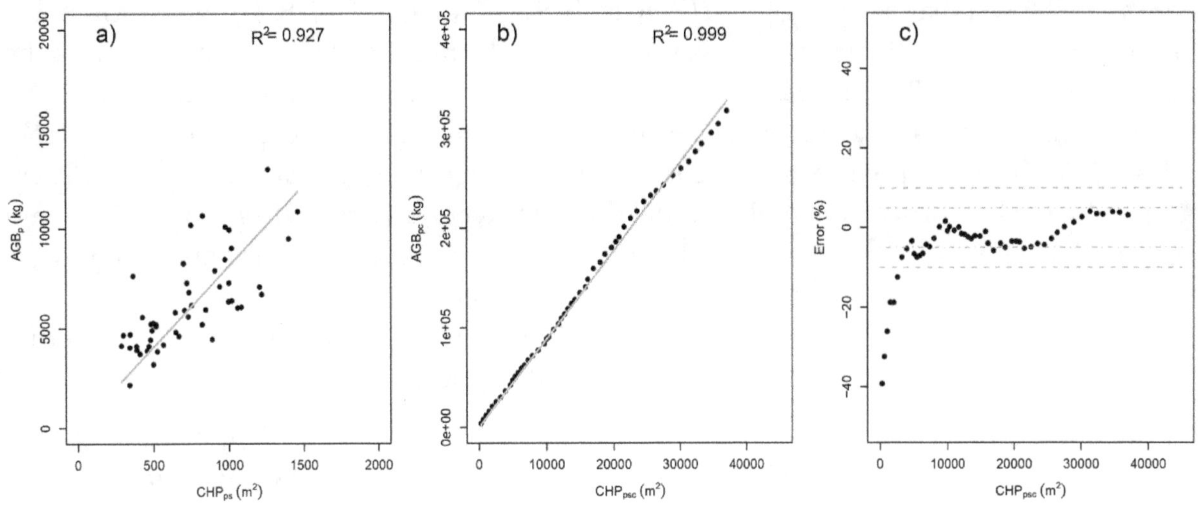

Figure 4. Crown horizontal projection derived from satellite image per plot (CHP_{ps}) versus above-ground biomass per plot (AGB_p) and equation (in grey) for model 9 (a) and their cumulative values, CHP_{psc} vs AGB_{pc}, and equation (in grey) (b) and error (c) for model 12.

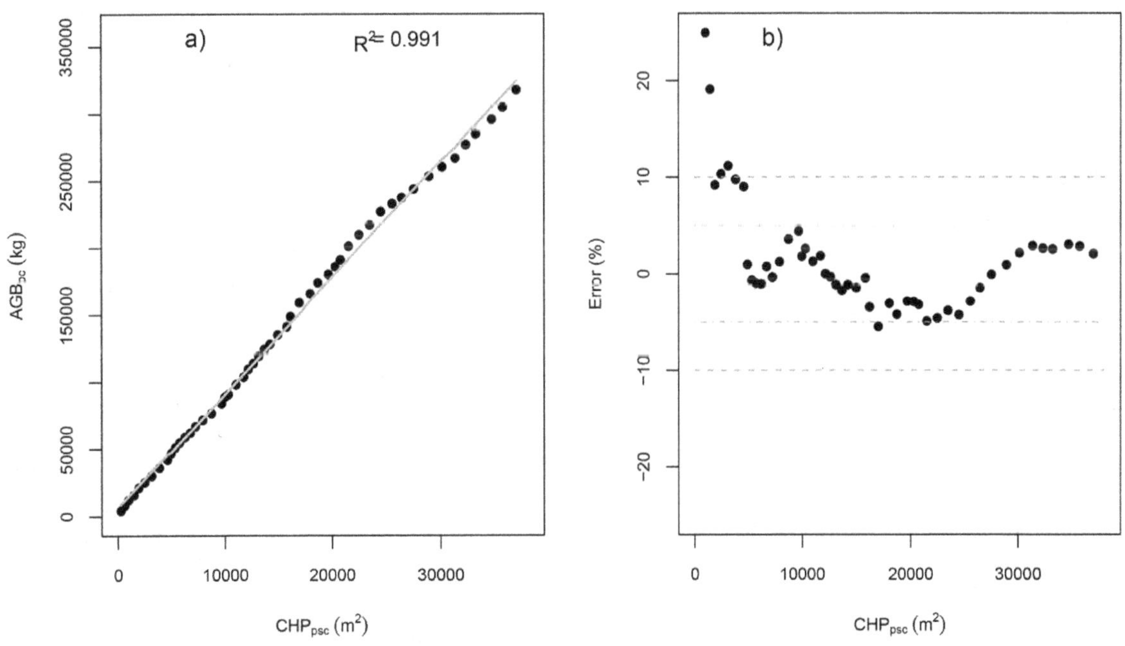

Figure 5. Cumulative values of crown horizontal projection derived from satellite image per plot (CHP_{psc}) versus cumulative values of above-ground biomass per plot (AGB_p) and equation (in grey, a), and error distribution (b) for model 14.

The inclusion of independent variables identifying the plot composition originated a generalised improvement in the statistical properties of the models (**Table 2**) for CHP_{ps} (model 13) and CHP_{psc} (model 14), when compared with those without composition variables (models 9 and 12). In fact, plot composition has revealed important as for both models the three dummy variables were included. Also, multicollinearity did not present any problem as VIF values were always lower than 10, being for *CHP*, *dQR*, *dQS* and *dQRQS*, 8.5, 3.8, 3.6 and 3.2 for model 13, and 4.6, 1. 8, 3.0 and 1.8 for model 14, respectively. The models accuracy increases from the models with only *CHP* (9 and 12) to those that include the dummy variables (models 13 and 14). Noteworthy is also the decrease in SQR of 36% from models 9 to 13, and of 39% from models 12 to 14, with the corresponding increase in the adjusted regression coefficients. There is a considerable improvement in the overall error from models 9 to 13, from 23.1 to 5.2% and from models 12 to 14, from 5.8 to 0.1%. The error for model 14 shows an irregular trend with error smaller than 10 and 5% for cumulative crown projection for areas larger than 1500 and 2000 m^2, corresponding to stand areas of 0.4 ha and 0.6 ha, respectively (**Figure 5**). Considering the threshold of 5% for model 14, -1.7 ha is needed than for model 12.

All the fitted models (models 7–14) have a high goodness of fit, with more than 90% of the variability explained. It seems that crown horizontal projection as better predictive abilities with linear functions than the shadow fraction processed from QuickBird images [49]. The models developed in this study show also better performances than multiple regression functions with crown diameter and total height as independent variables, derived from a panchromatic band of QuickBird satellite [50] or than exponential functions with reflectance of band 3 or NDVI for IKONOS image [47]. Likewise, biomass estimation functions derived from medium spatial resolution satellite image data do not show as good performance as those obtained in this study. This is probably related to dissimilarity between ground and satellite data [27]. The former was confirmed in a study with Landsat 5 TM and MODIS images for *Pinus pinaster* stands, whose independent variable is vegetation indices [17].

2.4. Conclusions

Remote sensing gives a good contribution to estimate the above-ground biomass. The high spatial resolution satellite image allows mapping with high accuracy the above-ground biomass at local and regional scale.

The overestimation of crown horizontal projection per plot using high spatial resolution satellite image, when compared with that calculated using inventory data, is related to the inclusion of mixed pixels in the boundary of tree crown delimitation in the multi-resolution segmentation and trees' spatial distribution. Though the cork and holm oak are visually similar, their spectral signature is sufficiently different to obtain a very good classification by forest specie.

In general, above-ground biomass allometric functions for plot and their cumulative values showed a good performance, though better for pure stands than for mixed stands, which might be explained by the larger variability observed in the latter. However, the inclusion of dummy variables, which reflect the differences between the species and the stand structure, originates a generalised improvement in the functions performance. The same threshold errors of 10 and 5% are attained by the latter with less 33 and 70% of the stand areas. Considering that to a 10% error correspond stand areas of 1.1 ha, it includes about 90% of the area occupied by these species in Portugal [65]. This method has the advantage of enabling the overall area evaluation, not requiring forest inventory or extrapolation procedures. It can be used when species composition is not (models 9 and 12) and is (models 13 and 14) differentiated.

Acknowledgements

The authors would like to thank the forest producers for allowing plot installation and to Paulo Mesquita for satellite image processing. The work was financed by Programa Operativo de Cooperação Transfronteiriço Espanha—Portugal (POCTEP); project Altercexa—Medidas de Adaptación y Mitigación del Cambio Climático a Través del Impulso de las Energías Alternativas en Centro, Alentejo y Extremadura (Ref_a 0317_Altercexa_I_4_E and 0406_ALTERCEXA_II_4_E); TrustEE—innovative market based Trust for Energy Efficiency investments in industry (Project ID: H2020—696140); and National Funds through FCT—Foundation for Science and Technology under the Project UID/AGR/00115/2013.

Abbreviations

AGB	Above-Ground Biomass per tree (kg)
AGB_t	tree Above-Ground Biomass per tree (kg)
AIC	Akaike Information Criterion
All	all plots
ALOS	Advanced Land Observing Satellite
APRESS	Absolute Predicted Residual Error Sum
ASTER	Advanced Spaceborne Thermal Emission and Reflection Radiometer
AVHRR	Advanced Very High Resolution Radiometer
c	cumulative values
CC	Crown Cover (%)
CHP	Crown Horizontal Projection (m^2)
CV	Coefficient of Variation
d	dummy variable
dbh	diameter at breast height, measured at 1.30 m height (cm)
dg	mean quadratic diameter (cm)
G	basal area per hectare (m^2ha^{-1})
GNSS	Global Navigation Satellite System
i	inventory data
ETM +	Enhanced Thematic Mapper +
LAI	Leaf Area Index
max	maximum value
mean	arithmetic mean value
min	minimum value
MODIS	Moderate Resolution Imaging Spectroradiometer
N	number of trees per hectare (tree ha^{-1})
NDVI	Normalized Difference Vegetation Index

NOAA	National Oceanic and Atmospheric Administration
p	plot
PALSAR	Phased Array type L-band Synthetic Aperture Radar
PRESS	Predicted Residual Error Sum of Squares
QR	holm oak (*Quercus rotundifolia*) pure plots
QRQS	mixed plots of holm and cork oaks (evergreen oaks)
QS	cork oak (*Quercus suber*) pure plots
R^2	coefficient of determination
R^2_{aj}	adjusted coefficient of determination
RADAR	RAdio Detection And Ranging
RMSE	Root Mean Square Error
s	satellite data images
SAR	Space-borne synthetic Aperture Radar
SD	Standard Deviation
SQR	Sum of Squares of the Residuals
t	tree
TM	Thematic Mapper
VIF	Variance Inflation Factor
WiFS	Wide Field Sensor
ww	wood biomass (kg)
wb	bark biomass (kg)
wc	crown biomass (kg)

Author details

Adélia M. O. Sousa*, Ana Cristina Gonçalves and José R. Marques da Silva

*Address all correspondence to: asousa@uevora.pt

Department of Rural Engineering, School of Sciences and Technology, Institute of Mediterranean Agricultural and Environmental Sciences (ICAAM), Institute of Research and Advanced Information (IIFA), University of Évora, Évora, Portugal

References

[1] McRoberts R, Tomppo E, Naesset E. Advanced and emerging issues on national forest inventories. Scandinavian Journal of Forest Research. 2010;25:368–381.

[2] Ter-Mikaelian MT, Korzukhin MD. Biomass equations for sixty-five North American tree species. Forest Ecology and Management. 1997;97:1–24.

[3] Eamus D, McGuinness K, William B. editors. Review of Allometric Relationships for Estimating Woody Biomass for Queensland, the Northern Territory and Western Australia, Technical report no. 5a. Australia: Australian Greenhouse Office. 2000. 56p.

[4] Keith H, Barrett D, Keenan R. editors. Review of Allometric Relationships for estimating Woody Biomass for New South Wales, the Australian Capital Territory, Victoria, Tasmania and South Australia, Technical report no. 5b. Australia: Australian Greenhouse Office. 2000. 112p.

[5] Jenkins JC, Chojnacky DC, Heath LS, Birdsey RA. National-scale biomass estimators for United States tree species. Forest Science. 2003;49(1):12–35.

[6] Zianis D, Muukkonen P, Mäkipää R, Mencuccini M. Biomass and Stem Volume equations for tree species in Europe. The Finnish Society of Forest Science, The Finnish Forest Research Institute. Tampere, Finland; 2005. 63p.

[7] Fehrmann L, Kleinn C. General considerations about the use of allometric equations for biomass estimation on the example of Norway spruce in central Europe. Forest Ecology and Management. 2006;236:412–421.

[8] Paulo JA, Tomé M. Equações para estimação do volume e biomassa de duas espécies de carvalhos: *Quercus suber* e *Quercus ilex*. [Equations for estimation of the volume and the biomass of two oak species: *Quercus suber* and *Quercus ilex*]. Publicações do GIMREF; RC1. Instituto Superior de Agronomia, Departamento de Engenharia Florestal. Lisboa. 2006.

[9] Miksys V, Varnagiryte-Kabasinskiene I, Stupak I, Armolaitis K, Kukkola M, Wójcik J. Above-ground biomass functions for Scots pine in Lithuania. Biomass and Bioenergy. 2007;31:685–692.

[10] Correia AC, Faias S, Tomé M, Evangelista M, Freire J, Ochoa P. Ajustamento simultâneo de equações de biomassa de pinheiro manso no Sul de Portugal. [Simultaneous fitting of biomass equations for stone pine in southern Portugal]. Silva Lusitana. 2008;16:197–205.

[11] Djomo AN, Ibrahima A, Saborowski J, Gravenhorst G. Allometric equations for biomass estimations in Cameroon and pan moist tropical equations including biomass data from Africa. Forest Ecology and Management. 2010;260: 1873–1885.

[12] Arias D, Calvo-Alvarado J, Richter DB, Dohrenbusch A. Productivity, aboveground biomass, nutrient uptake and carbon content in fast-growing tree plantations of native and introduced species in the Southern Region of Costa Rica. Biomass and Bioenergy. 2011;35:1779–1788.

[13] Somogyi Z, Cienciala E, Mäkipää R, Muukkonrn P, Lehtonen A, Weiss P. Indirect methods of large-scale forest biomass estimation. European Journal of Forest Research. 2007;126:197–207.

[14] Gail WB. Remote sensing in the coming decade: the vision and the reality. Journal of Application of Remote Sensing. 2007;1(1):012505.

[15] Boyd, DS, Danson FM. Satellite remote sensing of forest resources: three decades of research development. Progress in Physical Geography. 2005;29(1):1–26.

[16] Zhang X, Ni-meister W. Remote sensing of forest biomass. In: Hanes J.M. editor. Biophysical Applications of Satellite Remote Sensing. Springer-Verlag: Berlin Heidelberg; 2014. p. 63–98.

[17] Viana H, Aranha J, Lopes D, Cohen WB. Estimation of crown biomass of *Pinus pinaster* stands and shrubland above-ground biomass using forest inventory data, remotely sensed imagery and spatial prediction models. Ecological Modelling. 2012;226:22–35.

[18] Ahmed T, Tian L, Zhang Y, Ting KC. A review of remote sensing methods for biomass feedstock production. Biomass and Bioenergy. 2011;35:2455–2469.

[19] Pizaña JMG, Hernández JMN, Romero NC. Remote sensing-based biomass estimation. In: Marghny M. editor. Earth and Planetary Sciences, Geology and Geophysics, Environmental Applications of Remote Sensing. Croatia: Intech; 2016. p. 3–40.

[20] Woodcock CE, Collins JB, Jakabhazy VD, Li X, Macomber S, Wu Y. Inversion of the Li-Strahler canopy reflectance model for mapping forest structure. IEEE Transactions on Geoscience and Remote Sensing. 1997;35:405–414.

[21] Phua M, Saito H. Estimation of biomass of a mountainous tropical forest using Landsat TM data. Canadian Journal of Remote Sensing. 2003;29:429–440.

[22] Popescu SC, Wynne RH, Nelson, RF. Measuring individual tree crown diameter with LiDAR assessing its influence on estimating forest volume and biomass. Canadian Journal of Remote Sensing. 2003;29:564–577.

[23] Baccini A, Friedl MA, Woodcock CE, Warbington R. Forest biomass estimation over regional scales using multisource data. Geophysical Research Letters. 2004;31:L10501.

[24] Powell, SL, Warren B, Cohen WB, Healey SP, Kennedy RE, Moisen GG, Pierce KB, Ohmann JL. Quantification of live aboveground forest dynamics with Landsat time-series and field inventory data: a comparison of empirical modelling approaches. Remote Sensing of Environmental. 2010;114:1053–1068.

[25] Tomppo E, Nilsson M, Rosengren M, Aalto P, Kennedy P. Simultaneous use of Landsat-TM and IRS-1C WiFS data in estimating large area tree stem volume and aboveground biomass. Remote Sensing of Environment. 2002;82:156–171.

[26] Carreiras JM, Pereira JMC, Pereira JS. Estimation of tree canopy cover in evergreen oak woodlands using remote sensing. Forest Ecology and Management. 2006;223:45–53.

[27] Lu D. The potential and challenge of remote sensing-based biomass estimation. International Journal of Remote Sensing. 2006;27:1297–1328.

[28] Muukkonen P, Heiskanen J. Biomass estimation over a large area based on standwise forest inventory data and ASTER and MODIS satellite data: a possibility to verify carbon inventories. Remote Sensing of Environment. 2007;107:617–624.

[29] Steininger MK, Steininger, M.K., 2000. Satellite estimation of tropical secondary forest above ground biomass data from Brazil and Bolivia. International Journal of Remote Sensing. 2000;21:1139–1157.

[30] Foody GM, Boyd DS, Cutler MEJ. Predictive relations of tropical forest biomass from Landsat TM data and their transferability between regions. Remote Sensing of Environment. 2003;85:463–474.

[31] Pereira JMC, Oliveira TM, Paul JCP. Satellite-based estimation of Mediterranean shrubland structural parameters. EARSel Advances in Remote Sensing. 1995;4(3–XII): 14–20.

[32] Calvão T, Palmeirim JM. Mapping Mediterranean scrub with satellite imagery: biomass estimation and spectral behaviour. International Journal of Remote Sensing. 2004;25(16):3113–3126.

[33] Joffre R, Lacaze B. Estimating tree density in oak savanna-like 'dehesa' of southern Spain from SPOT data. International Journal of Remote Sensing. 1993;14:685–697.

[34] Salvador R, Pons X. On the applicability of Landsat TM images to Mediterranean forest inventories. Forest Ecology and Management. 1998;104:193–208.

[35] Maciel MNM, Bastos PCO, Carvalho JOP, Watrin OS. Uso de imagens orbitais na estimativa de parâmetros estruturais de uma floresta primária no município de Paragominas, Estado do Pará. [Use of orbital images in the estimation of structural parameters of a primary forest in the municipality of Paragominas, Pará state]. Revista Ciência Agrária. 2009;52:159–178.

[36] Asrar G, Fuchs M, Kanemasu ET, Hatfield JL. Estimating absorbed photosynthetic radiation and leaf area index from spectral reflectance in wheat. Agronomy Journal. 1984;76:300–306.

[37] Nelson RF, Kimes DS, Salas WA, Routhier M. Secondary forest age and tropical forest biomass estimation using Thematic Mapper imagery. BioScience. 2000;50:419–431.

[38] Foody GM, Cutler ME, McMorrow J, Pelz D, Tangki H, Boyd DS, Douglas I. Mapping the biomass of Bornean tropical rain forest from remotely sensed data. Global Ecology & Biogeography. 2001;10:379–387.

[39] Zheng D, Rademacher J, Chen J, Crow T, Bresee M, Le Moine J, Ryu S. Estimating aboveground biomass using Landsat 7 ETM+ data across a managed landscape in northern Wisconsin, USA. Remote Sensing of Environment 2004; 93:402–411

[40] Lu D, Chen Q, Wang G, Liu L, Li G, Moran EA. A survey of remote sensing-based aboveground biomass estimation methods in forest ecosystems. International Journal of Digital Earth. 2016;9(1):63–105.

[41] Fraser RH, Li Z. Estimating fire-related parameters in boreal forest using SPOT VEGETATION. Remote Sensing of Environment. 2002;82:95–110.

[42] Fernández-Manso O, Fernández-Manso A, Quintano C. Estimation pf aboveground biomass in Mediterranean forests by statistical modelling of ASTER fraction images. International Journal of Applied Earth Observation and Geoinformation. 2014;31:45–56.

[43] Tompkins S, John FM, Carld MP, Donald WF. Optimization of Endmembers Mixture Analysis for Spectral. Remote Sensing of Environment. 1997; 59:472–489.

[44] Baccini A, Laporte N, Goetz MS, Dong H. A first map of tropical Africa's above-ground biomass derived from satellite imagery. Environmental Research Letters. 2008;3:045011.

[45] Zhang X, Kondragunta S. Estimating forest biomass in the USA using generalized allometric models and Modis land products. Geophysical Research Letters. 2006;33:L09402.

[46] Häme T, Salli A, Andersson K, Lohi A. A new methodology for the estimation of biomass of conifer-dominated boreal forest using NOAA AVHRR data. International Journal of Remote Sensing. 1997;18(15):3211–3243.

[47] Thenkabail PS, Stucky N, Griscom BW, Ashton MS, Diels J, Meer B, Enclona E. Biomass estimations and carbon stock calculations in the oil palm plantations of African derived savannas using IKONOS data. International Journal of Remote Sensing. 2004;25(23): 5447–5472.

[48] Palace M, Keller M, Asner GP, Hagen S, Braswell B. Amazon forest structure from IKONOS satellite data and the automated characterization of forest canopy properties. Biotropica. 2008;40:141–150.

[49] Leboeuf A, Beaudoin A, Fournier RA, Guindon L, Luther JE, Lambert MC. A shadow fraction method for mapping biomass of northern boreal black spruce forests using QuickBird imagery. Remote Sensing of Environment. 2007;110:488–500.

[50] Upgupta S, Singh S, Tiwari PS. Estimation of aboveground phytomass of plantations using digital photogrammetry and high resolution remote sensing data. Journal of the Indian Society of Remote Sensing. 2015;43(2):311–323.

[51] Sousa AMO, Gonçalves AC, Mesquita P, Marques da Silva JR. Biomass estimation with high resolution satellite images: a case study of *Quercus rotundifolia*. ISPRS Journal of Photogrametric and Remote Sensing. 2015;101:69–79.

[52] Zhu X, Liu D. Improving forest aboveground biomass estimation using seasonal Landsat NDVI time-series. ISPRS Journal of Photogrammetry and Remote Sensing. 2015;102:222–231.

[53] Quegan, S. Recent advances in understanding SAR imagery. In: Danson FM, Plummer SE, editors. Advances in Environmental Remote Sensing. Chichester: Wiley; 1995. p. 89–104.

[54] Imhoff ML. A theoretical analysis of the effect of forest structure on synthetic aperture radar backscatter and the remote sensing of biomass. IEEE Transactions on Geoscience and Remote Sensing. 1995;33:341–52.

[55] Kuplich TM, Freitas CC, Soares JV. 2000. The study of ERS-1SAR and Landsat TMsynergism for landuse classification. International of Remote Sensing. 2000;21(10):2101–2111.

[56] Castel T, Guerra F, Caraglio Y, Hollier F. Retrieval biomass of a large Venezuelan pine plantation using JERS-1 SAR data. Analysis of forest structure impact on radar signature. Remote Sensing of Environment. 2002;79:30–41.

[57] Sun W, Heidt V, Gong P, Xu G. Information fusion for rural land-use classification with high resolution satellite imagery. IEEE Transactions on Geoscience and Remote Sensing. 2003;3(1):3–5.

[58] Santos JR. Airborne Pband SAR applied to the aboveground biomass studies in the Brazilian tropical rainforest. Remote Sensing of Environment. 2003;87:482–493.

[59] Treuhaft RN, Law BE, Asner GP. Forest attributes from Radar interferometric structure and its fusion with optical remote sensing. BioScience. 2004;54:561–71.

[60] Garcia M, Riaño D, Chuvieco E, Danson FM. Estimating biomass carbon stocks for Mediterranean forest in Spain using height and intensity LiDAR data. Remote Sensing of Environment. 2010;114:816–830.

[61] Ni-Meister W, Lee S, Strahler AH, Woodcock CE, Schaaf C, Yao T, Ranson KJ, Sun G, Blair JB. Assessing general relationships between aboveground biomass and vegetation structure parameters for improved carbon estimate from LiDAR remote sensing. Journal of Geophysical Research. 2010;115:1–12.

[62] He Q, Erxue C, An R, Li Y. Above-ground biomass and biomass components estimation using LiDAR data in a coniferous forest. Forests. 2013;4:984–1002.

[63] Tian S, Tanase MA, Panciera R, Hacker J, Lowell K. Forest biomass estimation using radar and LiDAR synergies. In: Proceedings of the IEEE International Geoscience and Remote Sensing Symposium – IGARSS. 21–26 July 2013; Melbourne, Australia: IEEE; 2013. p. 2145–2148.

[64] Correia AV, Oliveira AC. Principais espécies florestais com interesse para Portugal: zonas de influência mediterrânica. [Main forest species with interest for Portugal: zones of Mediterranean influence]. Direcção-Geral das Florestas. Estudos e Informação; 1999. 318p.

[65] IFN5, 2010. Inventário Florestal Nacional. IFN5 2005–2006. Portugal Continental. [National Forest Inventory. IFN5 2005–2006. Continental Portugal]. Autoridade Florestal Nacional, Lisboa.

[66] Envi. Reference Guide—Exelis Visual Information Solutions. Boulder, Colorado: Exelis Visual Information Solutions [Internet]. 2009. Available from: http://www.exelis-vis.com/portals/0/pdfs/envi/envi_zoom_user_guide.pdf. Accessed 2012-11-27.

[67] Chavez Jr. PS. An improved dark-object subtraction technique for atmospheric scattering correction of multispectral data. Remote Sensing of Environment. 1988;24(3): 459–479.

[68] Rouse JW, Haas RH, Schell JA, Deering DW. 1973. Monitoring vegetation systems in the Great Plains with ERTS. In 3rd ERTS Symposium, NASA SP-351 I. 309–317.

[69] Definiens Imaging. eCognition Developer 8.0.1 Reference Book [Internet]. 2010. Available from: http://www.definiens.com. Accessed 2012-10-23.

[70] Sousa AMO, Mesquita PA, Gonçalves AC, Marques da Silva JR. 2010. Segmentação e classificação de tipologias florestais a partir de imagens Quickbird. [Segmentation and classification of forest typologies using Quickbird images]. Ambiência (special edition). 2010;6:57–66.

[71] Esri. ArcGIS Desktop: Release 10. Redlands, CA: Environmental Systems Research Institute [Internet]. 2010. Available from: http://www.esri.com. Accessed 2013-01-23.

[72] Avery TE, Burkhart HE. editors. Forest Measurements. 4th ed. New York: McGraw-Hill Inc.; 1994. 480p.

[73] R Development Core Team. R: A Language and Environment for Statistical Computing. R Foundation for Statistical Computing [Internet]. 2012. Available from: http://www.R-project.org. Accessed 2013-02-21.

[74] Sheather SJ. editor. A Modern Approach to Regression with R. New York: Springer Texts in Statistics; 2009. 393 p.

[75] Legendre P, Legendre L. editors. Numerical Ecology. 3rd ed. Amsterdam: Elsevier Science BV. 2012; 24. 1006 p.

[76] Pretzsch H. editor. Forest Dynamics, Growth and Yield: From Measurement to Model. Berlin Heidelberg: Springer-Verlag; 2009. 664 p.

[77] Burkhart HE, Tomé M. Modelling Forest Trees and Stands. Dordrecht: Springer Science +Business Media; 2012. 457p.

[78] Clutter JL, Fortson JC, Pienaar LV, Briester GH, Bailey RL. editors. Timber Management: A Quantitative Approach. New York: John Wiley & Sons, Inc; 1983. 333p.

[79] Myers RH. editor. Classical and Modern Regression with Applications. Chicago: Duxbury Press; 1986. 488 p.

[80] Paulo JA, Palma JHN, Gomes AA, Faias SP, Tomé J, Tomé M. Predicting site index from climate and soil variables for cork oak (*Quercus suber* L.) stands in Portugal. New Forests. 2015;46:293–307.

[81] Congalton RG, Oderwald RG, Mead RA. Assessing Landsat classification accuracy using discrete multivariate statistical techniques. Photogrammetric Engineering and Remote Sensing. 1983;49:1671–1678.

[82] Stehman SV. Estimating the kappa coefficient and its variance under stratified random sampling. Photogrammetric Engineering & Remote Sensing. 1996;62:401–407.

[83] Oliver CD, Larson BC editors. Forest Stand Dynamics. Update editions. New York: John Wiley & Sons, Inc; 1996. 544p.

[84] Schütz JP. Sylviculture 2. La gestion des forêts irrégulières et mélangées. [Silviculture 2. The management of uneven aged and mixed forests]. Collection Gérer L'environement, no. 13. Lausanne: Presses Polytechniques et Universitaires Romandes; 1997. 178 p. [in French].

[85] Ke Y, Quackenbush LJ. A review of methods for automatic individual tree-crown detection and delineation from passive remote sensing. International Journal of Remote Sensing. 2011;32(17):4725–4747.

[86] Chemura A, Duren I, Leeuwen LM. Determination of the age of soil palm from crown projection area detected from WorldView-2 multispectral remote sensing data: the case of Ejisu-Juaben district, Ghana. ISPRS Journal of Photogrammetry and Remote Sensing. 2014;100:118–127.

10

Biomass Production on Reclaimed Areas Tailing Ponds

Martin Bosák

Abstract

This chapter presents the results of a multiannual systematic research and development of essentially new environmental safety technology of overlapping tailing pond modelled in terms of Vojany thermal power plant (EVO), Slovakia. Re-cultivated tailing area can be used to produce biomass (Swedish willow) and this biomass is used again for subsequent incineration with coal. Laboratory and small plot experiments conducted directly in the tailing pond area resulted in the development of another new dimension of environmental technology of decontamination of tailing ponds. This technology connects technical, safety, economical and environmental effects for biomass production.

Keywords: biomass, willow, tailing pond, re-cultivation, energy

1. Introduction

Nowadays the world is facing a global problem, which has an effect of plants on the environment. Slag (dross ashes mixture) and fly ash are formed by the combustion of coal. In the wider vicinity of power plants, it is a waste that burdens the environment. The problem of landfill sites must be effectively addressed because they are in terms with the landscape stability and severe environmental and safety problems.

There is a real danger of breaking up the dam especially, in the case of dumping large quantities of such wastes in an area of huge tailings ponds. Therefore, it can have serious consequences on the population, components of the environment and property, in general. The tailings ponds are stores of very fine waste containing significant amounts of water. Their mobility from the pond is high, if they are released. Subsequently, they can have trans-boundary impacts, also impact on the landscape areas and protected areas of European importance because they migrate long distances, particularly over the surface water flows.

Especially by using environmentally sound technologies, it is necessary to ensure adequate management of these wastes, for security and long-term stability of tailing ponds.

2. Tailing ponds

2.1. Tailing ponds in the Middle Europe

Industrial production generates some kind of waste (by-product) toxic substance, which contaminates the sites and often degrades the surroundings of human living including air, ground water and surface.

From the viewpoint of environmental safety of deposited materials, there are 56 tailing ponds of various levels and types in Slovakia (**Figure 1**). They are in different stages of their existence or life cycles (**Tables 1–3**). They contain mainly wastes from coal gangue, ore processing products (floating sludge), heating plants (slag, ash) and power plants. Safe closing or re-cultivation of tailing ponds is a current problem in the Slovak Republic and also for the environmental safety in Europe because generally the tailing ponds, as watery building constructions, are considered immensely dangerous objects, in terms of environment [1].

Types of tailing ponds in Slovakia:

- 15 with ash material,
- 27 with ore and
- 14 others (industrial).

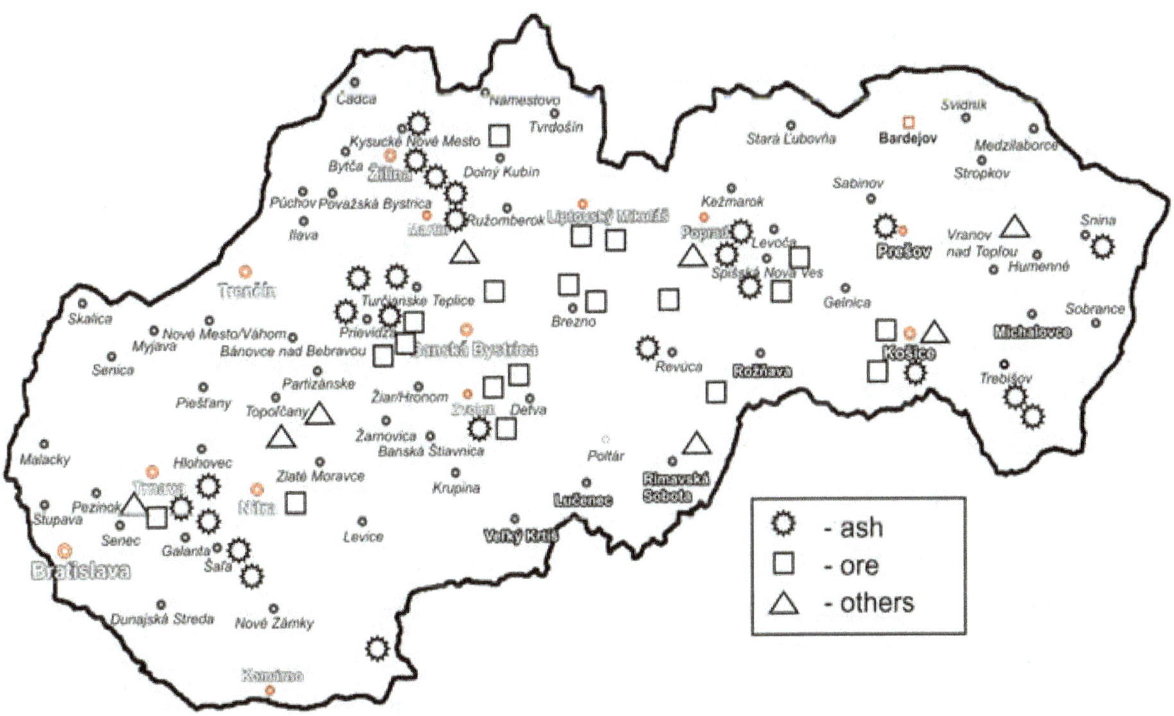

Figure 1. Registered tailing ponds in Slovakia.

No.	Name	Place	District	Category
1.	**Tailings pond ENO— new**	Zemianske Kostol'any	Prievidza	I
2.	**Tailings pond ENO —old**	Zemianske Kostol'any	Prievidza	II
3.	**Tailings pond ENO**	Bystričany —Chalmová	Prievidza	II
4.	**EVO Vojany**	Vojany - Drahnov	Mihalovce	II
5.	**Tailings pond KAPPAs.**	Štúrovo- časť Obid	Nové Zámky	II
6.	**Martin—old tailings pond**	Martin	Martin	II
7.	**Martin—new tailings pond**	Bystrička	Martin	II
8.	**Tailings pond Poša**	Poša-Nižný Hrabovec	Vranov n. T.	II
9.	**Tailings pond Snina**	Snina	Snina	II
10.	**Tailings pond DUSLO**	Tmovec n. Váhom,	Šal'a	II
11.	**Tailings pond Žilina**	Bytčica	Žilina	II
12.	**Tailings pond Košice**	Krásna nad Hornádom	Košice	III
13.	**Sered'**	Dolná Streda	Galanta	III
14.	**Zvolen**	Zvolen	Zvolen	III
15.	**Ash tailings pond**	Homé Opatovce	Žiar n. H.	III

Table 1. The list of ash tailing ponds in Slovakia.

No.	Name	Place	District	Category
1.	**Hačava**	Hačava	Rimavská Sobota	II
2.	**Hodruša Hámre**	Hodruša Hámre	Ziar nad Hronom	II
3.	**Jelšava**	Jelšava	Rožňava	II
4.	**Nižná Slaná**	Nižná Slaná	Rožňava	II
5.	**Rudňany**	Závadka	Spišská Nová Ves	II
6.	**Sedem žien**	Banská Belá	Žiar n. Hronom	II
7.	**Tailings pond Slovinky**	Slovinky	Spišská Nová Ves	II
8.	**Baňa Cígel' II.**	Sebedražie	Prievidza	III
9.	**Dúbrava 01**	Dúbrava	Liptovský Mikuláš	III
10.	**Dúbrava 02**	Dúbrava	Liptovský Mikuláš	III
11.	**Dúbrava 03**	Liptovský Mikuláš	Liptovský Mikuláš	III
12.	**Žiar nad Hrouom**	Žiar n. H.	Žiar n. H.	III
13.	**Košice – Bankov**	Košice	Košice	III
14.	**Lintych**	B. Štiavnica	B. Štiavnica	III
15.	**Pezinok**	Pezinok	Pezinok	III
16.	**Podrečanv**	Podrečany	Lučenec	III

No.	Name	Place	District	Category
17.	**Smolník**	Smolník	Spišská Nová Ves	III
18.	**Široká**	Široká	Dolný Kubín	IV
19.	**Baňa Cígel' I.**	Sebedražie	Prievidza	IV
20.	**Košice Bankov**	Košice	Košice	IV
21.	**Horná Ves (Kremnica)**	Horaá Ves	Žiar nad Hronom	IV
22.	**Hronský Beňadik,**	Hronský Beňadik	Nová Baňa	IV
23.	**Lubeník**	Jelšava	Rožňava	IV
24.	**Pezinok**	Pezinok	Pezinok	IV
25.	**Rožňava**	Rožňava	Rožňava	IV
26.	**Sered'**	Sered'	Galanta	IV
27.	Špania dolina	Špania dolina	B. Bystrica	IV

Table 2. The list of ore tailing ponds in Slovakia.

No.	Name	Place	District	Category
1.	**Čifáre**	Čifáre	Nitra	II
2.	**Bukocel**	Hencovce	Vranov n. T.	III
3.	**Dubová**	Dubová	B. Bystrica	III
4.	**Novácke odkalisko 7**	Nováky	Prievidza	III
5.	**Stabilizačný násyp Handlová**	Handlová	Prievidza	III
6.	**Odkalisko ČOV Sokol'any**	Sokol'any Bočiar	Košice	IV
7.	**Fámeš**	Pastuchov	Hlohovec	IV
8.	**Plešivec- Gemerská Hôrka**	Plešivec	Rožňava	IV
9.	**Nádrže oceliarenských kalov**	Vel'ká Ida	Košice	IV
10.	**Mokrá halda**	Vel'ká Ida	Košice	IV
11.	**Novácke odkalisko 6**	Nováky	Prievidza	IV
12.	**Sal'a RSTO**	Sal'a	Galanta	IV
13.	**Šulekovo**	Šulekovo	Trnava	IV
14.	**Veronika**	Dežerice	Topol'čany	IV

Table 3. The list of industrial tailing ponds in Slovakia.

According to the summary records of water cannons, 28 tailings of I–IV categories were located in the Czech Republic as on 1.1.2014, all listed in the following **Table 4**.

Not only red mud is produced in tailings, but there are also ash and uranium tailings in Hungary. Some of them are already being re-cultivated and prepared for liquidation. In Hungary, there are 20 tailing ponds as characterised below, in **Table 5**.

No.	Name	Place	Category
1.	**Hodějovice**	České Budějovice	III
2.	**Mydlovary**	České Budějovice	III
3.	**Zbrod North 1/4**	Hodonín	III
4.	**Nové Chalupy**	Karlovy Vary	III
5.	**Tailing ponds П.**	Ostrov	III
6.	**Dolní Radechová**	Náchod	III
7.	**Debrné**	Trutnov	III
8.	**Odkaliště TDK IV/3**	Trutnov	III
9.	**Stráž p. Ralskem**	Česká Lipa	II
10.	**Dřiteč**	Pardubice	III
11.	**Lhotka**	Pardubice	II
12..	**Semtín no. 7**	Pardubice	II
13.	**Chvaletice I.**	Přelouč	III
14.	**Božkov**	Plzeň	III
15.	**Panský les**	Mělník	II
16.	**Odkaliště Spolana**	Neratovice	III
17.	**Bvtíz**	Přibram	III
18.	**Rýzmburk**	Vlašim	III
19.	**Ušák**	Kadaň	II
20.	**SEPAP no. 4**	Litoměňce	III
21.	**Třískohipy**	Louny	III
22.	**Barbora III.**	Ústi nad Labem	III
23.	**Užin - old tailing ponds**	Ústi nad Labem	III
24.	**Dolni Rožinka**	Bystřie nad Pemštejnem	II
25.	**Zlatkov**	Bystřice nad Pemštejnem	II
26.	**Tailing ponds Synthesia a.s.**	Pardubice	IV
27.	**Tailing ponds**	Mladá Boleslav	IV
28.	**Ústí – new tailing ponds**	Ústi nad lab em	IV

Table 4. The list of registered tailing ponds in the Czech Republic.

On the presented maps (**Figures 1–3.**) are shown tailing ponds of middle European countries—Slovakia, the Czech Republic and Hungary. Even though Slovakia is the smallest country, it has the most tailing ponds (56) compared to the Czech Republic (28) and Hungary (20). Of course, this is also closely related with 'energy policy' of individual countries.

No.	Name	Type
1.	**Pellérd**	Uranium
2.	**Ajka**	Red mud
3.	**Almásfüzítő**	Red mud
4.	**Kurityán**	Red mud
5.	**Mosonmagyaróvár**	Red mud
6.	**Neszmély**	Red mud
7.	**Bokod**	Ash
8.	**Borsodnádasd**	Ash
9.	**Borsodszirák**	Ash
10.	**Estergom/Dorog**	Ash
11.	**Dunaújváros**	Ash
12.	**Gyöngyösoroszi**	Ash
13.	**Inota**	Ash
14.	**Kazincbarcika**	Ash
15.	**Múcsony**	Ash
16.	**Pécs**	Ash
17.	**Tatabány/Bánhida**	Ash
18.	**Tiszapalkonya**	Ash
19.	**Tiszaújváros**	Ash
20.	**Visonta**	Ash

Table 5. The list of tailing ponds in Hungary.

It is shown on the maps that locations of tailing ponds in Slovakia and the Czech Republic are distributed almost equally on the whole area of the countries. In Hungary, tailing ponds are placed mostly in the northern part of the country, near the border with the Slovakia.

The present tailing ponds are still considered environmentally dangerous and are also expensive to maintain; therefore, nowadays, a higher attention is given to them in Slovakia. This is demonstrated by the example. There was an accident in the Hungarian village Ajka, where on 4 October 2010, the dam pond broke after heavy rains. Subsequently, more than 0.7 million cubic meters of toxic mud struck in seven towns and villages and red sludge flooded the neighbourhood. The environmental disaster has claimed up to 10 human victims and over 150 people were injured and it destroyed dozens of homes. Other accidents at tailings ponds that are fatal and environmental devastation are shown in **Table 6**. Therefore, the discussion about closing these ponds is very important.

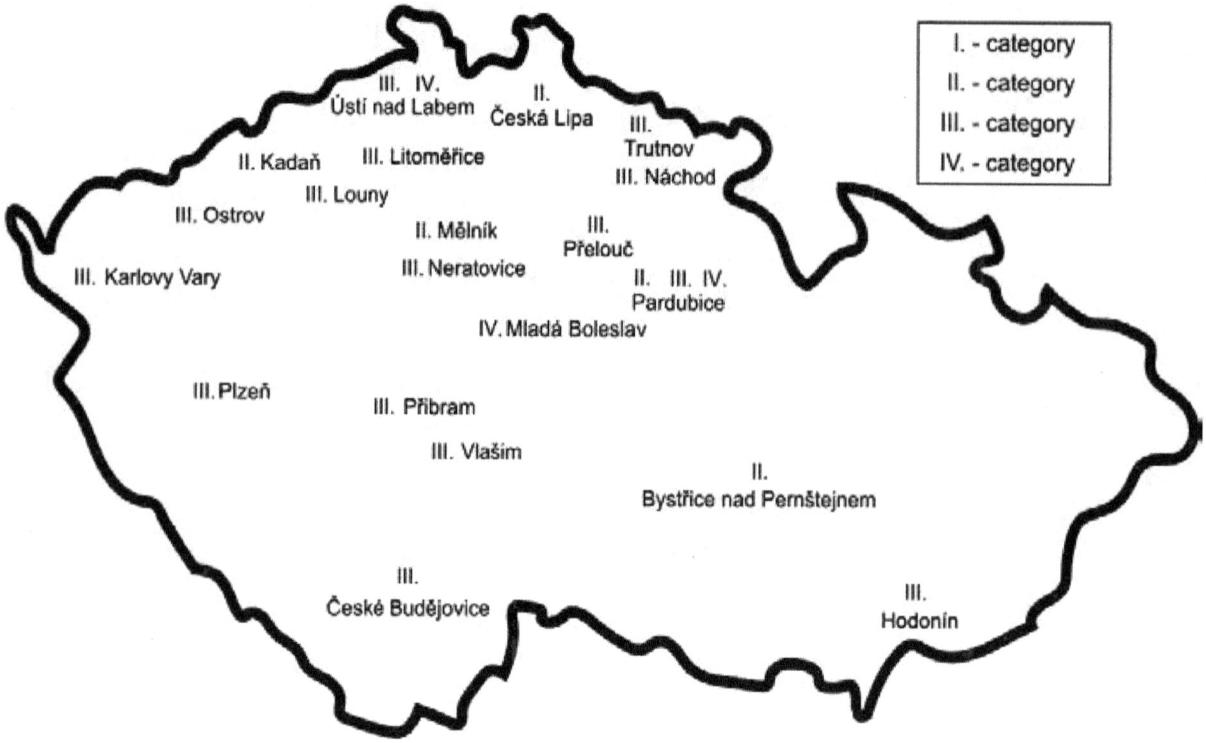

Figure 2. Registered tailing ponds in Czech Republic.

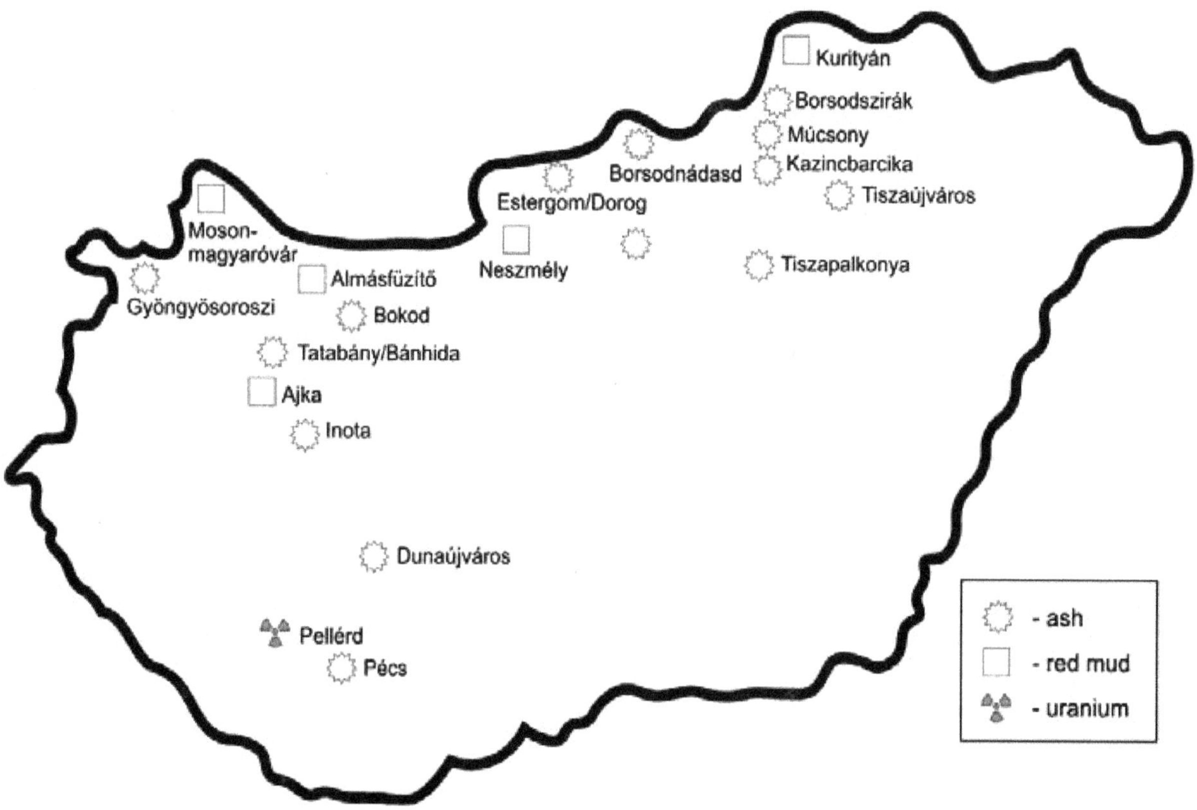

Figure 3. Registered tailing ponds in Hungary.

Town (Country)	Date	Number of death	Type of pond
Zemianske Kostol'any (Slovakia)	26.5.1965	4	Ashes from heat power plant
Stava (Italy)	19.7.1985	268	Fluorite sludge
Harmony (South Africa Republic)	6.2.1994	10	Cyanide pond
Placer (Philippines)	2.9.1995	12	Sludge
Ajka (Hungary)	4.10.2010	10	Red sludge

Table 6. Examples of tailing ponds from the world of accidents resulting in death.

Figure 4. Filling tailing pond.

For member countries, the European Union currently allocates huge funds in development projects. This is to prevent and rectify environmental damage; hence, the restoration and rehabilitation of tailing ponds dross ashes biological mixtures are important [1].

2.2. Tailing pond of EVO Vojany

The biggest fossil fuel plant in Slovakia is Vojany power plant, where mainly semi-anthracite coal from Ukraine and Russia is used as the fuel. Currently, for disposal of waste products from coal combustion the plant operates two facilities:

- dump with stabilisation material tailing,
- tailing ponds with dross ashes mixture.

Coal combustion products that are hydraulically transported are stored with stabilisation material in the tailing pond with two cassettes of dross ash mixtures (cassette no. 1 is already closed) (**Figure 4**). Stabilisation material is a by-product of the desulphurisation of power plant technology and combustion processes [2].

Safety and operation oversight within the relevant legislation of tailing pond of EVO plant Vojany are needed because it is a water work. On the verge of PLA Latorica, on the left bank of the river Laborec, it was built in 1965 to store dross ash mixture and is located in the administrative

Area of base	Cassette No. 1—47.2 ha
	Cassette No. 2—48.1 ha
Overall capacity of stock ash	Cassette No. 1—7,580,000 m^3
	Cassette No. 2—5,760,000 m^3
Dispozition capacity	Cassette No. 1—full
	Cassette No. 2—850.000 m^3

Table 7. Base parameters of tailing pond.

Figure 5. Place cassette no. 1 on pond in scale 1:10,000.

area of village Drahňov and Vojany. It is bounded on all sides by high grass-covered embankments. Two separate approximately similar cassettes create the tailing pond (**Table 7**):

- cassette No. 1–29 ha (with dam 47.2 ha), (**Figure 5**) and
- cassette No. 2–27 ha.

Cassettes are separated by dividing dam, which originally had the function of the peripheral dam cassette No. 1. This means that the area to be addressed after the final shutdown of the pond is about 56 ha [3].

3. Experiments and methods of research

This chapter presents development of original remediation technologies for unconventional tailing pond dross ashes mixture disposal and results of the experiments in the research. This technology uses structured layers of land, soil and stabilisation material. The reason is to replace previous legislative solution by the overlapping of hydrofilm material and drainage system.

3.1. Experiments of containers

The experiment simulating any large-scale use of this new non-traditional, uncertified practice or technology was based on the verification of replacement waterproofing properties of the stabiliser. The purpose of verification of experiments was to assess the possibility of using a stable material due to its potential to prevent solidification of penetrating rain water into the lower layers of the pond, with the risk of another accident.

A mixture of grass varieties that are resistant to typical and local conditions is used for covering of the energy crop with the future consideration of using pond grown plants as biomass for co-incineration with coal in power plants. The cultivation of fast growing Swedish willow with respect to its root system was experimentally verified. Therefore it was used in this experiment and the subsoil thickness was of 500 mm.

The laboratory experiment was set up with the following procedure:

- the stabilisation material of thickness of 0, 50, 100, 150, 200, 250, 300, 350, 400, 450 and 500 mm, was kept in the bottom of a 2 × 11 container with a size of 1000 mm × 1000 mm × 1000 mm,
- the second layer reflects the subsoil profile of the rehabilitated land and was then deposited in the soil with subsoil thicknesses of 300 mm for grass and 500 mm for Swedish willow (**Table 8**),
- like the subsoil, even topsoil profile describes the reclaimed area, the last build up layer of topsoil with a thickness of 200 mm is uniformly used for all variants,
- a part of containers was located in areas with variable weather conditions (the first half) and the other part is placed under the local climate conditions (the second half) (**Figures 6** and 7) [4].

	Subsoil	Topsoil	Soil together
Grass	300 mm	200 mm	500 mm
Willow	500 mm	200 mm	700 mm

Table 8. Requirements for growing.

Figure 6. Containers in terms of controllable.

3.1.1. *Stabilisation material: its analysis and usage*

We provide an analysis report of hazardous substances contained in the stabilised in the following tables and **Figure 8** show the structured layers in individual containers, which appeared in the experiment. The used stabilisation material has pH 8.45 and a conductivity of 38.7 (**Tables 9** and **10**).

Every container was filled with layers of stabilisation material, subsoil and topsoil according to their draft. The first container was filled without a layer of stabiliser, as a control container—nothing else has been done. Afterwards, sufficiently compacted individual layers were filled to bring them closer to actual conditions [4].

3.1.2. *Experimental method*

The half of containers were located in areas under local climatic conditions and the second half of containers were placed in areas where weather conditions were controlled.

Figure 7. Containers under conditions consistent with local climate.

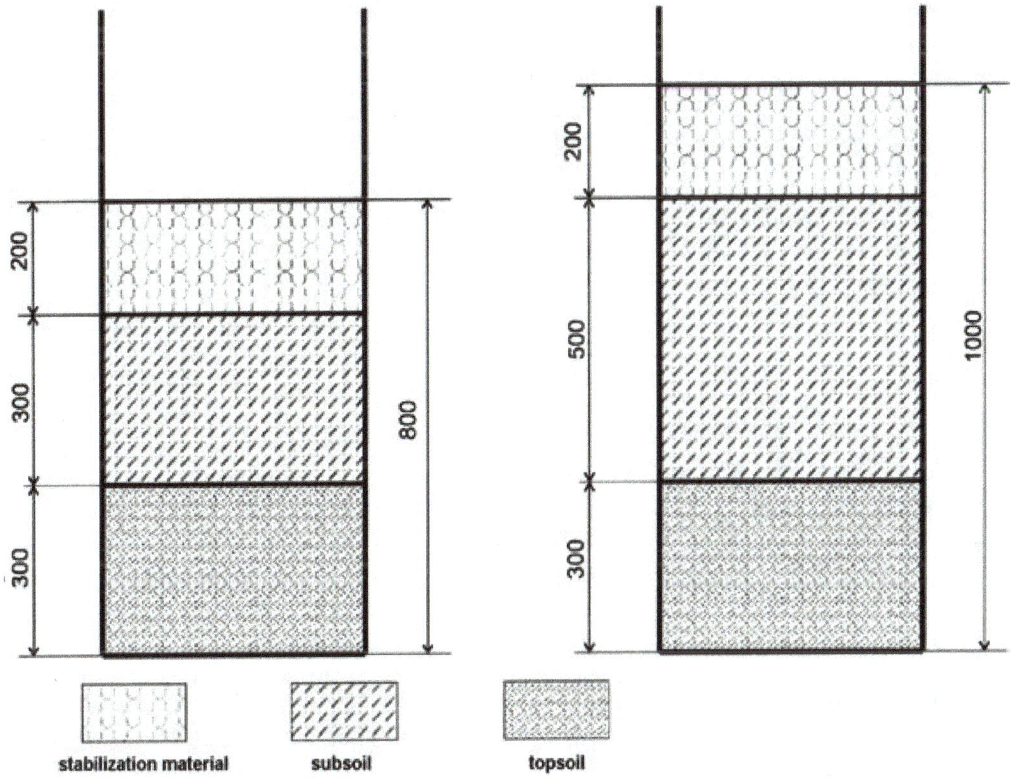

Figure 8. Diagram of the experimental variations of the experiment at 300 mm stabiliser.

	Indicator	Abbreviation	Value
1	Total organic carbon	TOC	**105,100**
2	Benzene, toluene, ethylbenzene and xylenes	BTEX	**<0.001**
3	Polychlorinated biphenyls, seven members	PCB	**<0.01**
4	Mineral oil (C10–C40)	NEL	**<1**
5	Polvcyclic aromatic hydrocarbons	PAH	**<0.05**

Table 9. Analysis of stabilisation material based on based on regulation of Ministry of Environment of SR No. 599/2005 Z.z.

	Indicator	Abbreviation	Value [mg]
1	Arsenic	As	**0.02**
2	Barium	Ba	**0.58**
3	Cadmium	Cd	**<0.003**
4	Total chromium	Cr	**0.03**
5	Copper	Cu	**0.03**
6	Mercury	Hg	**0.002**
7	Molybdenum	Mo	**0.05**
8	Nickel	Ni	**<0.02**
9	Lead	Pb	**0.05**
10	Antimony	Sb	**0.01**
11	Selenium	Se	**<0.01**
12	Zinc	Zn	**0.10**
13	Chlorides	cl^-	**31.7**
14	Fluorides	F^-	**2.2**
15	Sulphates	SO_4^{2-}	**1490**
16	Phenol index	FNI	**<0.3**
17	Dissolved organic carbon	DOC	**243**
18	Total solubles	CL	**3240**

Table 10. Analysis of water extract of stabilisation material based on regulation of Ministry of Environment of SR No. 599/2005 Z.z.

Reflecting the maximum monthly average for last 50 years indoor watering and simulation of precipitation in real outdoor was applied with water. Data on rainfall was used from the Slovak Hydrometeorological Institute, Regional Centre of Kosice from stations in Michalovce, Somotor and Milhostov ,which are the closest to the tailing pond.

Except for the rainy year 2010, when the monthly average rose to 85 mm/month, the long-term measurements of rainfall in the area show that the average monthly values range from 40 to 50 mm/month [4].

3.1.3. Results of experiment

According to different climatic conditions, the results of the experiment can be divided into two groups, with respect only to the quantity of water received:

3.1.3.1. Containers under conditions consistent with local climate–natural amount of rainwater

Rainfall was observed in containers on average 43 mm/month, which represents 1.43 mm of rainfall per day during the period January 2011–December 2011 (**Table 11**). The rainwater do not get through the layers, but in one container that contained stabilisation material, soil absorbed them.

Station	Month												Total	Average per month
	I	II	III	IV	V	VI	VII	VIII	IX	X	XI	XII		
Michalovce	40	6	30	17	30	92	180	27	51	16	1	77	567	47
Milhostov	28	4	31	14	46	112	166	11	41	14	1	58	525	44
Somotor	30	9	36	21	28	92	115	9	49	18	3	76	484	40
Average														**44**

Table 11. The rainfall [mm] of nearby stations in 2011.

3.1.3.2. Containers with controllable received amount of water (pouring)

The containers with variable conditions were simulated with extreme daily rainfall amounts of water, minimum 50 mm/day, that is to say, more than 7 times the maximum daily average precipitation. In May 2010, a long-term, extremely high quantity of rainwater were measured—the calculated average of monthly rainfall was 208 mm, it means that the average rainfall of the day is 6.9 mm (**Table 12**).

A part of the precipitation is absorbed by each layer in the container and seepage water accumulated the discharge outlet in the prepared containers and continuously measured.

Station	Year	Month	Rainfall [mm]
Michalovce	2011	VII	180
Milhostov	2010	V	219
Somotor	2010	V	226
Max. average			**208**

Table 12. Maximum of rainfall [mm] of nearby stations.

In order to determine the waterproofing ability of the individual layers for carrying out experiments, the maximum long-term nature of rainfall in the area of the tailing pond was taken as the basis of the daily rainfall amounts of water in the controlled.

In the following **Figures 9** and **10** are shown information depending on the amount of water tightness where the thickness of stabiliser and the subsoil are 300 and 500 mm, respectively.

Alternative use of 170 and 230 mm of stabilisation material, depending on the thickness of subsoil differences for willows and seed grass, resulted in implementation of laboratory experiments, leading to establish the maximum daily amount of rainwater at 100 mm in containers (70-times more than the average daily value). The root system of plants planted on the surface will gradually pump saturated water to the topsoil and subsoil.

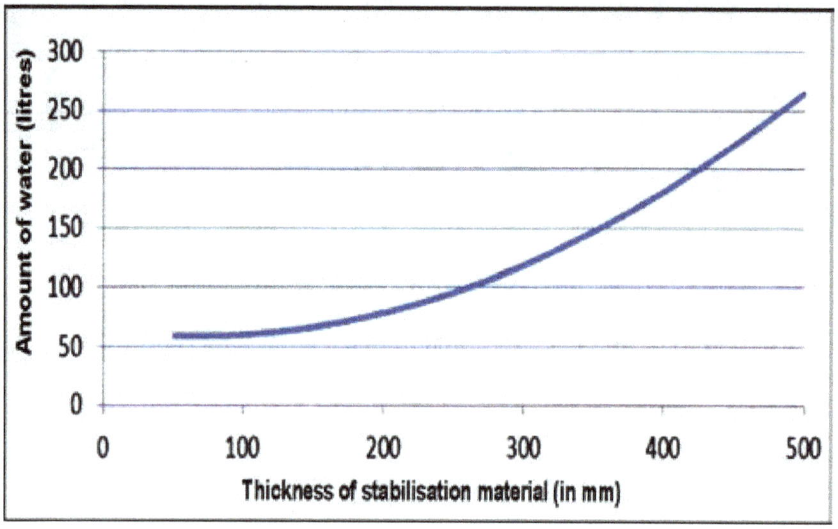

Figure 9. Dependence of the thickness of stabiliser and the amount of water tightness in thickness of the subsoil 300 mm.

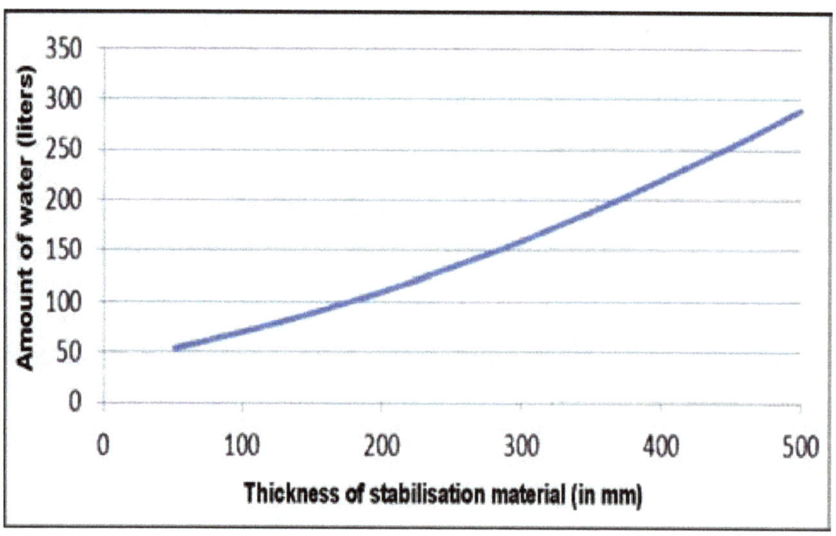

Figure 10. Dependence of the thickness of stabiliser and the amount of water tightness in thickness of the subsoil 500 mm.

3.2. Willow cultivation on tailings pond

3.2.1. Small plot trials on tailings pond

The second phase of experiment was to implement small plot trials on the surface of tailing where four large-sized parcels of 7 m × 20 m in a given structure have been built (**Tables 13** and **14**).

Parcel	Stabilisation material	Subsoil	Topsoil	Together
1st parcel	300 mm	500 mm	200 mm	1000 mm
2nd parcel	500 mm	500 mm	200 mm	1200 mm
3rd parcel	-	500 mm	200 mm	700 mm
4th parcel	-	500 mm	200 mm	700 mm

Table 13. Structure of parcels for willow.

Parcel	Stabilisation material	Subsoil	Topsoil	Together
1st parcel	300 mm	300 mm	200 mm	800 mm
2nd parcel	500 mm	300 mm	200 mm	1000 mm
3rd parcel	-	300 mm	200 mm	500 mm
4th parcel	-	300 mm	200 mm	500 mm

Table 14. Structure of parcels for grass.

The possibility cultivation of willow and grass cultivation in the closed tailing ponds in reality during the vegetation seasons are tested the proposed alternatives of the covering layer for the biological re-cultivation of the tailing ponds from the viewpoint of water permeability under natural conditions of atmospheric impacts (**Figures 11** and **12**). The results of experiment are

Figure 11. Experimental parcels on tailings pond.

Figure 12. Grass mixtures rapid and prosoil.

supplemented with 7 years of practical knowledge about the cultivation of Swedish willow on the territory of 15 ha in the nearby town of Kežmarok.

The average root striking of plant slips after the willow planting in two research stations of the Slovak University of Agriculture in Nitra was in the range between 66.91–89.51% and 45.67–91.35%. The planting was implemented in 2011.

Under suitable conditions, it is possible to achieve a root striking of more than 90% according to Dawson statement (2007). High root striking percentage of slips is inevitable for the optimal structure of vegetation and for optimal crops [5]. The number of rooted pieces can be influenced by the planting way of slips during the founding of commercial plantations. Better root striking in the case of planting of slips horizontally to the soil surface in comparison with the classical planting vertically to the soil surface was identified by Lowthe-Thomas et al. This planting method at the same time can significantly reduce the cost of planting [6].

We have achieved an average root striking of 91.76% (**Table 15**) in our experiment. The conditions for willow cultivation (structured layers of the stabilized stabilised waste, subsoil layer and arable layer) were suitably prepared. We can conclude from this result that, it is feasible to produce biomass directly in the power plant or in the tailing ponds which is 2 km away from the plant.

Plot	% of rooted plants
1.	88.72
2.	92.63
3.	91.14
4.	94.57
Average	**91.76**

Table 15. Number of rooted pieces of the Swedish willow in the experiment.

The optimal total quantity of precipitation in summer months should achieve 300 mm and during the total vegetation season it should achieve 550 mm [7]. Besides the extremely rainy year 2010, the measured values for the whole vegetation season were lower than 570 mm.

The precipitation data have been acquired from the three nearest monitoring stations to the tailing ponds (**Table 16**). The monitoring stations belong to relevant regional centre of the Slovak Hydro-Meteorological Institute.

Station	2009	2010	2011	2012	2013	2014	2015
Milhostov	585	892	526	496	529	567	556
Michalovce	636	929	567	635	578	625	593
Somotor	590	1030	486	481	522	520	537
Average	604	950	527	537	543	570	562

Table 16. The rainfall of nearby stations.

Swedish willow was planted in the spring of 2011. As most of the shoots were damaged, it was necessary to carry out planting of new shoots of Swedish willow in spring 2012. Rodents damaged willow root systems and most of them were completely destroyed by the spring of 2013. Subsequently the tree was replanted and fenced and rodent repellers were installed as well. Since the Swedish willow has been gradually replanted, various heights of tree shoots can be found on experimental fields. It is possible to achieve more than 90% of embeddedness of the Sweden tree shoots under appropriate conditions. High embeddedness of shoots is essential for achieving the optimal harvest. The method of planting Swedish willow cuttings might influence the number of in-rooted units. Worse embeddedness of shoots is monitored during planting cuttings vertically to the soil surface than planting the cuttings horizontally to the soil surface. We can conclude that conditions for growing Swedish willow were suitably prepared because in this experiment the average embeddedness of shoots was up to 92% (**Table 17, Figure 13**).

Approximately 580–600 mm of rainfall is the optimal precipitation value for the entire growing season. The willow can produce large amounts of biomass at this level. The rainiest year was 2010, as shown in **Table 13**, according to the atmospheric precipitation of individual

Season	Hight of willow
2012 – June	120 – 640 mm
2013 – April	1100 – 1750 mm
2014 – May	1900 – 2450 mm
2015 – April	2800 – 3200 mm

Table 17. Height of Swedish willow.

Figure 13. Swedish willow on 4th parcel.

periods. The total amount of rainfall was below the required level and did not exceed 550 mm in the next 3 years. As a result, Swedish willows have been drying out.

At Vojany thermal power plant through small plot trials continues the general process of experimental testing of the possibility of re-cultivating cinder/ash mixture tailings. The biomass yield of mowed grass is corresponding with the expected values which are increasing every year. It would be appropriate if the annual rainfall total was around 600 mm in terms of growing willow. We might conclude that total rainfall for the last period was below the long-term average. Because of lower-than-average precipitation during the year, the newly planted willow took a much lesser extent than initially expected.

3.2.2. *Results of experiment*

Upon analysing the individual components of experimental plots it was shown that at the edges of the experimental plots the formation of continual solidified layer of stabilisation material was not observed. The coherent layer of solidified stabilisation material was formatted in cuts, which was made closer to the centre of the plot. This situation could significantly affect the rainfall, whose intensity was at a time of experimentation very low. On the edges of the plots rainfalls withered quickly, more than the centre of the plots, in which the water was maintained for longer. In the edge stabilisation material was loose, but gradually towards the centre were produced visible larger chunks of hardened stabilisation material **Figure 14**.

From the plots were taken samples of manually cut root systems of Swedish willow (**Figure 15**). Selected root systems had different length, branching and direction depending on structured underlay of various plots. We can conclude, that in terms of length, the root system corresponds to the length of the part above the topsoil (mutually correlated). Another factor that cannot be ignored is, that the root system of Swedish willow did not crush the layer of stabilisation material in the vertical direction.

Figure 14. Solidified layer of stabilisation material.

Figure 15. Horizontal root system of Swedish willow above stabilisation material.

The experiments resulted in the application of new environmental recycling technology by new remediation technologies unconventional pond dross ashes mixture by using structured layers of stabilisation material, soil and land. Stabilisation material is the by-product of the desulphurisation process of power plant in combustion processes. The uniqueness new environmental recycling technology lies in the presented technology solutions, where a waste product of energy combustion processes will be used in another defused form.

During the growing season, the proposed real alternatives to the coating of bio-remediation of the pond in terms of water permeability under natural conditions and atmospheric effects were tested. It also verifies the best type of plants grown in the creation of sanitation, security conditions, with respect to the possibility of their further use as a co-incineration of biomass with coal in the same technology plant.

The final results showed that stabilisation layer prevented penetration of water into the lower layers of the tailing pond. The solution thus achieves a synergistic environmental-security effect [8].

3.3. Combustion of biomass

According to EU legislative from year 2008 EU is to provide 20% energy from renewable sources which is the energy sector goal for 2020 [8]. Co-incineration of biomass can be one of the ways to achieve this goal. Under the shared combustion of biomass and coal, there is a reduction, specifically, partial elimination of the environmental impact due to low content of nitrogen and sulphur in the biomass, resulting in a reduction of CO, NOx and SO_2, as well as reducing emissions of heavy metals [9].

One of the most promising methods for the provision of energy production in general is co-incineration of biomass energy willows and other biomass plant with coal in the near future and in the present [10]. It failed to apply in Slovakia and at global level, despite the availability, environmental and technological benefits announced by this system on a larger scale. Increased costs associated with the production and logistics of biomass assurance seems to be the biggest problem. Co-incineration of biomass with coal significantly increases the clean energy ratio. It is defined as the ratio of produced electricity to the total consumption of fossil energy. It is primarily reducing the greenhouse gas emissions from mining, transportation and combustion of coal when substituting a certain percentage of coal with biomass [11]. In practice, co-incineration of willow chips with wood wastes have been used in biomass power stations in Sweden. In the local energy supply, it plays an important role. Study of environmental impact of coal combustion from a power plant have been carried out in Poland and also in a number of countries [12].

The power plant Vojany started to implement them as well in collaboration with this author, in 2009 in line with the above trends. Project combustion of black coal with biomass in fluidised boilers was realised in the form of a scientific research complex. The research includes the provision of (growing) plant biomass in the pond area surrounding the facility gaining self-made slag-ash mixture. The first positive results in the reduction of emissions was produced by co-incineration of biomass, mainly wood chips mixed with black coal in a 4%

ratio by 40 kg per MWh produced, and operational savings associated with the consumption of limestone, creation and disposal of ash, water consumption and steam. The project's next phase was experimentally realised with the co-incineration of biomass with a value 7% and then 15%. The surroundings of the facility has good power potential of the fast-growing energy crops—willow was showed in research focused on the potential provision of biomass. In connection with this research specific purposeful cultivation of biomass was initiated in the large pond of about 56 ha that contained slag-ash mixture. One of most burdened areas in eastern Slovakia is the area of Vojany [13].

3.3.1. *Co-incineration of wood chips with coal*

In 2007, co-incineration of biomass in EVO boilers started. Forest biomass has been chosen for the first tests of wood chips. To achieve the same thermal input it was needed to deliver six cubic meters of biomass to the boiler, instead of one cubic of black coal, due to different heating value, calorific value and density of black coal and wood chips. It could not be replaced by any amount of coal with biomass because wood chips have a density of 0.3 t/m^3 and calorific value 10 MJ/kg and coal has a density of about 1 t/m^3 and heating value of 25 MJ/kg.

The added mixture of wood chips containing about 4% of the heat energy mix wood chips—coal, does not negatively affect dynamic characteristics of the power plant units—it was proved from calculations based on the time. It was also proved by tests in 2007 that it is possible to smoothly combust biomass (wood chip) in fluidised boilers in the power plant of Vojany. It turned out that wood chips have a positive impact on the boiler combustion mode, because of a higher proportion of volatile matter and a lower ignite temperature than coal. The resuls were a decrease in the concentration of carbon monoxide in the exhaust gas and more efficient combustion of irradiated fuel [14].

The launch of scientific-research activities aimed at complex solution for the issue of environmental energy-biomass combustion optimisation processes was initiated because of these partial positive results.

Replacement of a share of combusted black coal in thermal power EVO with biomass-based fuels was carried with priority to maximise the reduction of emissions especially carbon-sulphur oxides, by providing the required energy performance and therefore ultimately in increasing competitiveness improving economic indicators, manufacturing and energy-production companies.

The co-incineration of black semi-anthracite coal and wood biomass in fluidised FK5 boilers in the examined facility was performed as the experiment.

The wood chips ranged from 8.0 to 8.65 MJ/kg and the black coal average heating values in the experiment ranged from 25.4 to 28.1 MJ/kg. Graphically illustrated in **Figure 16** is the dependency of the heating value of wood chips on its moisture.

From different forms of concentrations of pollutants (the individual VOP) it is significant that:

- with increased performance there was a decrease in the concentration of carbon monoxide in co-incineration ratio of wood chips,

- during testing, legislative allowed emission limits for SO_2 were preserved and there has been an increase in performance and in values and
- other values of VOP were as well below the individual emission limits.

At the same time the efficiency of boiler and co-incineration of coal and the biomass, are presented in **Table 18.**

The co-incineration of a mixture of wood chips and coal led indeed into a slight reduction in boiler efficiency, which is negligible compared to the achieved environmental-safety effects—it can be stated when taking into account the partially different characteristics specifically, quality of supplied and combusted coal and wood chips.

The implementation of the I phase in plant biomass co-incineration in the power plant Vojany at the block no. 6 began in July 2009 based on these tests, and partial results of experimentation.

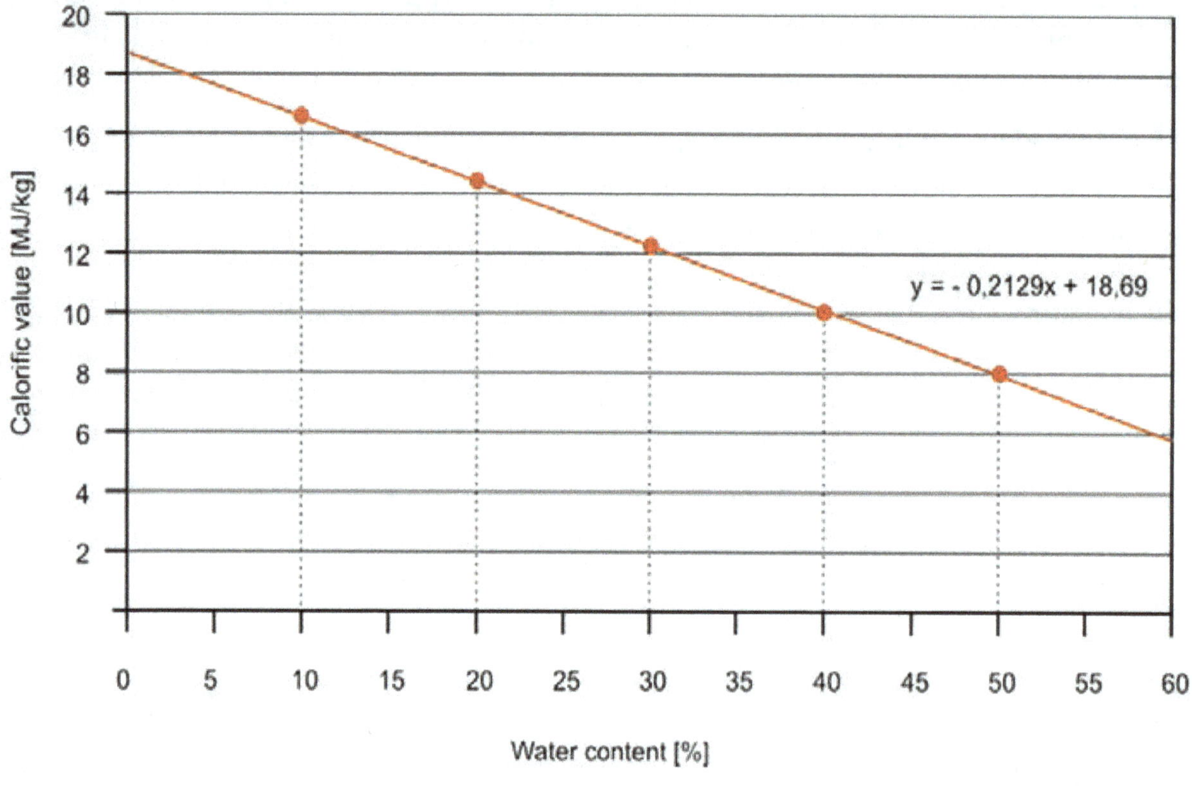

Figure 16. Dependency of the calorific value of wood chips on the water content.

Electrical performance	Black coal	Share of wood chip 1.91%	Share of wood chip 3.91%
[MW]	[%]	[%]	[%]
66	93.81	93.47	92.30
88	93.55	93.32	92.91
110	93.52	92.57	93.20

Table 18. The effectiveness of fluidised boiler K5 in Vojany.

A landfill with open capacity of 400 tons adjusted with transport (conveyor belts) and technology (crusher-sorter) biomass was built and customised to support experiments.

A rated power of conveyor belts was set to properly balance the mutual ratio and a mechanised system was used for wood chips, to transport them over conveyor belt and coal through the other one. The use of raw wood chip (hardwood and softwood) and its share was gradually increased to 5.3%.

Co-incineration of a better quality of the combusted wood chips with a higher calorific value than previously considered biomass was achieved with a share of 5.3%, not by increasing the weight. This means that projected calorific value changed from 9.5 MJ/kg to over 11 MJ/kg. This ratio has proved to be the best possible, to maintain the maximum dynamic properties of the boiler and in the execution of transporting the fuel mixture into the boiler. The implementation of II stage of biomass project co-incineration, which consisted of the construction of independent mechanised access to the boiler especially for biomass was determined in relation to the achieved results.

Table 19 shows that the combustion of 1 ton of biomass eliminated approximately one ton of carbon dioxide emissions and capacities of coal lines remained clear and the new path was used to transport higher volume of wood chips into the boiler (**Figure 17**).

3.3.2. Results of experiment

The production of biomass and co-incineration of coal in fluidised boilers, of thermal power stations SE, a.s. as well as other power plants is need in terms of environmental benefits. By burning coal and biomass is reflected in a significant reduction in emissions and production of solid waste.

The facility is obtaining its biomass currently from six local suppliers. EVO facility is a promising purchaser and will be their regular customer in the future as well because the surrounding wetlands around the facility provide good conditions for rapidly growing trees (poplar, willow). There is potential for increasing employment in the region for people with lower qualifications with the cultivation of biomass in the surrounding area of the plant or in the wetlands.

Year	Wood chip share [t]	CO_2 – eliminated [t]
2009	8,310	10,487
2010	21,443	27,061
2011	24,099	30,413
2012	26,917	33,969
2013	60,794	92,954
2014	48,752	84,899

Table 19. The amount of CO_2 saved by co-incineration of biomass.

Figure 17. Biomass—landfill and transport.

4. Summary

The research carried out in the tailing pond EVO Vojany showed that on the tailing pond can be planted Swedish willow, as a source of biomass, while the by-product (waste) of the desulphurisation of power plant technology, combustion processes, can be used as the stabilisation material, making it possible to be reused on reclaimed areas of tailing ponds.

Using biomass has a positive impact on the environment and it is environmentally adequate way of power generation. We developed our knowledge with this new technology with 7 years of experience in cooperation with the EVO. About 80 tons of wood chips are co-incinerated based on daily experiment based model. The European Union's commitment to continually increase the use of renewable energy of 20% by 2020—there are contributing implementations of research that results into practice.

Acknowledgements

In this chapter are the results of the project implementation VEGA No. 1/0936/15 Economics and Environmental Studies and experimental verification of the possibility of reclaiming tailings ash mixture in SE–EVO Vojany.

Author details

Martin Bosák

Address all correspondence to: martin.bosak@euke.sk

University of Economics in Bratislava, Košice, Slovak Republic

References

[1] Bosák M, Hajduová Z, Majerník M vodohospodárske dielo, Andrejovský P. Experimental-Energy Combustion of Biomass Combined with Coal in Thermal Power Plants. Polish Journal of Environmental Studies. Olsztyn, 2015, vol. 24, no. 4, pp. 1517–1523.

[2] Majerník M, Bosák M, Andrejkovič M, Hajduová Z, Turisová R. Prevention of Environmental Risks of Tailings Ponds, 12th International Multidisciplinary Scientific GeoConference SGEM 2012, Albena Resort, 2012, vol. 5, pp. 619–626.

[3] Hronec O, Vilček J, Majerník M, Tkáč M. Ecological stabilization tailings pond in EVO Vojany - study – štúdia, EU PHF Košice, 2009, 106 p.

[4] Tkáč M, Majerník M, Hronec O, Vilček J, Bosák M. Experimental verification options reclamation tailings pond in SE - EVO Vojany, EU PHF Košice, 2011, 53 p.

[5] Dawson WM. Short Rotation Coppice Willow – Best Practice Guidelines. Renew Project, 2007, 48 p.

[6] Lowthe-Thomas SC, Slater FM, Randerson PF. Reducing the Establishments Costs of Short Rotation Willow Coppice (SRC). Biomass and Bioenergy, 2010, vol. 34, no. 5, pp. 677–686.

[7] Hall RL. Short Rotation Coppice for Energy Production: Hydrological Guidelines. Report. Crown Copyright, 2003, 21 p.

[8] Bosák M, Tarča A, Majerník M, Tomčíková M. Application of New Environmental Recycling Technology. In: SGEM 2016, 16th International Multidisciplinary Scientific Geoconference, Albena, 2016,

[9] Heller M.C, Keoleian GA, Mann MK, Volk TA. Life Cycle Energy and Environmental Benefits of Generating Electricity from Willow Biomass. Renewable Energy, 2004, vol. 29, no. 7, pp. 1023–1042.

[10] McCormick K, Kåberger T. Key barriers for bioenergy in Europe: Economic conditions, Know-How and Institutional Capacity and Supply Chain Coordination. Biomass and Bioenergy, 2007, vol. 31, no. 7, pp. 443–452.

[11] Keoleian GA, Volk TA. Renewable Energy from Willow Biomass Crops. Life Cycle Energy, Environmental and Economics Performance. Critical Reviews in Plant Sciences, 2005, vol. 24, no. 5–6, pp. 386–406 .

[12] Kalembkiewicz J, Chmielarz U. Effects of Biomass Co-Combustion with Coal on Functional Speciation and Mobility of Heavy Metals in Industrial Ash. Polish Journal of Environmental Studies, 2013, vol. 22, no. 3, p. 741.

[13] Vilček J, Hronec O, Tomáš J. Risk Elements in Soils of Burdened Areas of Eastern Slovakia, Polish Journal of Environmental Studies, 2012, vol. 21, no. 5, 1429 p.

[14] Vaszily M. Three Years Experience. Slovenská energetika, Slovenské elektrárne, a. s., 2012, vol. 37, no. 6, pp. 3–5.

11

Progress Towards Engineering Microbial Surfaces to Degrade Biomass

Grace L. Huang and Robert T. Clubb

Abstract

Lignocellulosic biomass is a promising feedstock to sustainably produce useful biocommodities. However, its recalcitrance to hydrolysis limits its commercial utility. One attractive strategy to overcome this problem is to use consolidated bioprocessing (CBP) microbes to directly convert biomass into chemicals and biofuels. Several industrially useful microbes possess desirable consolidated bioprocessing characteristics, yet they lack the ability to degrade biomass. Engineering these microbes' surfaces to display cellulases and cellulosome-like structures could endow them with potent cellulolytic activity, enabling them to be used in CBP. In this chapter, we discuss recent progress in engineering the surfaces of *Saccharomyces cerevisiae, Escherichia coli, Bacillus subtilis, Corynebacterium glutamicum,* and lactic acid bacteria. We discuss the techniques used to display cellulases on their surfaces, their recombinantly achieved cellulolytic activities, and current obstacles that limit their utility.

Keywords: lignocellulose, consolidated bioprocessing, cellulase, minicellulosome, cell surface display

1. Introduction

As concerns over limited petroleum supplies rise, the momentum to produce renewable fuels, chemicals, and other materials from biomass has increased [1–3]. Second-generation biofuels derived from sustainable feedstocks such as lignocellulosic biomass are attractive, as plant biomass is cheap and highly abundant; over 1 billion tons of lignocellulosic biomass are produced annually in the United States, while an estimated 10–50 billion tons of waste lignocellulose are produced worldwide [4–6]. However, the resistance of lignocellulose to hydrolysis limits its use in biofuel production and has driven the search for new technologies to cost-effectively exploit this valuable resource [7, 8]. To produce fermentable sugars to generate cellulosic ethanol, conventional industrial processes utilize a multistep procedure to degrade lignocellulose (**Figure 1A**) [7, 9]. Typically, the biomass is first pretreated with strong acids and high temperatures to remove lignin and to partially degrade its cellulose and hemicellulose components [7]. It is then exposed to purified cellulase enzymes that hydrolyze the remaining polysaccharides into shorter polysaccharides and monosaccharides that can be fermented into ethanol by yeast. Significant effort has been put forth to optimize these steps, including the development of a range of new pretreatment approaches and enzyme cocktails [9–11]. It is generally believed that the cost of converting biomass into useful biocommodities could be greatly reduced by using a consolidated bioprocessing (CBP) microbe, a single microorganism that produces all of the necessary enzymes to degrade lignocellulose and then utilizes the resulting sugars to produce high levels of the biocommodity (**Figure 1B**) [12–14]. A CBP microbe would decrease costs by reducing the number of processing steps required to generate the final product and avoid the use of costly purified cellulase enzymes that are estimated to contribute $0.68–1.47 to the per gallon cost of cellulosic ethanol [15]. Given their great potential, a number of research groups have sought to develop a lignocellulose-utilizing CBP microbe using native and recombinant strategies [12]. In the native strategy, product production pathways are engineered into naturally cellulolytic microbes, while in the recombinant strategy, genetically well-studied microorganisms that may already be capable of producing a desired product are engineered to possess cellulolytic activity. Here, we describe progress towards creating recombinant cellulolytic microbes to convert biomass into useful commodities by engineering their surfaces to display cellulase enzymes.

Lignocellulose is recalcitrant to degradation and requires the action of many different types of enzymes to break it down [15]. It is composed of varying amounts of cellulose (25–55%), hemicellulose (8–30%), and lignin (18–35%) [7, 16]. Cellulose is a polymer of β-1,4-linked glucose molecules that can hydrogen-bond with other cellulose polymers to form both amorphous and crystalline regions [17]. It is synergistically degraded by three types of cellulases: endoglucanases, exoglucanases, and β-glucosidases [7]. Endoglucanases attack within a cellulose strand to hydrolyze the β-1,4-glucosidic bonds, producing new reducing and nonreducing ends that can be further broken down by exoglucanases [18]. The shorter cellodextrin chains that are produced by these enzymes, including the disaccharide cellobiose, are then degraded into glucose monomers by β-glucosidases [18]. Hemicellulose is a sugar polymer that is composed of a number of different types of pentose and hexose sugars [16]. Xylan is its main component and is the second most abundant polysaccharide in lignocellulose. As compared to cellulose, hemicellulose is more accessible to degradation by a range of enzymes with different substrate specificities, including among others: xylanases, arabinases, and mannanases. Finally, lignin surrounds and blocks enzyme access to cellulose and hemicellulose and is a complex polymer containing a mixture of phenolic compounds linked through radical coupling reactions [19]. A large number of enzymes are needed to degrade it, including peroxidases and laccases [20].

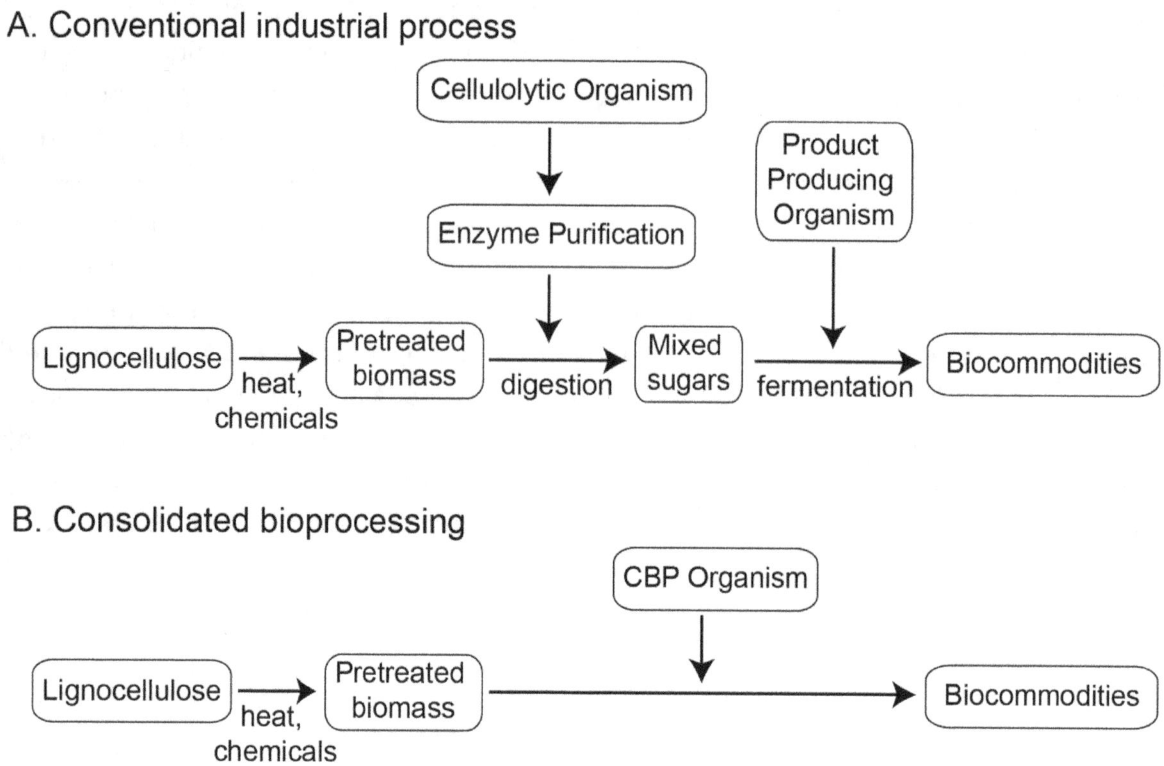

Figure 1. Conventional and consolidated bioprocessing (CBP) methods to convert biomass into biocommodities. (A) Many conventional industrial methods first pretreat biomass using heat and chemicals to remove lignin and to partially digest cellulose and hemicellulose. Complex enzyme mixtures are then added to the pretreated biomass to degrade its cellulose more fully. The resultant sugars are then fermented by a microorganism to generate the final product (currently only cellulosic ethanol). (B) A consolidated bioprocessing (CBP) microorganism could directly convert either pretreated or untreated lignocellulose into a biocommodity. It would bypass multiple steps in the conventional process, including the use of purified enzymes to degrade the biomass.

One promising recombinant strategy to create a useful CBP microbe is to engineer it to display a range of enzymes that degrade lignocellulose, thereby allowing it to produce sugars that can be further metabolized by the microorganism into useful chemicals [21–26]. In this approach, the goal was to engineer the microbial surface to effectively mimic the activity of naturally cellulolytic anaerobic bacteria. These microbes have the impressive capacity to adhere to, and degrade, untreated biomass and are typified by the cellulolytic anaerobic bacterium *Clostridium thermocellum* [27]. This eubacterium efficiently degrades biomass using a surface displayed complex called a cellulosome (**Figure 2A**) [27]. The cellulosome contains a central scaffolding protein that coordinates the binding of different enzymes. The primary scaffoldin, CipA, contains nine type-I cohesin modules that bind to cellulases harboring type-I dockerin modules [28]. CipA also contains a carbohydrate-binding module (CBM) that tethers the cellulosome complex to its cellulose substrate and a C-terminal type-II dockerin module [29]. The type-II dockerin module anchors the cellulosome complex to cell surface proteins by interacting with their type-II cohesin domains. As these surface proteins can contain multiple cohesin domains (1, 2, or 7 domains), large polycellulosomal structures can be displayed. Even more complex polycellulosomal structures exist in other species of bacteria and can contain over 100 enzymes [28].

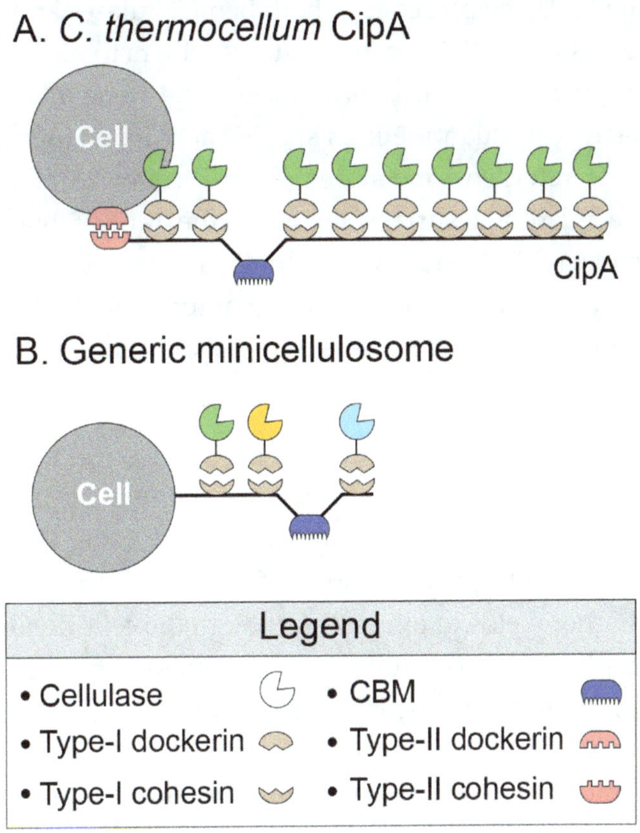

Figure 2. Natural cellulosomes and recombinant minicellulosomes. (A) The prototypical cellulosome from *C. thermocellum*. It contains the CipA scaffoldin that is capable of coordinating the binding of nine cellulases. CipA contains nine type-I cohesin domains that bind to cellulases containing type-I dockerin domains. It also contains a carbohydrate-binding module (CBM) that can bind to cellulose, holding the enzymes and the microorganism in close proximity to the substrate. CipA is anchored to the cell surface through its type-II dockerin domain that interacts with type-II cohesin modules present in the cell wall bound SdbA, Orf2, or OlpB proteins. (B) Basic structure of a minicellulosome that is displayed on a recombinant microbe. It is similar to the cellulosome, but contains fewer binding sites for enzymes. Color and symbol code: grey circles, cells; pink units, type-II cohesin-dockerin pairs; tan units, type-I cohesin-dockerin pairs; dark blue comb, CBM; and partial circular units: cellulases.

Microbes displaying cellulosomes are believed to degrade lignocellulose much more efficiently than microbes that degrade biomass by secreting cellulases [30]. There are three main reasons why increased efficiencies may be obtained. First, secreting enzymes presumably imposes a higher metabolic burden upon the microorganism as compared to displaying the enzymes on the cell surface. This is because the secreted enzymes can be lost to the environment with no guarantee that the sugars that they will produce will ultimately be accessible to the microbe for use as nutrients. As a result, larger quantities of the enzymes must be produced if they are to be secreted. For example, aerobic fungi (such as *Trichoderma reesei*, from which many purified industrial enzyme cocktails are derived) secrete 1–10 g of enzymes per liter of culture to degrade biomass, while cellulolytic anaerobic bacteria (such as *C. thermocellum*) need to produce only ~0.1 g of enzymes per liter of culture [31]. Second, colocalization of enzymes with different substrate preferences within a cellulosome promotes synergistic enzyme-enzyme and enzyme-proximity interactions, where the cellulolytic activity of the complexed enzymes is greater than that of individual enzymes due to their complementary activities and optimal enzyme spacing [32]. The presence of both hemicellulases and cellulases in a cellulosome complex also enables hemicellulose and cellulose fibers to be removed simultaneously, thereby overcoming potential physical hindrances. The benefits of the *C. thermocellum* cellulosomal system have been quantified: its specific activity against crystalline cellulose is 15-fold higher than the secreted enzyme system from *T. reesei* [33]. Finally, the placement of the cellulosome on the microbial surface increases the rate of hydrolysis by promoting cellulose–enzyme–microbe (CEM) synergy [34]. In this process, sugar uptake by the microbe presumably becomes more efficient by promoting import of the products into cells and removing potential enzyme inhibitors such as glucose and cellobiose from the environment [35].

Microbe-based CBP technologies are not currently used industrially to produce biocommodities from lignocellulosic biomass. However, a major step towards CBP of starch into ethanol has recently been demonstrated by Lallemand and Mascoma. These companies created TransFerm®, a genetically modified yeast strain that secretes a glucoamylase and that is optimized to produce higher yields of ethanol. Although the cells do not fully degrade starch on their own, they reduce by 20–45% the amount of exogenous enzymes that needs to be added to process starch. At present, microbe-based CBP of lignocellulose is not being performed industrially, and ongoing research is primarily focused on constructing and identifying microorganisms with optimal cellulolytic and biocommodity production capabilities. Furthermore, detailed cost analyses of CBP versus conventional pretreatment and saccharification approaches have not been reported [36]. This is because the specific costs associated with CBP will vary based on the biocommodity produced, and the microbe and biomass source that is employed. However, the greatest cost savings associated with CBP will likely be obtained by reducing the costs of saccharification. A detailed cost analysis has been performed for cellulosic ethanol production from corn stover using dilute acid pretreatment, enzymatic saccharification, and cofermentation [37]. In this analysis, on-site production of fungal enzymes was estimated to contribute $0.34 per gallon of ethanol (assuming enzyme loadings of 20 mg enzyme per gram of biomass), which could in principle be eliminated using a CBP microbe. Another analysis by Johnson explored the potential cost savings associated with altering the source of enzyme production from off-site to on-site cultivation, specifically on biomass as a primary substrate [38]. The estimated full cost of producing cellulosic ethanol was reduced by 19% if the enzymes were produced on-site because it eliminated the need for enzyme purification, formulation of the enzyme mixture to preserve stability, and transport. Similar substantial gains could be obtained using microbe-based CBP.

In this chapter, we review progress towards engineering microbes to display "minicellulosomes," smaller cellulosome-like complexes that can degrade biomass (**Figure 2B**). A list of the microorganisms engineered to display minicellulosomes discussed in this chapter is presented in **Table 1**. We discuss recent developments in displaying these structures on industrially useful microorganisms, including *Saccharomyces cerevisiae, Escherichia coli, Bacillus subtilis, Corynebacterium glutamicum,* and lactic acid bacteria. The mechanisms used to display enzymes, their cellulolytic activities, and current obstacles that limit their utility are discussed.

Anchor	Assembly (# enzymes/ complex)	Enzymes displayed		References
S. cerevisiae				
Aga2	*Ex vivo* (3)	*C. cellulolyticum*: *C. thermocellum* and *C. cellulolyticum* or *C. thermocellum*:	Exoglucanase (CelE) Endoglucanase (CelA) Endoglucanase (CelG) β-Glucosidase (BglA)	[73]
Aga2	*Ex vivo* consortium (3)	*C. thermocellum* *T. aurantiacus* and *C. cellulolyticum* or *T. reesei*:	Endoglucanase (CelA) β-Glucosidase (Bgl1) Exoglucanase (CelE) Cellobiohydrolase (CBHII)	[74]
α-Agglutinin	*Ex vivo* consortium (3)	*C. thermocellum*: *T. aurantiacus*: *T. reesei*:	Endoglucanase (CelA) β-Glucosidase (Bgl1) Cellobiohydrolase (CBHII)	[75]
Aga2	*Ex vivo* consortium (2)	*C. thermocellum*: *T. reesei*:	Endoglucanase (CelA) Exoglucanase (CBHII)	[76]
Aga2	*Ex vivo* adaptive assembly (4)	*C. cellulolyticum*: *C. thermocellum*:	Endoglucanase (CelG) β-Glucosidase (BglA)	[77]
Aga2	*In vivo* (3)	*A. aculeatus*: *T. reesei*:	β-Glucosidase 1 (BGL1) Cellobiohydrolase II (CBHII) Endoglucanase II (EGII)	[32]
Aga2	*In vivo* (5)	*A. aculeatus*: *T. reesei*: *H. insolens*: *T. aurantiacus*:	β-Glucosidase 1 (BGL1) Cellobiohydrolase II (CBHII) Endoglucanase II (EGII) Cellobiose dehydrogenase (CDH) Lytic polysaccharide monooxygenase (LPMO)	[78]
Aga2	*In vivo* adaptive assembly (12 max)	*C. cellulolyticum*:	Endoglucanase (CelCCA) Cellobiohydrolase (CelCCE) β-Glucosidase (Ccel_2454)	[79]
α-Agglutinin	*Ex vivo* (3)	*A. awamori*: *A. niger*: *T. lanuginosus*:	Acetylxylan esterase (AwAXEf) β-xylosidase (XlnDt) Endoxylanase (XynAc)	[87]
Aga2	*In vivo* (3)	*A. niger*:	Arabinofuranosidase (AbfB)	[88]

Anchor	Assembly (# enzymes/ complex)	Enzymes displayed		References
S. cerevisiae				
α-Agglutinin	*In vivo* (4)	*T. reesei*: *A. aculeatus*: *T. reesei*:	β-Xylosidase (XlnD) Endoxylanase (XynII) β-Glucosidase 1 (BGL1) Endoglucanase II (EGII)	[89]
Aga2	*Ex vivo* (N/A)	*C. cellulolyticum:*	Endoglucanase (CelA)	[90]
B. subtilis				
LysM	*Ex vivo* (3)	*B. subtilis*: *C. phytofermentans*: *C. thermocellum*:	Endoglucanase (Cel5) Cellobiohydrolase (Cel48) Endoglucanase (Cel9)	[31]
LPXTG from *S. aureus*	*Ex vivo* (3)	*C. cellulolyticum*: *C. thermocellum*:	Exoglucanase (Cel9E) Endoglucanase (Cel9G) Endoglucanase (Cel8A)	[118]
Lactic acid bacteria				
LPXTG from *S. pyogenes*	*Ex vivo* (1-2)	*E. coli*:	β-Glucuronidase (UidA)	[127]
LPXTG from *S. pyogenes*	*Ex vivo* (2)	*E. coli*:	β-Glucuronidase (UidA) β-Galactosidase (LacZ)	[128]
LPXTG from *L. plantarum*	*Ex vivo* consortium (2)	*T. fusca*:	Endoglucanase (Cel6A) Xylanase (Xyn11A)	[129]
C. glutamicum				
MscCG	*In vivo* (2)	*C. thermocellum*:	Endoglucanase (CelE) β-Glucosidase (BglA)	[133]

Table 1. Multi-enzyme display in minicellulosomes.

2. Displaying enzymes on *Saccharomyces cerevisiae*

Significant effort has been put forth to display cellulolytic enzymes on the surface of *S. cerevisiae* because of its established role in producing bioethanol [39]. The yeast cell surface is comprised of β-glucans, mannoproteins, and small amounts of chitin [40]. As β-glucan is the major constituent, displayed proteins are typically anchored to this fibrous scaffold. The most widely used approach to display cellulases and minicellulosomes employs a glycosylphosphatidylinositol (GPI)-anchoring system (**Figure 3A**). The protein of interest is expressed as a fusion with a polypeptide GPI attachment signal that is typically derived from the α-agglutinin protein or other cell wall proteins such as Sed1 and Cwp2 [41, 42]. After protein synthesis, GPI is added to the ω-site amino acid in the anchor signal sequence by the GPI transamidase complex in the endoplasmic reticulum [43, 44]. The GPI-linked protein is then directed towards the lipid bilayer and subsequently covalently linked to the cell wall β-1,6-glucan by the putative Dfg5 and Dcw1 cross-linkers [45–48]. Cellulase can also be displayed indirectly. In this approach, the enzyme is expressed as a fusion protein with the a-agglutinin Aga2 subunit, which in turn, forms disulfide bonds to the Aga1 subunit that is covalently linked to the cell wall via a GPI anchor. An estimated 1×10^4 to 1×10^5 proteins per cell are displayed using the Aga2 subunit [49]. The number of proteins displayed on the surface can be increased by genetically modifying the yeast strain. For example, deletion of a major endogenous GPI-anchored cell wall protein, SED1, can greatly improve heterologous protein display by reducing competition for cell wall-anchoring sites [50]. Display levels can also be improved by lowering the mannan content of the cell surface [51] and by employing the SED1 signal peptide, promoter and anchor [52, 53]. Reducing protein glycosylation also improves minicellulosome display [54]. The general strategies used to enhance enzyme display in yeast may also be useful in optimizing the other recombinant microbes that are discussed below.

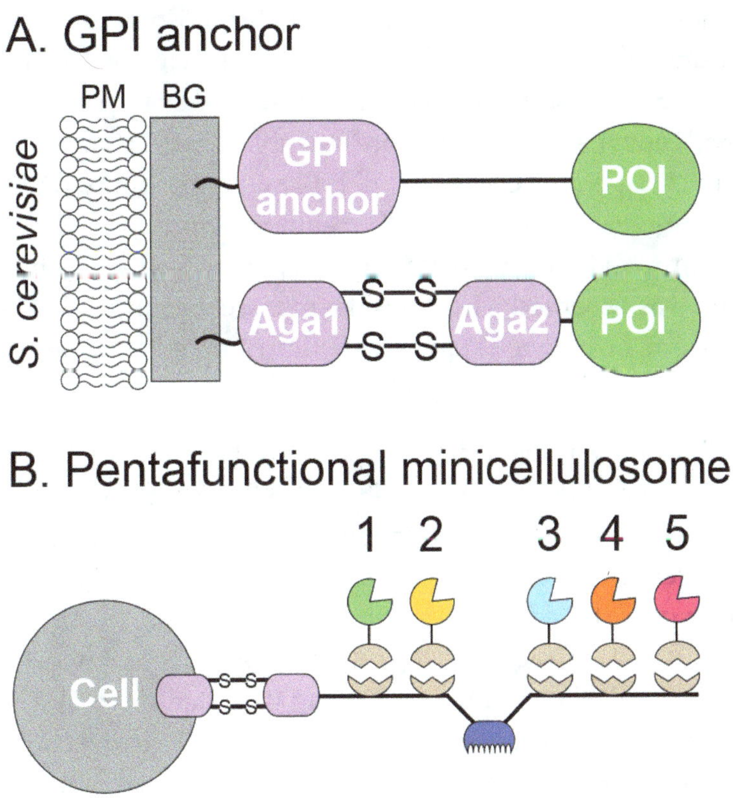

Figure 3. Cellulase and minicellulosome display in *S. cerevisiae*. (A) Methods used to display cellulases and their complexes on the cell surface. Proteins of interest (POI) are attached to the cell surface by a GPI anchor that is covalently linked to the β-glucan of the cell wall. The GPI anchor is attached to the displayed protein by expressing it as a fusion protein that contains the GPI attachment signal from either the α-agglutinin, SED1, or Cwp2 proteins. Proteins can also be displayed indirectly by expressing them as a fusion with the Aga2 protein that in turn interacts with the GPI anchored a-agglutinin Aga1 located on the cell surface. (B) One of the most active cellulolytic minicellulosomes that has been displayed on the surface of *S. cerevisiae*. It contains five enzyme functionalities and is displayed through fusion of the miniscaffoldin to the Aga2 protein. Symbols have been described in **Figure 2**. Enzymes 1–5 correspond to *T. reesei* endoglucanase (EGII), *T. reesei* cellobiohydrolase (CBHII), *A. aculeatus* β-glucosidase (BGL1), *T. aurantiacus* GH61a lytic polysaccharide monooxygenase (LPMO), and *H. insolens* cellobiose dehydrogenase (CDH), respectively. Abbreviations: POI, protein of interest; PM, plasma membrane; BG, β-glucan.

2.1. Individual cellulases

Individual cellulases were originally displayed on the cell surface by the Tanaka group [55]. They engineered cells to codisplay individual carboxymethylcellulase (CMCase) and β-glucosidase (BGL1) enzymes derived from *Aspergillus aculeatus* by fusing them to a C-terminal GPI anchor sequence from α-agglutinin. The cells could degrade cellodextrins, soluble glucose polymers that contain up to six glucose residues. Subsequently, the Kondo laboratory significantly advanced this technology by developing *S. cerevisiae* strains that display additional enzymatic activities. Initially, they constructed strains that codisplayed the *A. aculeatus* BGL1 and *T. reesei* endoglucanase II (EGII) enzymes, which when cultured in nutrient-rich media fermented β-glucan into ethanol [56]. Later, the inclusion of the *T. reesei* cellobiohydrolase II (CBHII) enzyme allowed the strains to ferment more complex amorphous phosphoric acid swollen cellulose (PASC) into ethanol [57]. Separately, the Kondo group also demonstrated that cellulose degradation can be further improved using a variety of approaches, including engineering a displayed cellulase to contain multiple cellulose-binding domains, integrating multiple copies of the cellulases genes into the yeast chromosome, and by strain diploidization [58–60]. Increasing the number and density of the displayed cellulases significantly improves the ability of *S. cerevisiae* to degrade cellulose. For example, a strain displaying the BGL1, EGII, and CBHII enzymes produces more ethanol from PASC than a strain that displays only BGL1 and secretes EGII and CBHII [61, 62]. Enzyme proximity on the cell surface also appears to have a major effect on activity, as the cellulolytic properties of cells displaying one or more cellulases improved as the enzyme density increased [63]. Interestingly, in these studies, good activity was observed when the enzymes were estimated to be separated on the cell surface by 10–100 nm, an enzyme spacing that is presumably similar to the spacing found in certain types of bacterial cellulosomes [64].

Yeast displaying cellulases may be industrially useful. For example, strains displaying the BGL1, CBHII, and EGII enzymes produced 43.1 g/L ethanol from 200 g/L liquid hot water pretreated rice straw in 72 h [65]. While supplementation with a purified cellulase cocktail at 10 filter paper units (FPU)/g-biomass was necessary to achieve this high ethanol yield, a control strain that did not display the enzymes required 10-fold more added cellulase to produce similar quantities of ethanol. Attractively, the cells displaying the enzymes could be reused in five fermentation cycles without significantly losing their activity [66]. Displaying enzymes also reduces the amount of purified cellulase lost through irreversible adsorption onto crystalline cellulose, facilitating more efficient cellulose degradation and higher ethanol yields [67]. Recently, studies using cellulase displaying cells have further reduced the amount of purified cellulase cocktail that needs to be added to convert biomass into ethanol [68]. These newer generation cells require 44% less commercial enzyme supplementation to degrade pretreated biomass by displaying four enzymes using the Sed1 anchor: *A. aculeatus* BGL1, *T. reesei* EGII, *Talaromyces emersonii* cellobiohydrolase I (CBHI), and *Chrysosporium lucknowense* cellobiohydrolase II (CBH2). When supplemented with 1 FPU of commercial enzyme cocktails, the cells yielded 18 g/L ethanol from 100 g/L of pretreated and milled rice straw in 96 h, obtaining 80% of the theoretical yield. In contrast, without enzymatic supplementation, only 7% of the theoretical yield was obtained using the same cells. Therefore, while the use of cellulase displaying yeast cells can significantly reduce the amount of commercial enzyme that needs to be added to degrade pretreated biomass, current generation cells still require enzyme supplementation.

Towards the goal of improving their cellulolytic activity, several studies have engineered cellulase displaying yeast cells to also produce complementary enzymes and transporters that improve cellulose solubilization and utilization. The Kondo group constructed cells that co-expressed three displayed cellulases, as well as the *Neurospora crassa* cellodextrin transporter (CTDI) and intracellular BGL1 enzyme [69]. This strain produces 1.7-fold more ethanol from PASC than a strain that only displays the cellulases, presumably by reducing the build-up of enzyme inhibitory products on the cell surface. Additionally, strains displaying cellulases as well as proteins that disrupt the structure of cellulose show improvements in ethanol production; cells codisplaying three cellulases and an *Aspergillus oryzae* expansin-like protein (AoelpI) produce 1.4-fold more ethanol from PASC than a control strain that only displayed the cellulases [70]. Finally, the Ueda group demonstrated that sequential exposure of biomass to engineered yeast cells could be beneficial. In this work, pretreatment of hydrothermally processed rice straw with yeast displaying the *Trametes* sp. Ha1. laccase I enzyme enabled 1.21-fold more ethanol to be produced from the biomass after it was subsequently exposed to yeast displaying three cellulases [71]. These studies highlight interesting enzyme functionalities that should be considered for inclusion in other recombinant microorganisms.

2.2. *Ex vivo* assembled minicellulosomes

Several research groups engineered *S. cerevisiae* to display minicellulosomes that are smaller than natural cellulosomes (**Figure 2B**). In the minicellulosome, a cell-surface displayed "miniscaffoldin" protein that contains cohesin domains noncovalently binds cellulases via their dockerin modules. These recombinant miniscaffoldins are either attached directly to the cell wall by fusing them to a GPI anchor sequence (from α-agglutinin or Cwp2) or indirectly by fusing them to the a-agglutinin Aga2 subunit. Two general methods are used to display minicellulosomes, an *ex vivo* approach in which cellulases need to be added to cells that display a miniscaffoldin (**Figure 4A**, **4B**), or an *in vivo* approach in which the microbe produces all of the protein components that are necessary to assemble and display the minicellulosome (**Figure 4C**).

In 2009, two groups demonstrated *ex vivo* minicellulosome assembly by creating cells that display a miniscaffoldin and then adding to the cells cellulases that contained a dockerin domain. The Volschenk group displayed a miniscaffoldin by fusing two cohesin modules and a CBM to the Cwp2 GPI anchor. They visually demonstrated binding of this yeast strain to filter paper via the CBM and constructed a minicellulosome by adding a purified endoglucanase–dockerin enzyme to the cells [72]. Concurrently, the Chen laboratory constructed an Aga2-fused recombinant minicellulosome that contained three cohesin modules derived from three different bacterial species [73]. This more elaborate minicellulosome used species-specific cohesin–dockerin interactions to target β-glucosidase, exoglucanase, and endoglucanase enzymes to specific sites within the complex. Quantitatively, the researchers demonstrated that incorporation into the minicellulosome enabled the enzymes to function synergistically to

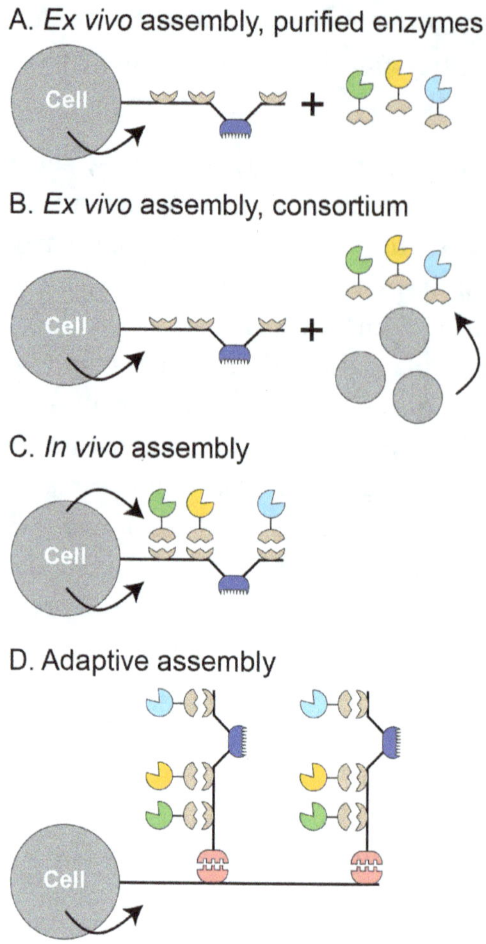

Figure 4. Summary of strategies used to assemble minicellulosomes on microbial surfaces. (A) *Ex vivo* assembly. Exogenous cellulase-dockerin fusion proteins (either purified or derived from cell lysates) are added to cells engineered to display a miniscaffoldin that binds to the cellulases. (B) *Ex vivo* assembly using a microbial consortium. Similar to (A), but miniscaffoldin displaying cells are cultured with a consortium of microbes that have been engineered to secrete cellulase-dockerin fusion proteins. (C) *In vivo* assembly. A single microorganism is engineered to express all of the components of the minicellulosome, enabling its spontaneous assembly on the cell surface. (D) Adaptive assembly method used to construct larger minicellulosomes. The miniscaffoldin contains two basic parts, a non-catalytic anchor-anchoring scaffoldin that attaches the structure to the cell wall and a primary scaffoldin that binds to the cellulase-dockerin fusion proteins. Color and symbol code as described in **Figure 2.**

produce ethanol from PASC. Later, the Chen laboratory bypassed the need to add purified enzymes or cell lysates to cells to produce the minicellulosome by using a consortium of yeast strains to produce the cellulases [74]. This was accomplished using four strains: one that displayed the miniscaffoldin and three strains that secreted each of the enzymes needed to form the complex. After determining the optimal ratio of strains, they demonstrated that ~1.9 g/L of ethanol could be produced from 10 g/L of PASC in 48 h. Additionally, they demonstrated that fusing the miniscaffoldin directly to the cell wall via an attached α-agglutinin GPI anchor instead of indirectly through the Aga2 subunit improved minicellulosome display [75]. Similarly, the Hahn laboratory utilized a consortium approach to build an *ex vivo* assembled minicellulosome that contained three identical cohesin–dockerin pairs and randomly incorporated enzymes [76]. Cells displaying this structure also obtained similar ethanol yields, 1.8 g/L of ethanol from 10 g/L of PASC in 94 h.

In order to build larger surface structures with higher enzyme densities, the Chen laboratory used an adaptive assembly approach [77]. In this procedure, the minicellulosome is built using two separate scaffoldin pieces: a primary scaffoldin that binds to the catalytic components and an anchoring scaffoldin that is attached to the cell surface and only binds to the primary scaffoldin (**Figure 4D**). They produced a yeast strain that displayed an Aga2-fused anchoring scaffoldin that contained two cohesin domains from *Acetivibrio cellulolyticus* and *Bacteroides cellulosolvens*. Two *E. coli* strains were then used to produce two primary scaffoldins that each contained the *C. thermocellum* and *R. flavefaciens* cohesins, and either the dockerin module from *A. cellulolyticus* or *B. cellulosolvens*. Thus, the final complex contained a total of four enzymes, with two copies each of the *C. cellulolyticum* endoglucanase (CelG) and *C. thermocellum* β-glucosidase (BglA) enzymes. This strain produced only 1.9 g/L of ethanol from 10 g/L of PASC in 72 h, presumably because it lacked exoglucanase activity.

2.3. *In vivo* assembled minicellulosome

The need to add enzymes to cells displaying a miniscaffoldin may make the use of *ex vivo* assembled minicellulosomes industrially impractical. To overcome this problem, two research groups engineered *S. cerevisiae* to express all of the components of the minicellulosome such that it could form spontaneously (*in vivo* assembly). The Zhao laboratory constructed a yeast strain that produced an Aga2-fused miniscaffoldin, *T. reesei* EGII and CBHII, and *A. aculeatus* BGL1 enzymes [32]. By comparing the display of a single enzyme to the simultaneous display of two or three noncomplexed enzymes, they demonstrated enzyme-enzyme synergy, which improved activity by 5.5-fold. They also demonstrated enzyme-proximity synergy, as the trifunctional minicellulosome degraded biomass better than cells containing three independently displayed enzymes (1.6-fold improvement). Quantitative measurements indicated that each cell displays 3×10^4 unifunctional miniscaffoldins, whereas 1.8×10^4 tri-functional minicellulosomes are displayed per cell. *In vivo* construction of increasingly more elaborate minicellulosomes decreased protein expression, likely due to metabolic burden. Compared to unifunctional minicellulosomes, the production of the miniscaffoldin and two enzymes caused the expression of all components to decrease slightly, and overexpression of an additional third enzyme caused protein expression levels of CBHII to drop significantly. This suggests that it will be increasingly more challenging to display larger, more complex enzyme structures on the cell surface. This is unfortunate, because adding additional enzyme functionalities is advantageous to cellulose solubilization. Although difficult to achieve, for example, the tri-functional displayed minicellulosome constructed by the Zhao group produces 1.8 g/L of ethanol from 10 g/L of PASC in 70 h. However, pentafunctional minicellulosomes are more effective, as they produce 2.7 g/L of ethanol from 10 g/L of PASC in 96 h and 1.8 g/L of ethanol from 10 g/L of Avicel in 96 h [78]. This larger pentafunctional complex contains two additional enzymes with new functionalities, the *Thermoascus aurantiacus* GH61a lytic polysaccharide monooxygenase (LPMO) and *Humicola insolens* cellobiose dehydrogenase (CDH) (**Figure 3B**). Even though fermentative growth was reduced because of the presence of oxygen needed for monooxygenase function, greater ethanol yields were still obtained.

The largest *in vivo* assembled minicellulosome reported thus far was constructed by the Tan group using an adaptive assembly approach [79]. This structure could display up to 12 enzymes and was formed using a cell surface-associated Aga2-fused anchoring scaffoldin that contained type-II cohesin domains that could bind four primary miniscaffoldins harboring type-II dockerin domains. Each primary scaffoldin also contained a CBM and three species-specific type-I cohesin modules for controlled cellulase binding. Combined, this strain displays in a single complex four copies each of the *C. cellulolyticum* endoglucanase, cellobiohydrolase, and β-glucosidase enzymes. However, the investigators found that as longer anchoring scaffoldins were used, a smaller percentage of cells displayed the scaffoldin. For this reason, they demonstrated the production of 1.4 g/L ethanol from 10 g/L Avicel in 96 h from a smaller six enzyme displaying strain.

2.4. Hemicellulases and hemicellulosomes

In order to develop xylose-fermenting strains of yeast, similar strategies have been employed to display hemicellulases and hemicellulosomes on the cell surface. The Kondo laboratory displayed an individual *T. reesei* endoxylanase II (XYNII) enzyme using the α-agglutinin anchor and demonstrated that the cells degraded birchwood xylan into xylobiose, xylotriose, and xylotetraose [80]. Subsequently, the cells were further engineered to produce ethanol from birchwood xylan by adding the *A. oryzae* β-xylosidase (XylA) to the cell surface and by expressing intracellular xylose utilization proteins (the *Pichia stipitis* xylose reductase (XYL1) and xylitol dehydrogenase (XYL2), as well as the *S. cerevisiae* xylulokinase (XKS1)) [81]. In a separate set of studies, the investigators later demonstrated cellodextrin degradation and xylose assimilation by yeast displaying *A. aculeatus* BGL1 enzyme, which enabled the microbe to produce ethanol from the sulfuric acid hydrolysate of wood chips [82]. Interestingly, the ability of this strain to co-utilize cellobiose and xylose avoided carbon catabolite repression [83]. Additional engineering created cells that displayed the XylA, XYNII, and BGL1 enzymes, enabling the xylose-utilizing strain to produce 8.2 g/L of ethanol from 80% (v/v) rice straw hydrolysate [84]. When *P. stipitis* xylose reductase (XR) was produced intracellularly in place of the three aforementioned xylose utilization proteins, xylitol accumulated, which is a starting material for chemical production of some pharmaceuticals [85]. Using an alternative pathway to utilize xylose, the Ueda group produced α-agglutinin anchored *Clostridium cellulovorans* xylose isomerase (XI) which converted xylose to xylulose extracellularly before fermenting it to ethanol [86].

Clustering hemicellulases within surface displayed complexes leads to improved enzymatic activity. The Silva group developed an *ex vivo* assembled xylanosome using the α-agglutinin anchor [87]. The xylanosome contained the *Thermomyces lanuginosus* endoxylanase (XynAC), *Aspergillus niger* β-xylosidase (XlnDt), *Aspergillus awamori* acetylxylan esterase (AwAXEf) enzymes, as well as the xylose-binding domain (XBD) from *Thermotoga maritima*. The xylanases in the complex functioned synergistically, with enzyme-enzyme synergy improving xylan hydrolysis by 1.6-fold, and enzyme-substrate synergy improving hydrolysis by 3.3-fold as compared to the free enzymes. Proximity of the endoxylanase to the XBD improved xylan hydrolysis by 2.5-fold, suggesting that placement of substrate binding domains can contribute significantly to hydrolysis. The Zhao laboratory created an *in vivo* assembled minihemicellulosome that is structurally related to their minicellulosome complex described above [88]. The minihemicellulosome contained the *A. niger* arabinofuranosidase (AbfB), β-xylosidase (XlnD), and *T. reesei* endoxylanase (XynII) enzymes. The yeast strain also possessed *P. stipitis* xylose utilization enzymes and could hydrolyze arabinoxylan better than the uni- and bifunctional yeast strains. Surprisingly, against birchwood xylan, a bifunctional minihemicellulosome containing XynII and XlnD exhibited a higher hydrolysis rate than the tri-functional complex, producing 0.95 g/L of ethanol from 10 g/L of birchwood xylan in 80 h.

2.5. Artificial cellulosome structures

In order to better control enzyme placement and increase the number of displayed enzymes on the cell surface, several groups have created artificial cellulosomes that are structurally distinct from naturally occurring cellulosomes. These structures use unique protein-protein interaction domains to tether the enzymes to the scaffoldin instead of naturally occurring cohesin-dockerin interactions. The Kondo laboratory created an *in vivo* assembled minicellulosome using a miniscaffoldin that contained two cohesin domains and two Z domains derived from *Staphylococcus aureus* protein A [89]. This enabled codisplay of the *A. aculeatus* BGL1—dockerin fusion protein and the *T. reesei* EGII protein fused to the Fc domain of human IgG. The group suggested that this artificial construct enables tighter regulation of the display ratio of the enzymes. Similarly, the use of type-I and type-II cohesin-dockerin pairs, or cohesin-dockerin pairs from different species, allows control over enzyme incorporation. The advantages of this noncellulosome-like structure were not extensively investigated, but activity against β-glucan was demonstrated. The Su group has displayed enzymes using a novel *ex vivo* assembled complex that contains an amyloid-like scaffoldin [90]. A multistep process was required to assemble the complex. Initially, yeast expressed an Aga2 anchored GFP-specific antibody fragment. Then, cells were incubated with a GFP-dockerin fusion protein to prime the cells to bind the protein scaffold. The protein scaffold itself was then created through the fibrillation of a recombinant-purified protein that contained a cohesin and a hydrophilic linker region that was fused to the N-terminus of Ure2, an amyloid-like yeast protein. In this way, multiple cohesin domains were incorporated into a large protein scaffold. As compared to cells displaying a single cohesin-dockerin pair harboring the *C. cellulolyticum* endoglucanase (CelA), cells displaying the amyloid-like scaffoldin had 8.5-fold greater activity against CMC. However, the process of assembly is complex, and certain cohesin domains may be rendered nonfunctional during fibrillation. The use of extremely large supramolecular scaffoldins to coordinate enzyme binding may also reduce the potential benefits of CEM synergy. Whether the enzymatic activities of these artificial cellulosomes are superior to complexes that more closely resemble natural cellulosomes remains to be determined, as their ability to degrade complex cellulose substrates to produce ethanol was not reported.

3. *Escherichia coli* surface display of individual cellulases

The model Gram-negative bacterium *E. coli* has several features that are attractive for CBP, including a robust genetic system that enables surface and metabolic engineering, as well as the microbe's innate ability to utilize the main components of lignocellulose, glucose, arabinose, and xylose [91, 92]. Its cell envelope is comprised of an inner membrane, a peptidoglycan layer, and an outer membrane. Only individual, noncomplexed cellulases and xylanases have been displayed on the surface of *E. coli* by fusing them to lipoproteins or integral membrane proteins (**Figure 5**). Lipoprotein anchors that have been used to display cellulases include the *E. coli*-derived Lpp-OmpA fusion protein [93, 94], the *P. syringae* ice nucleation protein (INP) [95–97], and the *E. coli* bacterial lipocalin (Blc) protein [98–100]. Cellulases have also been displayed by fusing them to outer membrane proteins such as the: *B. subtilis* poly-γ-glutamate synthase A (PgsA) [101, 102], *E. coli* outer membrane protein C (OmpC) [103], and the *E. coli* outer membrane protein X (OmpX) [104]. Cellulases can also be displayed through a unique mechanism in which they are fused to the *E. coli* autotransporter protein (AIDA-I) [105]. Other less commonly used approaches to display cellulases include fusing them to the *E. coli* inner membrane protein HdeD [100], or to the *B. anthracis* BclA exosporal protein whose mechanism of display in *E. coli* is not well understood [106]. In most cases, except for the work reported by the Kondo and the Karim groups, only single enzymes have been displayed on the surface, leading to cells that have limited cellulolytic activity [100, 101].

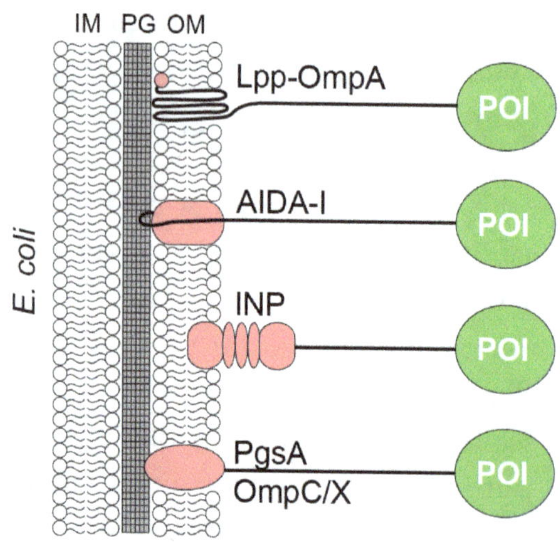

Figure 5. Methods used to display cellulases in *E. coli*. Proteins of interest are displayed as a result of fusion to lipoproteins (*E. coli* Lpp-OmpA, *P. syringae* INP, *E. coli* Blc), fusion to outer membrane proteins (*B. subtilis* PgsA, *E. coli* OmpC, *E. coli* OmpX), or fusion to the *E. coli* autotransporter AIDA-I. Not all anchoring methods are depicted here. At present, only individual enzymes have been displayed on the cell surface. Abbreviations: protein of interest (POI), inner membrane (IM), peptidoglycan (PG),and outer membrane (OM).

Pilot studies have shown that *E. coli* displaying cellulases can be used to produce ethanol or isopropanol from cellulosic substrates. Using the *E. coli* AIDA-I anchor, the display of the *T. fusca* β-glucosidase enzyme yielded cells that produced approximately 17–17.9 g/L of ethanol from 40 g/L of cellobiose in 72 h [105]. By utilizing the *E. coli* Blc anchor to display the *T. fusca* β-glucosidase, cells could produce 69.0 ± 11.6 mM of isopropanol in a 21 h fermentation containing 50 g/L of cellobiose [99]. The low yield of isopropanol, as compared to when glucose was supplied, suggested that the β-glucosidase activity was too low to efficiently produce glucose from cellobiose. The most complex *E. coli* display system thus far achieved displayed three enzymes using the *B. subtilis* PgsA anchor [101]. This strain displayed endocellulase, exocellulase, and β-glucosidase enzymes derived from *C. cellulolyticum*. From 10 g/L of PASC, it produced 3.59 ± 0.15 g/L of ethanol in 60 h. It could also make ethanol from dilute sulfuric acid pretreated corn stover hydrolysate (2 g/L), producing 0.71 ± 0.12 g/L of ethanol in 48 h. One notable study performed by Bokinsky et al. demonstrated that *E. coli* strains that secreted cellulases could produce advanced biofuels from ionic liquid pretreated biomass [107]. A consortium of two *E. coli* strains was used to produce either the *Bacillus* sp. D04 endocellulase and the *Cellvibrio japonicus* β-glucosidase, or the *Clostridium stercorarium* endoxylanase and the *C. japonicus* xylobiosidase. The consortia could grow on ionic liquid (IL)-treated biomass, albeit to half the cell density of strains grown on glucose. Additional engineering enabled cells to produce fatty-acid ethyl esters, butanol, and pinene from IL-pretreated switchgrass. However, only 5 and 6% of the available cellulose and hemicellulose were digested, respectively. Improved cellulolytic activity could be achieved by displaying cellulase containing complexes, but *E. coli* cells that display minicellulosomes have yet to be constructed.

4. Displaying cellulases and minicellulosomes in Gram-positive bacteria

Many species of Gram-positive bacteria are used industrially to produce biocommodities and have great promise as agents to produce second-generation biofuels. However, they are not naturally cellulolytic, prompting efforts designed to decorate their surfaces with cellulases. Below, we discuss progress towards creating cellulolytic strains of *B. subtilis*, *C. glutamicum*, and several industrially useful species of lactic acid bacteria. A common feature of these microbes is the absence of an extensive outer membrane, as they are surrounded by a thick peptidoglycan layer that in some instances is further surrounded by additional protective layers. Thus, the mechanisms used to display proteins on their surfaces are distinct from those employed to decorate Gram-negative bacteria.

4.1. *Bacillus subtilis*

B. subtilis is a model Gram-positive bacterium that is used industrially to produce commercial enzymes. It has several desirable traits that could enable its use in consolidated bioprocessing including established genetic tools to manipulate its genome as well as its generally recognized as safe status (GRAS) [108, 109]. It is also tolerant to high concentrations of salts and solvents and has the ability to utilize both pentose and hexose sugars produced from lignocellulose [110]. Finally, *B. subtilis* naturally secretes large quantities of extracellular enzymes (20–25 g of protein per liter of growth culture), suggesting that it should be capable of robustly exporting the enzymes needed to build cellulase containing complexes [111, 112]. The Gram-positive cell wall offers many sites for both covalent and noncovalent protein anchoring. Three mechanisms have been used to display cellulase on vegetative *B. subtilis* cells: membrane association via a lipoprotein anchor, noncovalent cell wall interactions using the LysM domain, and covalent attachment to the cell wall using sortase enzymes (**Figure 6A**). Covalent attachment occurs through sortase enzyme processing of a C-terminal cell wall sorting signal (CWSS) that contains a LPXTG motif and has been estimated to display ~2.4×10^5 proteins per cell [113, 114]. Noncovalent methods enable ~1.2×10^7 proteins to be displayed per cell when the protein of interest is fused to the LysM domain, a binding module that interacts with the N-acetyl-muramic acid and N-acetyl-D-glucosamine components of the peptidoglycan [115, 116]. Cellulases have also been displayed by expressing them as a fusion protein with the membrane-associated lipoprotein, PrsA [117]. However, this display mechanism requires lysozyme treatment of the cell to remove the peptidoglycan.

Two groups have displayed *ex vivo* assembled minicellulosomes on the surface of *B. subtilis*. The Zhang laboratory displayed a minicellulosome with a miniscaffoldin that contained three cohesin domains, a CMB, and three LysM domains [31]. An estimated 2×10^4 miniscaffoldins bound to the surface of each cell. Minicellulosomes were assembled by incubating the cells with purified *B. subtilis* endoglucanase, *C. thermocellum* endoglucanase, and *Clostridium phytofermentans* cellobiohydrolase enzymes. Against regenerated amorphous cellulose (RAC) and crystalline cellulose, respectively, cell-tethered minicellulosomes degraded substrate 2.3- and 4.5-fold better than free minicellulosomes. As compared to commercial fungal enzymes dosed at the same protein concentrations, minicellulosome-displaying cells degraded RAC to a similar extent after 72 h, but exhibited 30% greater hydrolytic activity on Avicel. The Clubb group also demonstrated *ex vivo* assembly of a minicellulosome complex using a LPXTG anchor that can be processed by a sortase enzyme [118]. Purified *C. cellulolyticum* endoglucanase and exoglucanase, and *C. thermocellum* endoglucanase enzymes associated with a covalently cell wall attached miniscaffoldin. A major challenge specific to *B. subtilis* is the large amount of proteases that this microbe produces that can degrade heterologous surface-exposed proteins [119]. The Zhang group addressed this problem by displaying the cellulosome in a *B. subtilis* strain in which six extracellular proteases had been deleted, while the Clubb group improved protein display by deleting WprA, a cell wall-associated protease. Display of an *in vivo* assembled minicellulosome has yet to be achieved. Notably, an *in vivo* assembled minicellulosome was reported that could degrade untreated biomass, but this work was later retracted.

B. subtilis sporulates to produce a highly stress-resistant, dormant spore cell that can be decorated with cellulases. During the process of spore formation within the mother cell, genomic DNA is encapsulated within multiple protective layers including a cortex, coat, and crust [120]. Attractively, proteins do not need to be translocated across the cytoplasmic membrane in order to be displayed on the spore. There has been a significant amount of research performed to optimize protein display on *B. subtilis* spores [121]. While spore coat proteins are typically used as carriers to display proteins, a recent study found that native, unmodified proteins could be overexpressed in the mother cell and absorbed to the spore surface for display (**Figure 6B**) [122]. Using this approach, monomeric *B. subtilis* carboxymethylcellulase (CelB) and multimeric *E. coli* β-galactosidase (LacZ) were successfully displayed on the spore surface, and vigorous physiochemical treatment was shown to be needed to

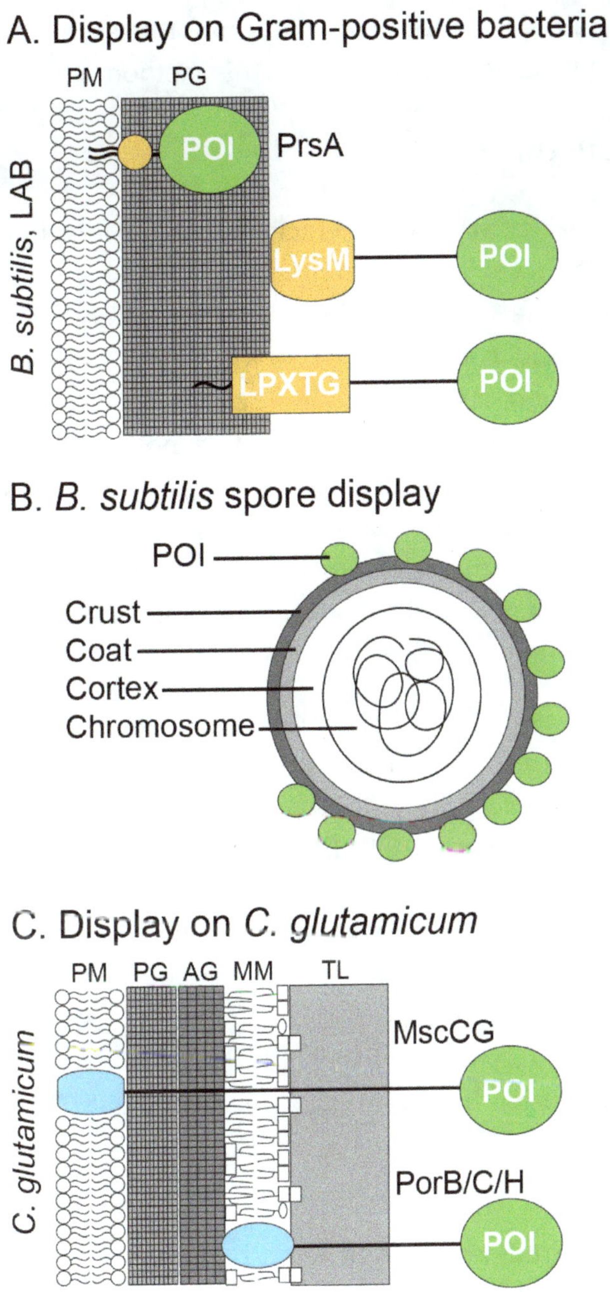

Figure 6. Cellulase and minicellulosome display in Gram-positive bacteria. (A) Methods used to display cellulases and their complexes. Proteins can be anchored to the plasma membrane through fusion to PrsA, but lysozyme treatment is necessary to expose this cellulase to the external environment. More conventionally, proteins are cell surface displayed either through noncovalent interactions with cell wall peptidoglycan by protein fusion to the LysM domain, or through covalent attachment to cell wall peptidoglycan by protein fusion to a cell wall sorting signal containing the LPXTG motif, which is processed by sortase transpeptidase. (B) Proteins can be displayed on the surface of *B. subtilis* spores. Native proteins, without fusion to anchor proteins, can be absorbed by the spore surface during spore formation. (C) Methods used to display cellulases and their complexes on the cell surface of *C. glutamicum*. Proteins of interest are displayed through fusion to membrane protein MscCG or porin proteins PorB, PorC, and PorH. Abbreviations: Protein of interest (POI), plasma membrane (PM), peptidoglycan (PG), arabinogalactan (AG), mycomembrane (MM), and top layer (TL).

remove the enzymes. Interestingly, proteins can also be directly adsorbed on the spore surface by incubating them with spores, resulting in 7.75×10^3 to 1.55×10^4 individual proteins being displayed per particle [123]. Thus, cellulases that cannot be secreted or expressed as fusion proteins can be readily displayed on spores using this method.

4.2. Lactic acid bacteria (LAB)

Gram-positive lactic acid bacteria (LAB) are widely used in the food industry to ferment sugars into lactic acid [124]. They have potential in biomass processing, as they can utilize pentose and hexose sugars, and some members of this group, namely *Lactobacillus plantarum,* are tolerant to low pH and ethanol concentrations up to 13% [26, 125]. Similar methods are used to display proteins in LAB and *B. subtilis,* as their cell wall envelopes are structurally related (**Figure 6A**). In particular, proteins can be displayed noncovalently using LysM-binding modules, or covalently using LPXTG anchors that are processed by sortase enzymes. Comparison studies using *Lactococcus lactis* have shown that covalent attachment using sortase enzymes leads to the largest number of functionally displayed proteins on this microbe's surface, but the relative efficiencies of the sortase and LysM display systems in other species of LAB have not been determined [126].

Two groups have used sortases to assemble minicellulosomes *ex vivo* in LAB. The Martin group anchored a variety of nuclease A-fused miniscaffoldin constructs onto the surface of a single protease-deficient strain of *L. lactis* [127]. These constructs either contained one cohesin, two cohesins, a cohesin and a CBM, or only a CBM. Using a purified *E. coli* β-glucuronidase (UidA)-dockerin fusion protein, they estimated that $\sim 10^4$ complexes are displayed per cell. Interestingly, unlike work in yeast, scaffoldin size was not a limiting factor for display. Instead, it was suggested that protein secretion might be limiting, as scaffoldins containing a CBM showed improved secretion, presumably due to more rapid folding facilitated by the CBM. Using similar approaches, this group later produced a bifunctional minicellulosome using specifically associating type-I and type-II cohesin-dockerin pairs [128]. They demonstrated that the order of enzyme docking affected the activities of the *E. coli* UidA and LacZ enzymes, due to steric factors that influenced enzyme binding.

In separate studies, the Mizrahi group engineered *L. plantarum* to display a sortase attached miniscaffoldin and then used a consortium of cellulase secreting strains to assemble the minicellulosome *ex vivo* [129]. In these studies, recombinant *T. fusca* endoglucanase (Cel6A) and xylanase (Xyn11A) activities were studied in their free form, individually bound to cells, or bound to surface displayed minicellulosomes. Against hypochlorite pretreated wheat straw, secreted enzymes had the best initial activity, followed by the minicellulosome. However, the sugar production rate of the surface displayed minicellulosome slowly increased over time, while the rate for the secreted enzymes decreased. Eventually, cellulosomal activity overtook that of the secreted enzymes, presumably because the cell-associated enzymes are more stable. Interestingly, cells displaying individually anchored enzymes had minimal activity. This is consistent with results obtained in yeast that have shown that enzymes with different substrate specificities are most active against cellulose when they are densely clustered with one another on the cell surface to promote synergistic interactions [63].

4.3. *Corynebacterium glutamicum*

C. glutamicum is an industrially important microbe that produces several tons of glutamate and lysine annually. Although it is a Gram-positive bacterium, its peptidoglycan layer is covered by an arabinogalactan layer, a mycolic acid bilayer, and a top layer composed of polysaccharides, glycolipids, and proteins [130]. The presence of the outer mycolic acid layer confers Gram-negative characteristics to the bacterium and, along with the cytoplasmic membrane, is the primary point to which cellulases are attached (**Figure 6C**). The Kondo laboratory demonstrated the feasibility of displaying heterologous proteins by attaching them to the mycolic acid layer [131]. This was achieved by expressing the cellulase as a fusion protein in which it was joined to the PorC porin. *C. glutamicum* displaying a *Saccharophagus degradans* β-glucosidase-PorC fusion produced 1.08 g/L of lysine from 20 g/L of cellobiose in 96 h [132]. The Han group also displayed an *in vivo* assembled bifunctional minicellulosome that associated with the cytoplasmic membrane. This was accomplished by expressing the scaffoldin as a fusion protein with the mechanosensitive channel (MscCG) [133]. The complex contained the *C. thermocellum* endoglucanase E (CelE) and β-glucosidase A (BglA) enzymes and could release sugars from pretreated rice straw, miscanthus, and rape stem. In the future, engineering these cells to more efficiently degrade complex lignocellulose may lead to more cost-effective ways to produce lysine and glutamate from inexpensive biomass.

5. Conclusions

Towards the goal of cost effectively converting biomass into useful biocommodities, several research groups have developed creative ways to display cellulases on microbes to endow them with cellulolytic activity. Comparing the biomass degradation efficiencies of different types of recombinant microorganisms is difficult. This is because investigators have measured their activities using a variety of cellulosic substrates that can vary dramatically in their resistance to enzymatic degradation as they differ in their solubility, enzyme accessibility, crystallinity, degree of polymerization, fraction of reducing ends, and the presence of hemicellulose and lignin [134]. Moreover, different methods are frequently used to pretreat biomass and to measure the extent of degradation, which can be reported as enzymatic activity, sugar released, biomass remaining, or product produced [17, 135, 136]. However, despite these qualifiers, the results of studies reported to date enable several major conclusions to be drawn. In particular, they provide convincing evidence that clustering enzymes on the surface within minicellulosomes leads to improved microbial cellulolytic activity by promoting synergistic interactions, and in some instances, by improving enzyme stability [32]. Interestingly, synergistic enzyme interactions can also be obtained by displaying different types of individual enzymes, as long as they are densely clustered and have complementary activities [63]. Because complexed enzyme systems require smaller amounts of enzymes to be produced than secreted enzyme systems to achieve efficient degradation of lignocellulose, the cellulosomal system may be optimal for CBP microorganisms, since conserved energy may be directed towards product production.

Displaying large, enzymatically diverse recombinant minicellulosomes remains a challenging problem, but progress has been made in *S. cerevisiae, B. subtilis, C. glutamicum,* and lactic acid bacteria (summarized in **Table 1**). Surface engineering in *S. cerevisiae* is the most advanced, with several groups developing strains that can assemble minicellulosomes *in vivo,* leading to significant improvements in biomass utilization [32, 78, 79, 89]. Complexes containing up to 12 enzymes have been displayed, but great challenges remain as expressing large structures with more enzymes reduces cellular display levels [79]. A variety of promising approaches may help to overcome this limitation, including adaptive assembly and host engineering methods. Furthermore, greater cellulolytic activity through the incorporation of new synergistic enzyme functionalities may offset the decreases associated with the production of larger proteins, as seen by the increased activity of a pentafunctional minicellulosome [78]. Approaches to display minicellulosome structures on eubacteria that have developed genetic systems are less advanced. Thus far, only *ex vivo* assembled complexes have been displayed in Gram-positive *B. subtilis* and LAB, and only individual enzymes have been displayed on the Gram-negative bacterium *E. coli*. The potential of displaying complexes on the cell surface of *E. coli* would seem to be bleak, as the presence of a second outer membrane in this microbe hampers protein secretion. However, studies of *C. glutamicum* are promising, as it is the only bacterium that has been engineered to display an *in vivo* assembled minicellulosome, although this complex contained only two enzymes. In *B. subtilis,* exogenous proteases appear to be a limiting factor, while in LAB protein expression levels are suboptimal [119, 137]. However, it would seem likely that these limitations can be overcome by optimizing display using genetic engineering and directed evolution approaches, as well as the use of cell consortiums to construct complexes *ex vivo*.

A great variety of surface engineering approaches have been developed to create ever more impressive recombinant cellulolytic organisms. However, it is clear that the lignocellulose hydrolysis rates of these recombinant microorganisms needs to be improved if they are to be used in CBP. Future studies may improve their cellulolytic activity by using directed evolution to enhance complex display and stability, by judiciously displaying cellulases that exhibit maximal enzyme synergy, and by devising new methods to stably attach proteins to the cell surface. Genetic engineering of the host will also be critical, enhancing expression, secretion, and display levels, and by eliminating proteins or factors affecting the stability and retention of anchored protein complexes over time. Displayed complexes will also have to be optimized to be maximally active against different types of biomass that have different sugar compositions and structures [138–140]. When these cells are further engineered to produce useful chemicals, their ability to cost-effectively produce biocommodities from biomass will be an important step towards reducing the world's dependency on oil [141–144].

Acknowledgements

This material is based on work supported by the U.S. Department of Energy Office of Science, Office of Biological and Environmental Research program under Award Number DE-FC02-

02ER63421. GL Huang was supported by a Ruth L. Kirschstein National Research Service Award GM007185.

Nomenclature

consolidated bioprocessing	CBP
cellulose–enzyme–microbe	CEM
phosphoric acid swollen cellulose	PASC
regenerated amorphous cellulose	RAC
ionic liquid	IL
glycosylphosphatidylinositol	GPI
cell wall sorting signal	CWS
cellulose-binding module	CBM

Author details

Grace L. Huang[1,2] and Robert T. Clubb[1,3*]

*Address all correspondence to: rclubb@mbi.ucla.edu

1 Department of Chemistry and Biochemistry, University of California, Los Angeles, Los Angeles, CA, USA

2 UCLA-DOE Institute of Genomics and Proteomics, University of California, Los Angeles, Los Angeles, CA, USA

3 Molecular Biology Institute, University of California, Los Angeles, Los Angeles, CA, USA

References

[1] Kerr, R.A., Energy. World oil crunch looming? Science, 2008. 322(5905): p. 1178–1179.

[2] Robertson, G.P., et al., Agriculture. Sustainable biofuels redux. Science, 2008. 322(5898): p. 49–50.

[3] Chu, S. and A. Majumdar, Opportunities and challenges for a sustainable energy future. Nature, 2012. 488(7411): p. 294–303.

[4] Perlack, R.D., et al. Biomass as feedstock for a bioenergy and bioproducts industry the technical feasibility of a billion-ton annual supply. 2005; Available from: http://www.ornl.gov/~webworks/cppr/y2001/rpt/123021.pdf.

[5] Lynd, L.R., et al., Fuel ethanol from cellulosic biomass. Science, 1991. 251(4999): p. 1318–1323.

[6] Taha, M., et al., Commercial feasibility of lignocellulose biodegradation: possibilities and challenges. Curr Opin Biol, 2016. 38: p. 190–197.

[7] Himmel, M.E., et al., Biomass recalcitrance: engineering plants and enzymes for biofuels production. Science, 2007. 315(5813): p. 804–807.

[8] Chundawat, S.P.S., et al., Deconstruction of lignocellulosic biomass to fuels and chemicals. Annu Rev Chem Biomol Eng, 2011. 2: p. 121–145.

[9] Menon, V. and M. Rao, Trends in bioconversion of lignocellulose: biofuels, platform chemicals & biorefinery concept. Prog Energy Combust Sci, 2012. 38(4): p. 522–550.

[10] Zhao, X., K. Cheng, and D. Liu, Organosolv pretreatment of lignocellulosic biomass for enzymatic hydrolysis. Appl Microbiol Biotechnol, 2009. 82(5): p. 815–27.

[11] Hendriks, A.T. and G. Zeeman, Pretreatments to enhance the digestibility of lignocellulosic biomass. Bioresour Technol, 2009. 100(1): p. 10–8.

[12] Olson, D.G., et al., Recent progress in consolidated bioprocessing. Curr Opin Biotechnol, 2012. 23(3): p. 396–405.

[13] la Grange, D.C., R. den Haan, and W.H. van Zyl, Engineering cellulolytic ability into bioprocessing organisms. Appl Microbiol Biotechnol, 2010. 87(4): p. 1195–1208.

[14] Lynd, L.R., et al., Consolidated bioprocessing of cellulosic biomass: an update. Curr Opin Biotechnol, 2005. 16(5): p. 577–583.

[15] Klein-Marcuschamer, D., et al., The challenge of enzyme cost in the production of lignocellulosic biofuels. Biotechnol Bioeng, 2012. 109(4): p. 1083–1087.

[16] Zhao, X., L. Zhang, and D. Liu, Biomass recalcitrance. Part I: the chemical compositions and physical structures affecting the enzymatic hydrolysis of lignocellulose. Biofuels Bioprod Biorefin, 2012. 6(4): p. 465–482.

[17] Zhao, X.B., L.H. Zhang, and D.H. Liu, Biomass recalcitrance. Part II: fundamentals of different pre-treatments to increase the enzymatic digestibility of lignocellulose. Biofuels Bioprod Bioref, 2012. 6(5): p. 561–579.

[18] Ghose, T., Cellulase biosynthesis and hydrolysis of cellulosic substances, in Advances in Biochemical Engineering, T. Ghose, Editor. 1977, Springer Berlin/Heidelberg.

[19] Boerjan, W., J. Ralph, and M. Baucher, Lignin biosynthesis. Annu Rev Plant Biol, 2003. 54: p. 519–546.

[20] Bugg, T.D., et al., Pathways for degradation of lignin in bacteria and fungi. Nat Prod Rep, 2011. 28(12): p. 1883–1896.

[21] Wieczorek, A.S., D. Biot-Pelletier, and V. J.J, Recombinant Cellulase and Cellulosome Systems, in Cellulose - Biomass Conversion, J. Kadla, Editor. 2013: InTech.

[22] Huang, G.L., T.D. Anderson, and R.T. Clubb, Engineering microbial surfaces to degrade lignocellulosic biomass. Bioengineered, 2014. 5(2): p. 96–106.

[23] Tanaka, T. and A. Kondo, Cell surface engineering of industrial microorganisms for biorefining applications. Biotechnol Adv, 2015. 33(7): p. 1403–1411.

[24] Liu, Z., et al., Recent advances in yeast cell-surface display technologies for waste biorefineries. Bioresour Technol, 2016. 215: p. 324–333.

[25] Munoz-Gutierrez, I. and A. Martinez, Polysaccharide hydrolysis with engineered Escherichia coli for the production of biocommodities. J Ind Microbiol Biotechnol, 2013. 40(5): p. 401–410.

[26] Michon, C., et al., Display of recombinant proteins at the surface of lactic acid bacteria: strategies and applications. Microbial Cell Factor, 2016. 15.

[27] Bayer, E.A., et al., Cellulosomes-structure and ultrastructure. J Struct Biol, 1998. 124(2–3): p. 221–234.

[28] Fontes, C.M. and H.J. Gilbert, Cellulosomes: highly efficient nanomachines designed to deconstruct plant cell wall complex carbohydrates. Annu Rev Biochem, 2010. 79: p. 655–681.

[29] Leibovitz, E., et al., Characterization and subcellular localization of the Clostridium thermocellum scaffoldin dockerin binding protein SdbA. J Bacteriol, 1997. 179(8): p. 2519–2523.

[30] Ding, S.Y., et al., A biophysical perspective on the cellulosome: new opportunities for biomass conversion. Curr Opin Biotechnol, 2008. 19(3): p. 218–227.

[31] You, C., et al., Enhanced microbial utilization of recalcitrant cellulose by an ex vivo cellulosome-microbe complex. Appl Environ Microbiol, 2012. 78(5): p. 1437–1444.

[32] Wen, F., J. Sun, and H. Zhao, Yeast surface display of trifunctional minicellulosomes for simultaneous saccharification and fermentation of cellulose to ethanol. Appl Environ Microbiol, 2010. 76(4): p. 1251–1260.

[33] Lynd, L.R., et al., Microbial cellulose utilization: fundamentals and biotechnology. Microbiol Mol Biol Rev, 2002. 66(3): p. 506–577, table of contents.

[34] Lu, Y., Y.H. Zhang, and L.R. Lynd, Enzyme-microbe synergy during cellulose hydrolysis by Clostridium thermocellum. Proc Natl Acad Sci U S A, 2006. 103(44): p. 16165–16169.

[35] Demain, A.L., M. Newcomb, and J.H. Wu, Cellulase, clostridia, and ethanol. Microbiol Mol Biol Rev, 2005. 69(1): p. 124–154.

[36] Klein-Marcuschamer, D. and H.W. Blanch, Renewable fuels from biomass: technical hurdles and economic assessment of biological routes. AIChE J, 2015. 61(9): p. 2689–2701.

[37] Humbird, D., National Renewable Energy Laboratory (U.S.), and Harris Group Inc., Process design and economics for biochemical conversion of lignocellulosic biomass to ethanol dilute-acid pretreatment and enzymatic hydrolysis of corn stover, in Nrel/Tp 5100-47764. 2011, National Renewable Energy Laboratory: Golden, CO. p. 1 online resource (ix, 136 p.) ill.

[38] Johnson, E., Integrated enzyme production lowers the cost of cellulosic ethanol. Biofuels Bioprod Bioref, 2016. 10(2): p. 164–174.

[39] Tanaka, T. and A. Kondo, Cell-surface display of enzymes by the yeast Saccharomyces cerevisiae for synthetic biology. FEMS Yeast Res, 2014.

[40] Lipke, P.N. and R. Ovalle, Cell wall architecture in yeast: new structure and new challenges. J Bacteriol, 1998. 180(15): p. 37353740.

[41] Van der Vaart, J.M., et al., Comparison of cell wall proteins of Saccharomyces cerevisiae as anchors for cell surface expression of heterologous proteins. Appl Environ Microbiol, 1997. 63(2): p. 615–620.

[42] van der Vaart, J.M., et al., The retention mechanism of cell wall proteins in Saccharomyces cerevisiae. Wall-bound Cwp2p is beta-1,6-glucosylated. Biochim Biophys Acta, 1996. 1291(3): p. 206–214.

[43] Fujita, M. and T. Kinoshita, Structural remodeling of GPI anchors during biosynthesis and after attachment to proteins. FEBS Lett, 2010. 584(9): p. 1670–1677.

[44] Ohishi, K., N. Inoue, and T. Kinoshita, PIG-S and PIG-T, essential for GPI anchor attachment to proteins, form a complex with GAA1 and GPI8. EMBO J, 2001. 20(15): p. 4088–4098.

[45] Orlean, P. and A.K. Menon, Thematic review series: lipid posttranslational modifications. GPI anchoring of protein in yeast and mammalian cells, or: how we learned to stop worrying and love glycophospholipids. J Lipid Res, 2007. 48(5): p. 993–1011.

[46] Lu, C.F., J. Kurjan, and P.N. Lipke, A pathway for cell wall anchorage of Saccharomyces cerevisiae alpha-agglutinin. Mol Cell Biol, 1994. 14(7): p. 4825–4833.

[47] Lu, C.F., et al., Glycosyl phosphatidylinositol-dependent cross-linking of alpha-agglutinin and beta 1,6-glucan in the Saccharomyces cerevisiae cell wall. J Cell Biol, 1995. 128(3): p. 333–340.

[48] Orlean, P., Architecture and biosynthesis of the Saccharomyces cerevisiae cell wall. Genetics, 2012. 192(3): p. 775–818.

[49] Chao, G., et al., Isolating and engineering human antibodies using yeast surface display. Nat Protoc, 2006. 1(2): p. 755–768.

[50] Kotaka, A., et al., Enhancement of beta-glucosidase activity on the cell-surface of sake yeast by disruption of SED1. J Biosci Bioeng, 2010. 109(5): p. 442–446.

[51] Matsuoka, H., et al., Cell wall structure suitable for surface display of proteins in Saccharomyces cerevisiae. Yeast, 2014. 31(2): p. 67-76.

[52] Inokuma, K., T. Hasunuma, and A. Kondo, Efficient yeast cell-surface display of exo— and endo-cellulase using the SED1 anchoring region and its original promoter. Biotechnol Biofuels, 2014. 7.

[53] Inokuma, K., et al., Enhanced cell-surface display and secretory production of cellulolytic enzymes with Saccharomyces cerevisiae Sed1 signal peptide. Biotechnol Bioeng, 2016.

[54] Suzuki, H., et al., Deglycosylation of cellulosomal enzyme enhances cellulosome assembly in Saccharomyces cerevisiae. J Biotechnol, 2012. 157(1): p. 64–70.

[55] Murai, T., et al., Assimilation of cellooligosaccharides by a cell surface-engineered yeast expressing beta-glucosidase and carboxymethylcellulase from aspergillus aculeatus. Appl Environ Microbiol, 1998. 64(12): p. 4857–4861.

[56] Fujita, Y., et al., Direct and efficient production of ethanol from cellulosic material with a yeast strain displaying cellulolytic enzymes. Appl Environ Microbiol, 2002. 68(10): p. 5136–5141.

[57] Fujita, Y., et al., Synergistic saccharification, and direct fermentation to ethanol, of amorphous cellulose by use of an engineered yeast strain codisplaying three types of cellulolytic enzyme. Appl Environ Microbiol, 2004. 70(2): p. 1207–1212.

[58] Ito, J., et al., Improvement of cellulose-degrading ability of a yeast strain displaying Trichoderma reesei endoglucanase II by recombination of cellulose-binding domains. Biotechnol Prog, 2004. 20(3): p. 688–691.

[59] Yamada, R., et al., Cocktail delta-integration: a novel method to construct cellulolytic enzyme expression ratio-optimized yeast strains. Microb Cell Fact, 2010. 9: p. 32.

[60] Yamada, R., et al., Direct ethanol production from cellulosic materials using a diploid strain of Saccharomyces cerevisiae with optimized cellulase expression. Biotechnol Biofuels, 2011. 4: p. 8.

[61] Yanase, S., et al., Ethanol production from cellulosic materials using cellulase-expressing yeast. Biotechnol J, 2010. 5(5): p. 449–455.

[62] Liu, Z., et al., Combined cell-surface display- and secretion-based strategies for production of cellulosic ethanol with Saccharomyces cerevisiae. Biotechnol Biofuels, 2015. 8.

[63] Bae, J., K. Kuroda, and M. Ueda, Proximity effect among cellulose-degrading enzymes displayed on the Saccharomyces cerevisiae cell surface. Appl Environ Microbiol, 2015. 81(1): p. 59–66.

[64] Vazana, Y., et al., A synthetic biology approach for evaluating the functional contribution of designer cellulosome components to deconstruction of cellulosic substrates. Biotechnol Biofuels, 2013. 6(1): p. 182.

[65] Matano, Y., T. Hasunuma, and A. Kondo, Display of cellulases on the cell surface of Saccharomyces cerevisiae for high yield ethanol production from high-solid lignocellulosic biomass. Bioresour Technol, 2012. 108: p. 128–33.

[66] Matano, Y., T. Hasunuma, and A. Kondo, Cell recycle batch fermentation of high-solid lignocellulose using a recombinant cellulase-displaying yeast strain for high yield ethanol production in consolidated bioprocessing. Bioresour Technol, 2013. 135: p. 403–9.

[67] Matano, Y., T. Hasunuma, and A. Kondo, Simultaneous improvement of saccharification and ethanol production from crystalline cellulose by alleviation of irreversible adsorption of cellulase with a cell surface-engineered yeast strain. Appl Microbiol Biotechnol, 2013. 97(5): p. 2231–2237.

[68] Liu, Z., et al., Engineering of a novel cellulose-adherent cellulolytic Saccharomyces cerevisiae for cellulosic biofuel production. Sci Rep, 2016. 6: p. 24550.

[69] Yamada, R., et al., Efficient direct ethanol production from cellulose by cellulase—and cellodextrin transporter-co-expressing Saccharomyces cerevisiae. AMB Express, 2013. 3(1): p. 34.

[70] Nakatani, Y., et al., Synergetic effect of yeast cell-surface expression of cellulase and expansin-like protein on direct ethanol production from cellulose. Microb Cell Factor, 2013. 12.

[71] Nakanishi, A., et al., Effect of pretreatment of hydrothermally processed rice straw with laccase-displaying yeast on ethanol fermentation. Appl Microbiol Biotechnol, 2012. 94(4): p. 939–948.

[72] Lilly, M., et al., Heterologous expression of a Clostridium minicellulosome in Saccharomyces cerevisiae. FEMS Yeast Res, 2009. 9(8): p. 1236–1249.

[73] Tsai, S.L., et al., Functional assembly of minicellulosomes on the Saccharomyces cerevisiae cell surface for cellulose hydrolysis and ethanol production. Appl Environ Microbiol, 2009. 75(19): p. 6087–6093.

[74] Tsai, S.L., G. Goyal, and W. Chen, Surface display of a functional minicellulosome by intracellular complementation using a synthetic yeast consortium and its application to cellulose hydrolysis and ethanol production. Appl Environ Microbiol, 2010. 76(22): p. 7514–7520.

[75] Goyal, G., et al., Simultaneous cell growth and ethanol production from cellulose by an engineered yeast consortium displaying a functional mini-cellulosome. Microb Cell Fact, 2011. 10: p. 89.

[76] Kim, S., et al., Cellulosic ethanol production using a yeast consortium displaying a minicellulosome and beta-glucosidase. Microb Cell Fact, 2013. 12: p. 14.

[77] Tsai, S.L., N.A. DaSilva, and W. Chen, Functional display of complex cellulosomes on the yeast surface via adaptive assembly. ACS Synth Biol, 2013. 2(1): p. 14–21.

[78] Liang, Y., et al., Engineered pentafunctional minicellulosome for simultaneous saccharification and ethanol fermentation in Saccharomyces cerevisiae. Appl Environ Microbiol, 2014. 80(21): p. 6677–6684.

[79] Fan, L.H., et al., Self-surface assembly of cellulosomes with two miniscaffoldins on Saccharomyces cerevisiae for cellulosic ethanol production. Proc Natl Acad Sci U S A, 2012. 109(33): p. 13260–13265.

[80] Fujita, Y., et al., Construction of whole-cell biocatalyst for xylan degradation through cell-surface xylanase display in Saccharomyces cerevisiae. J Mol Catal B Enzym, 2002. 17(3–5): p. 189–195.

[81] Katahira, S., et al., Construction of a xylan-fermenting yeast strain through codisplay of xylanolytic enzymes on the surface of xylose-utilizing Saccharomyces cerevisiae cells. Appl Environ Microbiol, 2004. 70(9): p. 5407–5414.

[82] Katahira, S., et al., Ethanol fermentation from lignocellulosic hydrolysate by a recombinant xylose- and cellooligosaccharide-assimilating yeast strain. Appl Microbiol Biotechnol, 2006. 72(6): p. 1136–1143.

[83] Nakamura, N., et al., Effective xylose/cellobiose co-fermentation and ethanol production by xylose-assimilating S-cerevisiae via expression of beta-glucosidase on its cell surface. Enzym Microb Technol, 2008. 43(3): p. 233–236.

[84] Sakamoto, T., et al., Direct ethanol production from hemicellulosic materials of rice straw by use of an engineered yeast strain codisplaying three types of hemicellulolytic enzymes on the surface of xylose-utilizing Saccharomyces cerevisiae cells. J Biotechnol, 2012. 158(4): p. 203–210.

[85] Guirimand, G., et al., Cell surface engineering of Saccharomyces cerevisiae combined with membrane separation technology for xylitol production from rice straw hydrolysate. Appl Microbiol Biotechnol, 2016. 100(8): p. 3477–3487.

[86] Ota, M., et al., Display of Clostridium cellulovorans xylose isomerase on the cell surface of Saccharomyces cerevisiae and its direct application to xylose fermentation. Biotechnol Prog, 2013. 29(2): p. 346–51.

[87] Srikrishnan, S., W. Chen, and N.A. Da Silva, Functional assembly and characterization of a modular xylanosome for hemicellulose hydrolysis in yeast. Biotechnol Bioeng, 2013. 110(1): p. 275–85.

[88] Sun, J., et al., Direct conversion of xylan to ethanol by recombinant Saccharomyces cerevisiae strains displaying an engineered minihemicellulosome. Appl Environ Microbiol, 2012. 78(11): p. 3837–3845.

[89] Ito, J., et al., Regulation of the display ratio of enzymes on the saccharomyces cerevisiae cell surface by the immunoglobulin g and cellulosomal enzyme binding domains. Appl Environ Microbiol, 2009. 75(12): p. 4149–4154.

[90] Han, Z.L., et al., self-assembled amyloid-like oligomeric-cohesin scaffoldin for augmented protein display on the saccharomyces cerevisiae cell surface. Appl Environ Microbiol, 2012. 78(9): p. 3249–3255.

[91] van Bloois, E., et al., Decorating microbes: surface display of proteins on Escherichia coli. Trends Biotechnol, 2011. 29(2): p. 79–86.

[92] Clomburg, J.M. and R. Gonzalez, Biofuel production in Escherichia coli: the role of metabolic engineering and synthetic biology. Appl Microbiol Biotechnol, 2010. 86(2): p. 419–434.

[93] Francisco, J.A., et al., Specific adhesion and hydrolysis of cellulose by intact Escherichia coli expressing surface anchored cellulase or cellulose binding domains. Biotechnology (NY), 1993. 11(4): p. 491–495.

[94] Qu, W., Y. Xue, and Q. Ding, Display of fungi xylanase on Escherichia coli cell surface and use of the enzyme in xylan biodegradation. Curr Microbiol, 2015. 70(6): p. 779–785.

[95] Jung, H.C., et al., Expression of carboxymethylcellulase on the surface of Escherichia coli using Pseudomonas syringae ice nucleation protein. Enzym Microb Technol, 1998. 22(5): p. 348–354.

[96] Kim, Y.S., H.C. Jung, and J.G. Pan, Bacterial cell surface display of an enzyme library for selective screening of improved cellulase variants. Appl Environ Microbiol, 2000. 66(2): p. 788–793.

[97] Liu, W., et al., Engineering of Clostridium phytofermentans endoglucanase Cel5A for improved thermostability. Appl Environ Microbiol, 2010. 76(14): p. 4914–4917.

[98] Tanaka, T., et al., Creation of a cellooligosaccharide-assimilating Escherichia coli strain by displaying active beta-glucosidase on the cell surface via a novel anchor protein. Appl Environ Microbiol, 2011. 77(17): p. 6265–6270.

[99] Soma, Y., et al., Direct isopropanol production from cellobiose by engineered Escherichia coli using a synthetic pathway and a cell surface display system. J Biosci Bioeng, 2012. 114(1): p. 80–85.

[100] Tanaka, T., et al., Creation of cellobiose and xylooligosaccharides-coutilizing Escherichia coli displaying both beta-glucosidase and beta-xylosidase on its cell surface. ACS Synth Biol, 2014. 3(7): p. 446–453.

[101] Ryu, S. and M.N. Karim, A whole cell biocatalyst for cellulosic ethanol production from dilute acid-pretreated corn stover hydrolyzates. Appl Microbiol Biotechnol, 2011. 91(3): p. 529–542.

[102] Chen, Y.P., et al., Enhancing the stability of xylanase from Cellulomonas fimi by cell-surface display on Escherichia coli. J Appl Microbiol, 2012. 112(3): p. 455–463.

[103] Ko, K.C., et al., Bacterial cell surface display of a multifunctional cellulolytic enzyme screened from a bovine rumen metagenomic resource. J Microbiol Biotechnol, 2015. 25(11): p. 1835–1841.

[104] Yim, S.S., et al., Isolation of a potential anchoring motif based on proteome analysis of Escherichia coli and its use for cell surface display. Appl Biochem Biotechnol, 2013. 170(4): p. 787–804.

[105] Munoz-Gutierrez, I., et al., Cell surface display of a beta-glucosidase employing the type V secretion system on ethanologenic Escherichia coli for the fermentation of cellobiose to ethanol. J Ind Microbiol Biotechnol, 2012. 39(8): p. 1141–1152.

[106] Park, T.J., et al., Surface display of recombinant proteins on Escherichia coli by BclA exosporium of Bacillus anthracis. Microb Cell Fact, 2013. 12: p. 81.

[107] Bokinsky, G., et al., Synthesis of three advanced biofuels from ionic liquid-pretreated switchgrass using engineered Escherichia coli. Proc Natl Acad Sci U S A, 2011. 108(50): p. 19949–19954.

[108] Schallmey, M., A. Singh, and O.P. Ward, Developments in the use of Bacillus species for industrial production. Can J Microbiol, 2004. 50(1): p. 1–17.

[109] Liu, L., et al., Developing Bacillus spp. as a cell factory for production of microbial enzymes and industrially important biochemicals in the context of systems and synthetic biology. Appl Biochem Biotechnol, 2013. 97(14): p. 6113–6127.

[110] Zhang, X.-Z. and Y.-H.P. Zhang, One-step production of biocommodities from lignocellulosic biomass by recombinant cellulolytic Bacillus subtilis: Opportunities and challenges. Eng Life Sci, 2010. 10(5): p. 398–406.

[111] van Dijl, J.M. and M. Hecker, Bacillus subtilis: from soil bacterium to super-secreting cell factory. Microbial Cell Factories, 2013. 12(3).

[112] Kang, Z., et al., Molecular engineering of secretory machinery components for high-level secretion of proteins in Bacillus species. J Ind Microbiol Biotechnol, 2014. 41(11): p. 1599–1607.

[113] Spirig, T., E.M. Weiner, and R.T. Clubb, Sortase enzymes in Gram-positive bacteria. Mol Microbiol, 2011. 82(5): p. 1044–1059.

[114] Nguyen, H.D. and W. Schumann, Establishment of an experimental system allowing immobilization of proteins on the surface of Bacillus subtilis cells. J Biotechnol, 2006. 122(4): p. 473–482.

[115] Buist, G., et al., LysM, a widely distributed protein motif for binding to (peptido)glycans. Mol Microbiol, 2008. 68(4): p. 838–847.

[116] Chen, C.L., et al., Development of a LytE-based high-density surface display system in Bacillus subtilis. Microb Biotechnol, 2008. 1(2): p. 177–90.

[117] Kim, J.H., I.S. Park, and B.G. Kim, Development and characterization of membrane surface display system using molecular chaperon, prsA, of Bacillus subtilis. Biochem Biophys Res Commun, 2005. 334(4): p. 1248–1253.

[118] Anderson, T.D., et al., Assembly of minicellulosomes on the surface of Bacillus subtilis. Appl Environ Microbiol, 2011. 77(14): p. 4849–4858.

[119] Westers, L., H. Westers, and W.J. Quax, Bacillus subtilis as cell factory for pharmaceutical proteins: a biotechnological approach to optimize the host organism. Biochim Biophys Acta, 2004. 1694(1–3): p. 299–310.

[120] McKenney, P.T., A. Driks, and P. Eichenberger, The Bacillus subtilis endospore: assembly and functions of the multilayered coat. Nat Rev Microbiol, 2013. 11(1): p. 33–44.

[121] Isticato, R. and E. Ricca, Spore Surface Display. Microbiol Spectr, 2014. 2(5).

[122] Pan, J.G., et al., Display of native proteins on Bacillus subtilis spores. FEMS Microbiol Lett, 2014. 358(2): p. 209–217.

[123] Sirec, T., et al., Adsorption of beta-galactosidase of Alicyclobacillus acidocaldarius on wild type and mutants spores of Bacillus subtilis. Microbial Cell Factories, 2012. 11(100).

[124] Chapot-Chartier, M.P. and S. Kulakauskas, Cell wall structure and function in lactic acid bacteria. Microbial Cell Factories, 2014. 13(Suppl 1: S9).

[125] G-Alegria, E., et al., High tolerance of wild Lactobacillus plantarum and Oenococcus oeni strains to lyophilisation and stress environmental conditions of acid pH and ethanol. FEMS Microbiol Lett, 2004. 230(1): p. 53–61.

[126] Kyla-Nikkila, K., U. Alakuijala, and P.E.J. Saris, Immobilization of Lactococcus lactis to cellulosic material by cellulose-binding domain of Cellvibrio japonicus. J Appl Microbiol, 2010. 109(4): p. 1274–1283.

[127] Wieczorek, A.S. and V.J. Martin, Engineering the cell surface display of cohesins for assembly of cellulosome-inspired enzyme complexes on Lactococcus lactis. Microb Cell Fact, 2010. 9: p. 69.

[128] Wieczorek, A.S. and V.J.J. Martin, Effects of synthetic cohesin-containing scaffold protein architecture on binding dockerin-enzyme fusions on the surface of Lactococcus lactis. Microbial Cell Factories, 2012. 11(160).

[129] Morais, S., et al., A combined cell-consortium approach for lignocellulose degradation by specialized Lactobacillus plantarum cells. Biotechnology for Biofuels, 2014. 7(112).

[130] Burkovski, A., Cell envelope of corynebacteria: structure and influence on pathogenicity. ISRN Microbiol, 2013. 2013: p. 935736.

[131] Tateno, T., et al., Development of novel cell surface display in Corynebacterium glutamicum using porin. Appl Microbiol Biotechnol, 2009. 84(4): p. 733–739.

[132] Adachi, N., et al., Direct L-lysine production from cellobiose by Corynebacterium glutamicum displaying beta-glucosidase on its cell surface. Appl Microbiol Biotechnol, 2013. 97(16): p. 7165–7172.

[133] Kim, S.J., et al., Bi-functional cellulases complexes displayed on the cell surface of Corynebacterium glutamicum increase hydrolysis of lignocelluloses at elevated temperature. Enzym Microb Technol, 2014. 66: p. 67–73.

[134] Zhang, P.Y.H., M.E. Himmel, and J.R. Mielenz, Outlook for cellulase improvement: screening and selection strategies. Biotechnol Adv, 2006. 24(5): p. 452–481.

[135] Yang, B. and C.E. Wyman, Pretreatment: the key to unlocking low-cost cellulosic ethanol. Biofuels Bioprod Bioref, 2008. 2(1): p. 26–40.

[136] Dashtban, M., et al., Cellulase activities in biomass conversion: measurement methods and comparison. Crit Rev Biotechnol, 2010. 30(4): p. 302–309.

[137] Le Loir, Y., et al., Protein secretion in Lactococcus lactis: an efficient way to increase the overall heterologous protcin production. Microbial Cell Factories, 2005. 4(2).

[138] Banerjee, G., et al., Rapid optimization of enzyme mixtures for deconstruction of diverse pretreatment/biomass feedstock combinations. Biotechnology for Biofuels, 2010. 3(22).

[139] Engel, P., et al., Rational approach to optimize cellulase mixtures for hydrolysis of regenerated cellulose containing residual ionic liquid. Bioresour Technol, 2012. 115: p. 27–34.

[140] Kim, I.J., et al., Customized optimization of cellulase mixtures for differently pretreated rice straw. Bioprocess Biosyst Eng, 2015. 38(5): p. 929–937.

[141] Dellomonaco, C., F. Fava, and R. Gonzalez, The path to next generation biofuels: successes and challenges in the era of synthetic biology. Microb Cell Fact, 2010. 9: p. 3.

[142] Jang, Y.S., et al., Engineering of microorganisms for the production of biofuels and perspectives based on systems metabolic engineering approaches. Biotechnol Adv, 2012. 30(5): p. 989–1000.

[143] Rabinovitch-Deere, C.A., et al., Synthetic biology and metabolic engineering approaches to produce biofuels. Chem Rev, 2013. 113(7): p. 4611–4632.

[144] Li, H., A.F. Cann, and J.C. Liao, Biofuels: biomolecular engineering fundamentals and advances. Annu Rev Chem Biomol Eng, 2010. 1: p. 19–36.

Permissions

All chapters in this book were first published in BVEVE, by InTech Open; hereby published with permission under the Creative Commons Attribution License or equivalent. Every chapter published in this book has been scrutinized by our experts. Their significance has been extensively debated. The topics covered herein carry significant findings which will fuel the growth of the discipline. They may even be implemented as practical applications or may be referred to as a beginning point for another development.

The contributors of this book come from diverse backgrounds, making this book a truly international effort. This book will bring forth new frontiers with its revolutionizing research information and detailed analysis of the nascent developments around the world.

We would like to thank all the contributing authors for lending their expertise to make the book truly unique. They have played a crucial role in the development of this book. Without their invaluable contributions this book wouldn't have been possible. They have made vital efforts to compile up to date information on the varied aspects of this subject to make this book a valuable addition to the collection of many professionals and students.

This book was conceptualized with the vision of imparting up-to-date information and advanced data in this field. To ensure the same, a matchless editorial board was set up. Every individual on the board went through rigorous rounds of assessment to prove their worth. After which they invested a large part of their time researching and compiling the most relevant data for our readers.

The editorial board has been involved in producing this book since its inception. They have spent rigorous hours researching and exploring the diverse topics which have resulted in the successful publishing of this book. They have passed on their knowledge of decades through this book. To expedite this challenging task, the publisher supported the team at every step. A small team of assistant editors was also appointed to further simplify the editing procedure and attain best results for the readers.

Apart from the editorial board, the designing team has also invested a significant amount of their time in understanding the subject and creating the most relevant covers. They scrutinized every image to scout for the most suitable representation of the subject and create an appropriate cover for the book.

The publishing team has been an ardent support to the editorial, designing and production team. Their endless efforts to recruit the best for this project, has resulted in the accomplishment of this book. They are a veteran in the field of academics and their pool of knowledge is as vast as their experience in printing. Their expertise and guidance has proved useful at every step. Their uncompromising quality standards have made this book an exceptional effort. Their encouragement from time to time has been an inspiration for everyone.

The publisher and the editorial board hope that this book will prove to be a valuable piece of knowledge for researchers, students, practitioners and scholars across the globe.

List of Contributors

Thorsten Ahrens and Silvia Drescher-Hartung
Institute for Biotechnology and Environmental Engineering, Ostfalia University of Applied Sciences, Wolfenbüttel, Germany

Olga Anne
Department of Natural Science, Klaipeda University, Klaipeda, Lithuania

Claudia Ortiz-Calderon, Héctor Cid Silva and Daniel Barros Vásquez
Faculty of Chemistry and Biology, Department of Biology, University of Santiago of Chile, Santiago, Chile

Lei Shi
State Forestry Administration Key Laboratory for Science & Technology of Bamboo & Rattan, International Center for Bamboo and Rattan, Beijing, China

Shirong Liu
Institute of Forest Ecology, Environment and Protection, Chinese Academy of Forestry, Beijing, China

Mohammed Abed Fattah Hamad, Aly Moustafa Radwan and Ashraf Amin
National Research Centre, Cairo, Egypt

Tran Dang Xuan
Graduate School for International Development and Cooperation (IDEC), Hiroshima University, Higashi Hiroshima, Japan

Nguyen Thi Phuong
Vietnam Academy of Agricultural Science (VASS), Hanoi, Vietnam

Tran Dang Khanh
Agricultural Genetics Institute (AGI), Hanoi, Vietnam

Aiguo G. Wang, Danielle Austin and Hua Song
University of Calgary, Calgary, Canada

Leticia Saiz Rodríguez and José M. Bermejo Muñoz
Signus Ecovalor S.L, Madrid, Spain

Adrien Zambon and Jean P. Faure
Aliapur, Lyon, France

Borja Velázquez-Martí
Department of Rural and Food Engineering, Universitat Politècnica de València, Valencia, Spain

Isabel López-Cortés and Domingo Salazar-Hernández
Department of Vegetal Production, Universitat Politècnica de València, Valencia, Spain

Ángel Jesús Callejón-Ferre
Department of Engineering, University of Almeria, Almería, Spain

Adélia M. O. Sousa, Ana Cristina Gonçalves and José R. Marques da Silva
Department of Rural Engineering, School of Sciences and Technology, Portugal
Institute of diterranean Agricultural and Environmental Sciences (ICAAM), Portugal
Institute of Research and Advanced Information (IIFA), University of Évora, Évora, Portugal

Martin Bosák
University of Economics in Bratislava, Košice, Slovak Republic

Grace L. Huang
Department of Chemistry and Biochemistry, University of California, Los Angeles, Los Angeles, CA, USA
UCLA-DOE Institute of Genomics and Proteomics, University of California, Los Angeles, Los Angeles, CA, USA

Robert T. Clubb
Department of Chemistry and Biochemistry, University of California, Los Angeles, Los Angeles, CA, USA
Molecular Biology Institute, University of California, Los Angeles, Los Angeles, CA, USA

Index

www.ingramcontent.com/pod-product-compliance
Lightning Source LLC
LaVergne TN
LVHW081014210926
848485LV00004B/864